Pi Technology Libr

Bought from www.ama

by Charlie Watraby 31 October 2000

Karl Kordesch, Günter Simader

Fuel Cells

and Their Applications

VCH

© VCH Verlagsgesellschaft mbH, D-69451 Weinheim, Federal Republic of Germany, 1996

Distribution:

VCH, P.O. Box 10 11 61, D-69451 Weinheim, Federal Republic of Germany

Switzerland: VCH, P.O. Box, CH-4020 Basel, Switzerland

United Kingdom and Ireland: VCH, 8 Wellington Court, Cambridge CB1 1HZ, United Kingdom

USA and Canada: VCH, 220 East 23rd Street, New York, NY 10010–4606, USA

Japan: VCH Eikow Building, 10-9 Hongo 1-chome, Bunkyo-Ku, Tokyo 113, Japan

ISBN 3-527-28579-2

Karl Kordesch, Günter Simader

Fuel Cells

and Their Applications

VCH Weinheim · New York · Basel · Cambridge · Tokyo

Prof. Dr. Dr. h. c. K. Kordesch
Stemayrgasse 16, III
A-8010 Graz
Austria

Dr. Günter Simader
Gaishorn 125
A-8783 Gaishorn/See
Austria

Published jointly by
VCH Verlagsgesellschaft mbH, Weinheim (Federal Republic of Germany)
VCH Publishers, Inc., New York, NY (USA)

Editorial Director: Dr. Barbara Böck
Production Manager: Peter J. Biel
Cover Illustration: Susanne Baum

Library of Congress Card No. applied for
A CIP catalogue record for this book is available from the British Library

Die Deutsche Bibliothek – CIP-Einheitsaufnahme

Kordesch, Karl:
Fuel cells and their applications/Karl Kordesch; Günter Simader. –
Weinheim; New York; Basel; Cambridge; Tokyo: VCH, 1996
ISBN 3-527-28579-2
NE: Simader, Günter:

Composition, Printing and Bookbinding: Druckhaus "Thomas Müntzer" GmbH, D-99947 Bad Langensalza
Indexing: Borkowski & Borkowski Publishing Service

Printed in the Federal Republic of Germany

Preface

Fuel cells have excited the minds of theoretical scientists and practical technologists for over 100 years, ever since the first experiments of W.R. Grove showed that the electrolysis of water was a reversible process, and W. Oswald provided the basic thermodynamic equations showing the definite advantages of "Low Temperature Electrochemical Oxidation" over "High Temperature Combustion" of fuels.

Fuel cell system developers are now in the fourth or fifth cycle of attempts to turn this knowledge into realistic systems which can compete with the simple burning of fuels in power plants and combustion engines. The question is: Why does it take so long to verify these theories in actual and practical devices?

The authors have tried to give rational answers to this question. These are based on the stepwise progress and historic advances in the development of fuel cells and their competition with the existing, highly-developed combustion engine technology. Unfortunately, there is no single answer. The sometimes gradual, sometimes stepwise improvement in the availability of suitable construction materials was of great importance. However, the need for fuel cells was not always clearly established: in times of a surplus of cheap fossil fuels, the research and development efforts (especially in the USA) stopped. Unquestionably, the pressure exerted by environmental groups has had much to do with the renaissance in fuel cell research and development we are experiencing today. Interest, at least in the Western world, has rocketed.

Dispersed, efficient power production and the extremely low emission electric vehicle are the main points dominating the present applicability scenario of fuel cells. The opponents have serious, negative key replies: a long life is needed for power plants and affordable systems are required for electric automobiles.

There are many excellent books about fuel cell technologies available. However, the general aspects are often missed in an attempt to supply design details to the reader. The comparison with other systems is sometimes missing, or enthusiasm for one type of fuel cell or a specific application makes the author overlook the advantages of others. In fuel cell technology, as in other promising new fields, fashions and trends develop quickly and adverse results tend to be minimized. The authors have tried to avoid such pitfalls.

However, it shall not be hidden from the readers that Karl Kordesch has a deep conviction that low-cost, alkaline hydron-air fuel cell systems with circulating electrolyte combined with a rechargeable battery in a hybrid configuration will play a more important role in the future of electric vehicles than is generally believed at the present time. The reasoning will be extensively discussed in the appropriate chapter, but the basis is clearly actual experience with his own vehicle which was operated in regular traffic and service from 1970 to 1973 – a surprisingly long time.

The "secret" of endurance may be simply explained: there is a considerable difference in life expectancy if a fuel cell is operated intermittently (like the combustion engine in a car), or is continuously activated and subject to deterioration during standing periods.

After 25 years, this point is now again under investigation and, if confirmed, a new era for fuel cell electric vehicles can be expected. It may turn out that the tremendous efforts made to immobilize electrolytes and avoid circulating pumps was just the wrong way to go.

The co-author, Dr. Günter R. Simader, chose the fuel cell technology as one of the energy innovation possibilities open to the Austrian utility companies, and published his doctoral thesis on that subject in 1994. He studied the thermodynamics of fuel cell systems involving heat processes (exergy considerations included) and has in addition extensively investigated the application of fuel cells for dispersed power applications. One of the surprising conclusions from his involvement in fuel cell technologies was the opinion that the real bottleneck in the progress of fuel cell applications is the fuel to be used. Hydrogen, as a gas, has its disadvantages and a more convenient and far less costly energy carrier must be looked for. Alcohols, ammonia and metal hydrides are reasonable choices, but not ideal for the distribution on a large consumer market scale. His studies of sponge iron as a hydrogen source on demand (by reaction with steam) yielded in a promising evaluation depending on the kind of application. At present he is working in IEA related PEMFC activities at ESTCO Energy Inc. in Ottawa (Canada).

A certain emphasis has been put on the combination of renewable energy and fuel cells. These aspects of long-time energy storage are important for the use of fuel cells in smaller stationary applications, remote locations and households. The combination of fuel cells with secondary batteries in hybrid-photovoltaic power supplies is described in connection with the Zn-Br Flow Battery and the recently available, rechargeable alkaline manganese dioxide – zinc battery (known in the USA under the trade name "Renewal"), which also has an excellent shelf-life at elevated temperatures (in contrast to e.g. Ni-Cd batteries). These systems have also been researched and further developed at the Technical University Graz by teams, supervised by Karl Kordesch.

The authors are extremely grateful for cooperation of Dr. Julio T. Oliveira from ESTCO Energy Inc., a former department of the University of Ottawa, who was responsible for the extensive coverage of membrane systems. Financial support for dissertations was received from the Funds for the Support of Scientific Research in Austria (FWF) and from Battery Technologies, Inc. (BTI) in Canada. The travel grants received from Austrian electricity producing and natural gas distribution companies, especially the "Grazer Stadtwerke" made it possible to obtain a detailed picture of the phosphoric acid fuel cell programs and (near commercial) power plant efforts in Japan. The manuscript was produced at the facilities of the Institute for Inorganic Chemical Technology of the Technical University Graz. Our special appreciation is due to Mrs. Karin Grand and Ms. Karin Wukounig for their help with copies, drawings, photographs.

Graz, February 1996   Dr. Karl V. Kordesch
                                                             Dr. Günter R. Simader

Acknowledgement

The authors are indebted to several organisations and to many colleagues in the Fuel Cell and Battery field. However we must credit also many personal friends and non-scientific people for their help during work on this book, which does represent a more than two years effort, competing with our other technological commitments and administrative duties.

For a continuous generous financial support we want to thank Battery Technologies Inc., in Richmond Hill, Ontario, Canada. Payments for projects related to Fuel Cell research we also obtained from The Austrian Funds for the Support of Scientific Research (FWF), especially by granting the Erwin Schrödinger Stipend to one of us, who could then pursue his scientific and publishing efforts at the University of Ottawa within the ESTCO Energy Inc. grouping.

For our publishing effort we required the library facilities and the help of the personnel of the Institute of Inorganic Technology of the Technical University Graz and we wish to acknowledge the excellent and untiring work of Mrs. Karin Grand, who provided nearly all the graphs and drawings in computer formats. Related costs were carried by the Institute administration and by private sponsors.

Last but not least we are very grateful for the efforts of the Publisher, the VCH-Verlagsgesellschaft and its personnel, insisting on a very high standard of quality, editing and proof-reading the manuscripts, with our special thanks to Dr. Barbara Böck for her patience.

Karl Kordesch
Günter Simader

Contents

Abbreviations

ABS	acrylonitrile butadiene styrene
AFC	alkaline fuel cell
BAM	Ballard advanced materials
CAAA	Clean Air Act Amendments
CEC	Commission of the European Communities
CRIEPI	Central Research Institute of Electric Power Industry
DEMS	differential electrochemical mass spectroscopy
DMFC	direct methanol fuel cell
DOE	US Department of Energy
DSM	Demand Side management
ECTDMS	electrochemical thermal desorption mass spectroscopy
EMF	electromotoric force
EMIRS	electro-modulated infrared reflectance spectroscopy
EPRI	Electric Power Research Institute
EPSI	Ergenics Power Systems Inc.
ERC	Energy Research Corporation
FCV	fuel cell vehicles
FFFC	alkaline falling film fuel cell
HER	hydrogen evolution reaction
HFCP	HERMES fuel cell power plant
ICE	internal combustion engine
IEM	ion exchange membrane
IFC	International Fuel Cells Corporation
IGFC	integrated fuel cell/coal gasifiers
IMHEX	internally manifolded heat exchanger
IRMCFC	internal reforming molten carbonate fuel cell
IRP	integrated resource planning
KTI	Kinetics Technology International
LCP	least cost planning
MCFC	molten carbonate fuel cell
MELCO	Mitsubishi Electric Corporation
NECAR	new electric car (Daimler-Benz)
PAFC	phosphoric acid fuel cell
PEFC	polymer electrolyte fuel cell
PEM	proton exchange membrane
PEMFC	proton exchange membrane fuel cell
PSA	pressure swing adsorption
PTFE	polytetrafluoroethylene
SOFC	solid oxide fuel cell
SPE	solid polymer electrolyte
UTC	United Technology Corporation
ZEV	zero emission vehicle

1. Introduction

1.1. Fuel Cell Technology: a Dream, Challenge or a Necessity?

The last two decades of the 20th Century can be considered as a transition time for the methods of energy production, storage and conversion. The fossil fuels, coal, oil and natural gas, which were responsible for the – now in retrospect – nearly unbelievable development of technology in the Western world and increase of mobility of mankind, are now considered as dangerous for the survival of the natural environment as we know it. At the same time, the fear is often expressed that the rapid usage and the resulting disappearance of fossil fuels will stop the future progress of technology at a time when the increasing world population needs far more food, better housing, improved industrial products and extended means of transportation and communication.

The dilemma is caused by the realization that the previous concept of unlimited energy available through nuclear power generation turned out to be a very dangerous proposition. The situation is worsened by the fact that the hopes for a substantial global energy fraction supplied by solar energy, or other even smaller renewable sources, are also unrealistic.

Several authors have described the future of the world as critically energy dependent. The pessimistic scenarios predict human catastrophes and solutions based on forced energy savings by "going back to the basic life styles". More optimistic views consider the effect of new technologies leading to a better utilization of fossil fuels in addition to the use of solar and other renewable energies, including the proper use of atomic energy.

1.2. The Historic Trends in Fuel Cell System Developments

The wish to convert the chemical energy of fossile fuels directly into electricity already existed around 1900, and resulted in surprisingly large scale experiments trying to oxidize coal and coal gas electrochemically in "piles". We can also find hydrogen fuel cell systems described in the literature of that time which look very much like the H_2–O_2 space power cells of the 1960's. The demise of all these attempts was caused by material problems. The systems were poorly designed and their operating life-time was short.

After 1920, the gas-diffusion electrode was recognized as the key for successful low temperature operations. A. Schmid was the pioneer who built the first platinum-catalyzed, porous carbon-hydrogen electrodes in tubular form. Paired with air-electrodes of a similar design, practical fuel cells were built. The second WW stopped most of the research in the area of direct energy conversion. The internal combustion engine ruled the transportation field, followed by jet-engines in the air and gas turbines as power plants.

In the UK, however, F.T. Bacon worked on an alkaline fuel cell system with porous metal electrodes, and after the war his work was published in the still "closed" literature. His fuel cell system was the first prototype of the later NASA Space Fuel Cells, which enabled men to fly to the moon in 1968. Research in the fuel cell field became exciting again, porous carbon as substrate with low catalyst loadings proved to be a low cost solution for hydrogen-air fuel cells on earth and the interest in electric automobiles propelled by fuel cells became widespread, leading to several prototypes. In 1970, K. Kordesch built a hydrogen fuel cell/battery hybrid vehicle for 4 passengers, which was operated for 3 years in city traffic.

The mid-1970s showed an interesting change in the direction of fuel cell technology. The alkaline system, which had reached the highest level of development in the space programs, was replaced in world-wide research and development efforts by the phosphoric-acid system, which was seemingly better suited for stationary power plants. Parallel to these efforts, the development of reformers became necessary as hydrocarbons would be the preferred fuels. This trend to large-scale power plants was especially noticeable in Japan after a loss of interest in the USA. Power plants in the 50–100 kW and up to 10 Megawatt sizes achieved acceptable life times for prototypes.

However, due to their obviously better overall efficiency together with the heat from a high temperature plant, the development of molten carbonate fuel cell systems in the 1980s and of solid oxide fuel cells in the 1990s accelerated. Unfortunately, the life expectancy problems have still to be solved for the high temperature fuel cells.

Another surprising turn in technology occurred in the 1990s. The membrane fuel cell system appeared as the most attractive object for development. This system was already in existence in the 1960s, but did not perform reliable in the space fuel cell projects and its importance fell behind the alkaline systems. High power densities were obtained as a result of new membrane types and catalyst research. Operating life expectancies improved also considerably. One large draw-back remained: the high cost of the membranes and the expensive auxiliary system for heat and water removal.

1.3. Current Projects and Applications for Fuel Cells

1.3.1. Fuel Cell Technology Programs in the USA and Canada

The US-Department of Energy (DOE) has expended nearly $ 300 million to date for the phosphoric acid fuel cell (PAFC) technology, including pilot plant demonstrations up to the 11 Megawatt size. This amount was actually matched in cost sharing programs with private corporations, who then commercialized this system. For example, ONSI Corp. began manufacturing 200 kW on-site co-generating systems in 1992. By 1993, nearly 60 units had been sold to electric and gas utility companies in the USA, Japan and Europe.

As a result, only an additional $ 18 million were authorized for further work programs on PAFC stationary systems in 1993. Preferential attention is now being given to the higher temperature fuel cell systems, which have a higher overal efficiency.

Molten carbonate fuel cell (MCFC) units are presently being produced on a pilot-plant scale by the Energy Research Corp. (ERC). 2000 hours of cumulative operating time had been already achieved in 1992. In 1993, ERC completed a 120 kW stack (using internal reforming of natural gas) and operated it for 250 hours at an efficiency of 50%. This stack was a prototype component for the 2 Megawatt power plant to be installed in 1995 in Santa Clara (CA) (total cost $ 46 million operational in 1998). Commercial production totaling 100 MW of units is also expected then. The capital cost projections for 2 MW size MCFC power plants are now $ 3000/kW and are estimated to drop to $ 1000/kW by the end of the century

In the future, coal-gas fueled MCFC systems with a special gasifier will also be built. M-C Power Corp., with a production capacity of 2 to 3 MW of units per year, is engaged in several projects in California involving 250 kW stacks for prototype testing. The Electric Power Research Institute (EPRI) and the Gas Research Institute (GRI) are co-sponsors.

Despite these efforts, significant improvements are needed to reach a commercially viable state, and the DOE issued a $150 million procurement program for complete power plant developments in 1993. The Department of Defense (DOD) also joined the program. The target in the USA is to achieve commercial availability of MCFC power plants for operation on natural gas by the year 2000, and on coal-derived gas by 2010.

Solid Oxide Fuel Cells (SOFC) in the USA follow essentially three paths. The tubular design of Westinghouse Electric Corp. with 1 to 2 m long tubes, fuel gas feed (anode) outside, cathode inside and yttrium-stabilised zirconium forming the solid electrolyte structure. A monolithic design (a corrugated sequence of electrodes and electrolyte) is used by Allied Signal Corp. Other companies follow with a thin planar disk arrangement. The US-DOE has spent $ 64 million, Westinghouse and Japanese partners $ 76 million since 1991 in a 5-year cooperative agreement. After first testing 25 kW modules (cell bundles), the results should be some 100 kW and a Megawatt-scale prototype in 1996. Some small units have achieved 35000 hours of operating time. DOE supports also the monolithic SOFC program. 50000 hours of operation are required for commercial plants.

Alkaline fuel cell systems are not considered to be suitable for power plants, at present and no efforts are being made in the USA. The development of alkaline fuel cells in the 0.1 to 0.5 kW range (portable units for demonstrations) is being pursued by Astris, Inc. in Mississauga, Canada.

Solid Polymer Electrolyte (SPE)[1] Fuel Cells, also known as Proton Exchange Membrane (PEM) fuel cells, Ion Exchange Membrane (IEM) fuel cells or simply Polymer-Electrolyte Fuel Cells (PEFC's), have a long history going back to the Gemini space programs. Due to the development of new membrane materials (DOW Chemical Co.), they are now capable of high current densities at low temperatures. They are best suited for smaller applications, such as buildings, vehicles etc. In Canada, Ballard Power System built a prototype

---

1 SPE is a TM of General Electric Co.

transit bus which was demon-strated in Vancouver in 1994 ($ 10 million were available for that project; see Section 1.5. Fuel Cell Operated Electric Vehicles).

Some remarkable progress has also been made recently with fuel cells utilizing methanol directly. The cell voltage "barrier" of 0.4 V has been overcome with UTC membrane cells producing over 0.7 V. Fuel utilization is still low, but a milestone was set in 1994.

1.3.2. Fuel Cell Activities in Japan

The phosphoric acid system is still dominating fuel cell development efforts. Toshiba and the US-based International Fuel Cell Co. (IFC), with its subsidiary ONSI, manufactured and installed an 11 Megawatt PAFC plant at the Tokyo Electric Power Co. plant in Goi (1991). 56 ONSI PC-25 plants of 200 kW size have been ordered by 26 customers in 10 countries. 36 units had been installed by 1994. Newer ONSI units will have an improved, higher heat level output, therefore a better overall efficiency. Several Japanese Power Companies are installing test units in the 50 to 500 kW size. These have been built by Japanese manufacturers, with the Government providing one third of the cost. Fuji Electric Corp. alone is said to be capable of producing 15 MW of fuel cell units per year. A 5 MW Fuji PAFC power plant (cost shared with the PAFC Technology Research Association) is scheduled for delivery to a group of electricity and gas companies in 1995. Mitsubishi Electric Co. will have three 200 kW plants installed in 1994/95. Life-times have passed 6800 cumulative hours (reported by ONSI). This is not long enough for commercial plants and therefore prototype testing is being continued.

Molten Carbonate Fuel Cells are now also receiving more attention in Japan and Hitachi Ltd., Mitsubishi and other companies are obtaining Government support for 100 kW modules. Toshiba researched one-meter square units and has developed its own MCFC system. The Fuel Cell Development Information Center listed about $ 50 Million for MCFC support in 1993. This is about 10 times the support for SOFC, and 25 times the support for the Solid Polymer Fuel Cell. The factual situation, however, is changing rapidly and is following the 1994 outlook, recognizing new and abandoning old trends and seeking more international cooperation.

Solid Oxide Fuel Cell Research and Development in Japan in 1993 still concerned material studies and basic design work, and systems of only several 10 kW capacity were projected for 1996/97. This schedule will certainly be changed in view of US and European efforts.

1.3.3. Fuel Cell Programs in Europe

The only Alkaline Fuel Cell (AFC) system considered for commercialization was built by ELENCO N.V. in Belgium (Hugo v.d. Broeck announced by the end of April 1995 that ELENCO N.V. discontinued its fuel cell business). Some research and development efforts are going on in Sweden at the Royal Institute of Technology in Stockholm. Work at the national IFP in France stopped in the 1980's. In Germany, the efforts of the Siemens AG in connection with the European Space Shuttle HERMES (abandoned in 1993) have been discontinued. The development of AFCs in the Eastern countries (especially in the former USSR) have also been stopped. Some work on oxygen electrodes is continuing for Zn-Air cells.

PAFC development work on a larger scale was never carried out in Europe. Only AEG worked on acidic fuel cells but discontinued their efforts. However, during the last 10 years large licensing and marketing efforts have been apparent. IFC and the Italian Co. Ansaldo have joined to market 200 kW PC-25 PAFC plants, and presently about 10 units have been installed. Two stacks, forming a 1 MW PAFC plant, are scheduled to be tested in Milan. A 50 kW Fuji stack is also being tested there. Other 25 kW Fuji-built PAFC units are to be operated at the ENEA Research Center in Rome.

Extending and shifting their interests, Ansaldo is now also active in MCFC system developments. 2 or 3 MW of molten carbonate fuel cell stacks per year are to be fabricated. New MCFC installations in Spain are also being planned by Ansaldo.

The MCFC system has a long history in the Netherlands. During the 1950s and 60s, Broers and Ketelaar operated the first molten electrolyte cells at 650 °C. In the mid-80s, a national program was initiated by the Dutch Government. An agreement between the Chicago-based Institute of Gas Technology (IGT) and ENC, the Netherlands Energy Research Foundation, was reached in 1985. A lifetime of 5000 hours was reached in 1992 with a 2 kW stack consisting of 20 cells. Industrial partners joined ENC to form Brandstofcel Netherland (BNC) with an initial budget of about $ 30 million. Natural gas-fed and coal gas-fed demonstration units of 10 to 25 and up to 250 kW are planned and, after proof of production feasibility commercialization is scheduled to be started in 1996.

Solid Oxide Fuel Cell development was carried out very actively in Europe. However, BBC and Dornier in Germany discontinued their efforts about 1980 and only some high temperature electrolysis work continued. The Commission of the European Communities (CEC) had to start again with basic programs in 1985–87 in Germany, Greece, the Netherlands, Spain and Portugal. A direct methanol/ethanol fuel cell program was also started in 1988 (probably now superseded by the recent US-results). At the present time (1994), SOFC development in Europe is gaining speed and funding by the European Community Energy Commission is rapidly increasing in the face of the US and Japanese efforts.

The research and development work on PEM Fuel Cells has reached a sudden peak, also as a result of the successful demonstrations in Canada and the R&D fundings in the USA. Siemens AG in Germany is building fuel cell stacks for submarines on a government contract. The license has been obtained from International Fuel Cells (IFC), Canada. (See also Section 1.5. Fuel Cell Operated Electric Vehicles).

1.4. Fuels for Fuel Cells: More Reformer Development

Hydrogen plays an important role in many of the realistic, industrially-oriented scenarios – not as an energy source, but as an energy carrier, a means for short or long-time energy storage and as an intermediate product of chemical or electrolytic processes. Pure hydrogen is such a desirable, widely-usable substance because it has a high energy content and reacts with air or oxygen forming pure water.

With respect to safety, hydrogen is no different to the previously widely-used city gas (household gas with 40–60 % H_2 and up to 8 % CO) as far as flammability is concerned – which is in the same range as low fractions of gasoline, methane and propane. The general draw-back is that hydrogen must be transferred in pipes, stored as a compressed gas or as a cryogenic liquid. The alternative choice is the electrolysis of water on-site or the chemical production from other H_2-containing "fuels" by "reforming". Such fuel carriers can be natural gas, alcohols or ammonia. New technologies provide solid metal hydrides – in the future they may be preferred for electric vehicles powered by fuel cells.

PAFC systems use steam reforming for the conversion of hydrocarbon fuels and alcohols. The reformer unit is integrated into the fuel cell system to make the best possible use of heat exchanging possibilities. The reformer process usually goes through several catalytic steps, the final ones shifting the gas composition to a mixture of H_2 and CO_2 with a minimum of CO. The CO_2 can be separated to enrich the gas supplied to the fuel cell electrodes, and it can be disposed of (as some concepts suggest, as liquid or solid CO_2) or at least partly re-utilized for the manufacture of organic chemical products. Electrolytic cells for that purpose have been designed. If high temperature heat is available (e.g. from nuclear reactors), a re-conversion with steam to CO containing fuel gases (or even liquid fuels) is a conceptual possibility, starting another useful oxidation cycle.

The important advantage of reforming hydrocarbon fuels to hydrogen, instead of just burning them in a heat engine or turbine, is the higher efficiency of the fuel cell cycle. The result is simply less CO_2 per produced kWh.

MCFC power plants do not need separate steam reformers due to the high temperature at which the conversion can be carried out internally at the electrodes. SOFC systems can have an even higher overall efficiency because the off-heat may be used in co-generation systems. This statement would seem to oppose the thermodynamic rule that a low temperature fuel cell has the highest efficiency. However, for the end purpose all practical applications must be considered. If only electric power is the desired product, the low temperature fuel cells (the AFC or the PEFC) are still the most useful ones, for example, for electric vehicles.

1.5. Fuel Cell Operated Electric Vehicles

In the 1970s only alkaline fuel cell systems were considered for use in electric vehicles. The only AFC vehicle program of the 90's was the City Bus project in Belgium, using ELENCO hydrogen-air fuel cells. The presentation of a prototype bus was scheduled for summer 1994 in Amsterdam. The project was discontinued, despite of the earlier participation of large European companies. The prospects for more efficient and longer-living power plants for vehicles have shifted to the PEFC systems.

The PAFC system, which was for a certain time the favorite choice in the USA because it can use reformed methanol as fuel, has lost its attraction. It took several years to bring the prototype to perform with a Fuji PAFC. The bus was finally presented in 1994, but is only one of the originally scheduled vehicle fleet. The system is a hybrid design, using a secondary battery for peak performance. The future of the project is uncertain, despite claims that the performance and economy of a diesel bus were obtained. The cost of the PAFC city bus was calculated to approach the cost of a regular transit bus over a period of 20 years. This fact may save the PAFC as a possible vehicle propulsion system.

One very serious obstacle to the application of membrane systems at the present time is the extremely high cost estimates (up to \$ 20000 per installed kW). It is expected that cooperation between PEM fuel cell manufacturers and membrane producers (Ballard, DOW-Chemical Corp. signed a contract in 1993) can be of help in this respect. Many companies experimenting with PEMFC systems are still using the NAFION™ membranes of DuPont and lower resistance membranes from Asahi (Japan) and recognize that the membrane alone is not the decisive component. An improvement in the water removal system or in the pressurized air loop can have very large effects on the overall cost, efficiency and operating life of the units. Hybrid designs using secondary batteries for peak performance are generally considered to be unavoidable.

Ballard Power Systems "Zero Emission Transit Bus" is presently the most sought after prototype for a PEMFC vehicle. It does not use a hybrid system. It uses compressed hydrogen gas and can carry 21 passengers over a distance of 160 km. It looks like a diesel transit bus and is supposed to match diesel performance at a lower electricity cost than an electric trolley bus. In the development program, which is supported by the British Columbia Transit Co., the BC Government, the Canadian Government and California's South Coast Air Quality Management District, the use of a methanol reforming system is also being considered.

A current highly-commented PEMFC project is the Energy Partners "Green Car"™.

A combined pool of industry grants is said to be able to provide the capital for equipment and research to take the technology from the prototype stage to a commercial product. The program started in 1991, and its goal is a (lead-acid) hybrid fuel cell vehicle. The three, 7 kW PEM hydrogen-air fuel cell stacks are to be built by Energy Partners, Inc. Compressed pure hydrogen is the selected fuel for the 125 V, 15 kW fuel cell system. The weight of the prototype is 1.5 tons and the city driving range is 100 km. Motor power is 35 kW.

In Europe the Mercedes Benz AG has started a large electric vehicle project on the basis of the licensed PEM Ballard fuel cell concept, but with new design goals aiming at principal improvements of the system (heat exchange, water removal, etc.). Research efforts concern the anodic degradation of the membrane materials, gas cross-leakage and, essentially, operating life problems due to catalyst poisoning. A reduction of cost by minimizing the amount of platinum catalyst is of course essential. At the Paul Scherrer Institute in Switzerland intensive basic studies of ion exchange membrane problems are being performed.

1.6. Outlook for Fuel Cell Technology

Fuel cell systems are important components of a rapidly emerging "Energy Technology", which has many industrial supporters but also an established competitive technology to match in cost and performance. The chances to become an essential part of the environmentally-benign energy supply technology of the next century are good, particularly for the dispersed and local fuel cell power plants which can produce electricity and heat. Co-generation will become routine planning for electricity networks. The types of fuel cell systems to be applied for specific uses will depend on economic considerations.

The competition from large conventional power plants will be strong and the clean combustion of fossil fuels will be a major goal worldwide, with the Western countries leading. Nuclear energy will be needed to keep global carbon dioxide within manageable limits. The combination with fuel cells can be a success by providing cheap electricity for electrolysis. Solar energy will have the same capability to provide hydrogen, but the cost question is still open. On the other hand, the conversion of all fuel types (also biomass as an example) to hydrogen by reforming is the key to a higher efficiency of universal usage – in interconnected networks, in large and dispersed power plants and for fuel cell vehicles.

The operational life of fuel cell systems will strongly influence the cost question, and therefore decide the competition between fuel cell systems. The concept of intermittent operation of low temperature fuel cells could ultimately change the whole vehicle fuel cell scenario. This is, at least, the personal opinion of the authors. Considering the turbulent history of fuel cells, such a "surprise" is definitely conceivable. The principle could easily be tested and extended to other fuel cell types.

Electric vehicles operated by fuel cells are visualized as the way to meet the growing human mobility requirements worldwide, replacing fossil fuels in the distant future. However, we should not overlook the fact that natural gas is a relatively clean fuel, and its abundance is guaranteed for centuries. A hybrid vehicle using a clean combustion engine and a secondary battery in parallel is undoubtedly a concept which is affordable and which requires no substantially new technologies to implement. It would also be accepted by existing motor vehicle lobbies. The improvement or replacement of the lead-acid battery by a more suitable rechargeable battery would tremendously help all hybrid concepts.

2. General Aspects of Fuel Cell Systems

2.1. Introduction "Fuel Cells as Energy Converter"

A fuel cell is an electrochemical cell which can continuously convert the chemical energy of a fuel and an oxidant to electrical energy by a process involving an essentially invariant electrode-electrolyte system [1]. Fuel cells work at high efficiency with emission levels far below the most strict standards.

Fuel cell systems have the advantage of being modular and can therefore be built in a wide range of power requirements from a few hundred Watts up to multi-kilowatt and megawatt sizes. This range allows the construction of highly efficient power plants at specific locations. Long term planning, based on predictions of expected increased needs, can be more flexible and the early investment in over-sized systems can be avoided. Because of their very low emission level, fuel cell power plants can be installed on-site, where energy is consumed, even in densely populated areas. As a result, power transmission lines are more economic and transmission losses are reduced.

The basic principles of a fuel cell are those of the well-known electrochemical batteries, which are involved in many activities of our everyday life. The big difference is that, in the case of batteries, the chemical energy is stored in substances located inside them. When this energy has been converted to electrical energy, the battery must be thrown away (primary batteries) or recharged appropriately (secondary batteries). In a fuel cell, the chemical energy is provided by a fuel and an oxidant stored outside the cell in which the chemical reactions take place. As long as the cell is supplied with fuel and oxidant, electrical power can be obtained (Figure 2-1).

Fuel Cell = Electrochemical energy generator (Chemical reaction producing electricity)
Primary Cell = Electrochemical energy producing device (one-way chemical reaction producing electricity)
Rechargeable or secondary battery = Electrochemical energy storage device (reversible chemical reaction producing or using electricity)

Despite their relatively long history and the progress attained in developments during the past 30 years, fuel cell have experienced very few applications up to now. The most important ones have been their use as main electrical energy sources for the manned spacecrafts of the Gemini, Apollo and Space Shuttle programs of the National Aeronautics and Space Administration (NASA) in the USA. Attempts to develop systems on earth in the mid to upper power range (50 kW–20 MW) were started in the 1980s. They are aimed at applications such as load leveling for energy utilities, for remote power plants, co-generation, industrial waste utilization or emergency power supply. In the

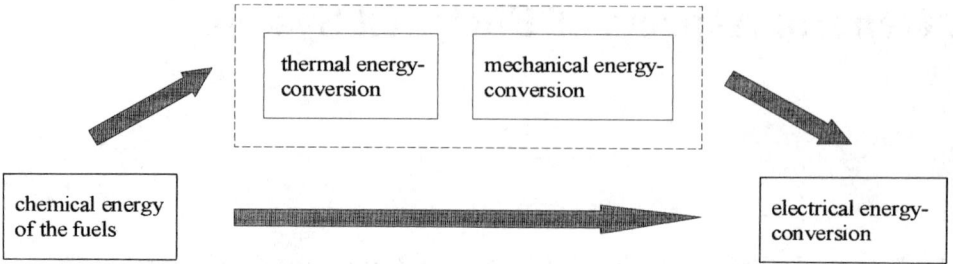

Figure 2-1. Direct Energy Conversion with Fuel Cells in comparison to Conventional Indirect Technology

lower power (kilowatt) range, studies focused on the development of fuel cell systems for electric vehicle propulsion and military purposes.

Fuel cells will probably play a key role in the future world energy scenario. Their most important characteristics, namely high efficiency and very low emission and noise levels, will be mandatory in the next generation of power plants. Should hydrogen be the main energy carrier of the 21st century, as many technical studies point out, then fuel cells will have undisputed advantages over all other energy conversion devices [2].

2.2. Classification of Fuel Cell Systems

There are several types of fuel cells and nearly as many types of classifications have appeared in the literature. An attempt has been made in Figure 2-2 to give a fairly general type of classification. All of the different types can be covered with the descriptions direct, indirect or regenerative fuel cells. In a direct fuel cell, the products of the cell reactions are discarded, whereas in a regenerative fuel cell the spent reactants are regenerated from the products by one of several methods as indicated in the table. The two types – direct and regenerative fuel cells – are comparable to primary and secondary batteries. A third type is the indirect fuel cell, and an example of this type is the reformer fuel cell using organic fuels which must be converted to hydrogen. A biochemical fuel cell is another example: a biochemical substance is decomposed by means of an enzyme in solution (sometimes supplied by adding bacteria) to produce hydrogen.

The bulk of fuel cell research is concerned with the development of primary fuel cells and converters. A further subdivision according to the temperature range in which the fuel cell operates would appear appropriate, and for this purpose the following classification has been adopted: low, intermediate, high, and very-high temperature fuel cells, operating in the ranges 25 to 100 °C, 100 to 500 °C, 500 to 1000 °C and over 1000 °C, respectively.

In each of these temperature ranges, the different types of fuel cells can be further subdivided depending on the type of fuel used (See Figure 2-2). Some of the fuels may be immediately available, like natural gas, or may be easily produced, like hydrogen. Among organic compounds, a variety of potential fuels are conceivable, e.g. hydrocarbons, alco-

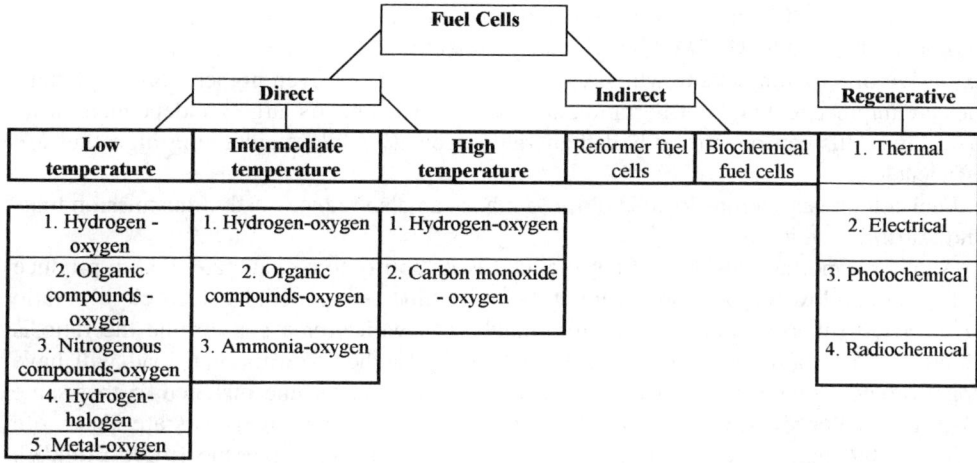

Figure 2-2. Classification of Fuel Cells

hols. Carbon (or graphite) can also be considered as a fuel. The nitrogen-containing fuels which have been used are ammonia, hydrazine and the methyl-substituted hydrazines. Oxygen, either pure or as air, is used in practically all fuel cells as the oxidant.

Another subdivision is possible from the point of view of the nature of the electrolyte used. This classification is not presented in Figure 2-2. From the temperature ranges, it can be seen that aqueous electrolytes are used mostly in the low-temperature range and (pressurized) at intermediate temperatures. Molten electroytes are used occasionally at intermediate temperatures, but usually at high temperatures; solid electrolytes (e.g. mixed oxides, where the oxide ion is the current carrier in the electrolyte) are used at very high temperature.

Much of the current research concerns regenerative fuel cells with electrical regeneration (H_2/O_2-cells). In this case, the electrochemical energy converter can also be used as an electrochemical energy-storage device (e.g. using compressed gases).

2.3. Characteristics of Fuel Cell Systems

Fuel cells offer advantages in efficiency, reliability, economy, cleanliness, unique operating characteristics, planning flexibility, and future development potential.

2.3.1. High Efficiency and Reliability

The fuel cell can convert up to 90% of the energy contained in its fuel into usable electric power and heat. Current phosphoric acid fuel cell (PAFC) designs offer 42% electrical conversion efficiency on a high-heat value basis, with 46% electrical conversion efficiencies for PAFC possible in the near-term through already well known science and engineering. The Electric Power Research Institute has estimated that advanced molten carbonate fuel cells (MCFC) may achieve electrical efficiencies greater than 60%, exclusive of a topping cycle, which would raise efficiencies even higher. "Such efficiencies are unprecedented." Furthermore, a fuel cell's efficiency is largely independent of its size. Fuel cells can operate at half their rated capacity while maintaining high fuel-use efficiencies.

Fuel cell power stations located close to loads can also reduce costly transmission lines and transmission losses.

Another important attribute of the fuel cell is its ability to co-generate; i.e., to produce hot water and low-temperature steam at the same time as it generates electricity. Its ratio of electric to thermal output is approximately 1.0, while for a gas turbine the ratio is about 0.5. This advantage means that a fuel cell matched to a thermal load will have approximately twice the electric output of a combustion turbine matched to that same load. In smaller sizes, and up to public utility power systems, fuel cells are also more efficient (by about a factor of two) when compared to, for example, the 15 000 Btu/kWh heat rate of a 2 MW combined cycle system. The fuel cell's load-following capability (while maintaining high efficiency) may also give it an advantage in co-generation markets with varying heat demands.

As they contain fewer moving parts, fuel cell systems should have higher reliability than combustion turbine combined-cycle systems or internal combustion engines. Fuel cells will not experience a catastrophic breakdown, such as occurs with a combustion turbine or internal combustion engine when rotating parts fail; they experience only a gradual loss of efficiency. Mature fuel cell systems are expected to have higher on-line availabilities than competing technologies.

2.3.2. Unparalleled Environmental Performance

Substitution of fuel cells for conventional power plants should improve air quality and reduce water consumption and waste water discharge. The conventional generation of electricity produces more particulates, sulphur oxides and nitrogen oxides than all other stationary industrial sources combined. Fuel cell power plant emissions are ten times lower than those specified by the most stringent environmental regulations. Fuel cells also produce lower carbon dioxide (CO_2) emissions than conventional generating plants, a question of increasing concern due to the so-called "greenhouse effect".

Because the electrochemical reaction of the fuel cell produces water as a by-product, little if any external water is required for power plant operation. This low water usage is in marked contrast to large steam power plants that require massive quantities of water for cooling. The quantities of waste water discharged from fuel cell systems are also lower, and the quality is superior compared with conventional fossil-fueled power plants, requiring no pretreatment prior to disposal in many communities. Fuel cells eliminate or reduce water quality problems associated with thermal discharge, power plant site run-off and the disposal of wastes from air emission control systems.

The quiet, electrochemical nature of fuel cells eliminates many of the sources of noise associated with conventional steam-powered systems and can easily comply with OSHA

(Occupational Health and Safety Administration) standards. No ash or large volume wastes are produced from fuel cell operation. Land requirements are acceptable, and connecting transmission corridors are not required as is the case with outside power sources. Fuel cells are among the least hazardous methods of energy conversion due to their comparatively small size, the absence of a combustion cycle, their state-of-the-art safety systems and low pollutant emissions.

Table 2-1. Test Results of NO_x Emission Levels[1]

Power Plants	Capacity [kW][2]	Fuel	NO_x [ppm]
NEDO (Kansai EPC)	1000	LNG	10
Tohoku EPC	50	LNG LPG	4,5 null[3]
Tokyo Gas	50	LNG	2
Okinawa EPC	200	Methanol	< 2
Shikoku EPC	4	Methanol	null
Forklift Truck	5	Methanol	null
Bus (DOE)	25	Methanol	null

2.3.3. Unique Operating Characteristics

Fuel cells have beneficial operating characteristics matched by no other technology. These characteristics save costs in meeting system operating requirements. Dynamic operating benefits include load following, power factor correction, quick response to generating unit outages, control of distribution line voltage and quality control.

The power conditioning unit of the fuel cell plant can be used to control real and re-active power independently. Control of power factor, line voltage and frequency can minimise transmission losses, and reduce requirements for reserve capacity and auxiliary electrical equipment such as capacitors, tap-changing transformers and voltage regulators.

When new generating capacity is added to an electric power system, substation equipment often has to be upgraded because of the expectation of increased fault current (which lowers the reliability of the electric system). However, with fuel cell power units it is not necessary to upgrade the fault-current interrupting capability of existing substation equipment, because the fuel cell power units have a short-circuit ratio of about 1:1. The fuel cell also controls a disconnect if necessary.

---

1 based on 7% O_2
2 All levels are measured under 100% load operation
3 Null means less than detection limit

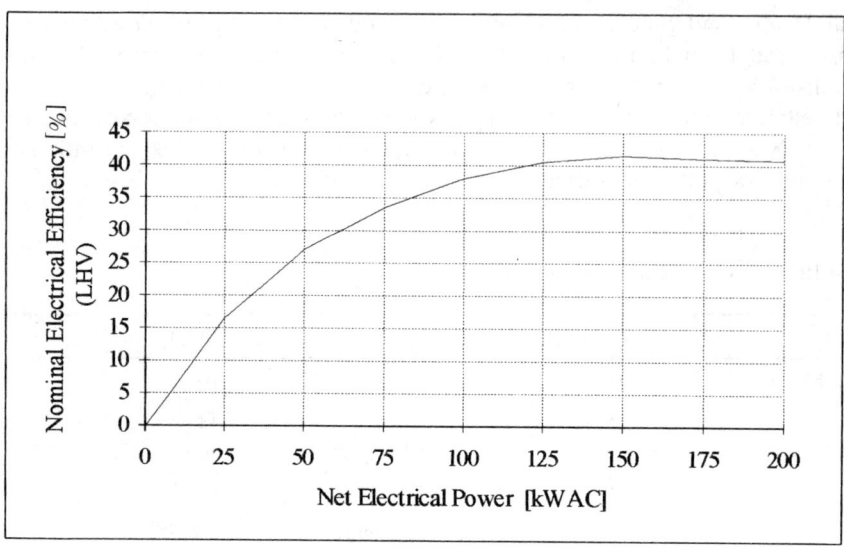

Figure 2-3. Part Load Characteristics

Fuel cells can easily regulate any deviations from system frequency on an area-by-area basis. This benefit leads to an increase in overall system stability, and overcomes the present fuel penalties which result from the need to switch to other (often oil-fired) generating units to meet the deviations in frequency caused by localised variations in system load.

Fuel cell units have an excellent part-load heat rate and can respond rapidly to transient loads (see Figure 2-3). For example, the heat rate of a phosphoric acid demonstration unit is anticipated to be 8300 Btu/kWh at rated power and to increase slightly at 50% of rated capacity. It is also expected to be able to go from 30% of rated power to rated power in only 7 seconds. Spinning reserve requirements can thus be lessened when fuel cells are used.

One of the recurring problems facing energy control dispatchers is how to operate the system during periods of high load pickup, while still maintaining a reserve or regulating margin. Dispatchers typically meet this "regulatory margin" by loading (more expensive) units above their economic assignment level early in the pickup period. The cost penalty for these inefficiencies can be overcome by keeping fuel cells on-line longer and taking advantage of their high part-load efficiencies. Fuel cell power plants are designed for unattended operation.

2.3.4. Planning Flexibility

Planning flexibility, including modularity, results in strategic and financial benefits to the utility and its customers. Because fuel cell power plants may be built within two years of order, and because performance is largely independent of plant size, they can be used to increase utility system capacity by small increments in response to customer needs.

By better matching increases in electric demand, long periods of overcapacity are avoided and average fixed costs can be lowered. If demand growth is uncertain, the fuel cell's short lead time becomes even more valuable. A utility can slow down or accelerate its response to growth. In addition, as experience is gained with fuel cells, utilities may be able to reduce required reserve margins while maintaining the same reliability – again resulting in lower fixed costs.

The community benefits from the increased reliability of smaller units distributed throughout the grid. Placing generation at the distribution level reduces the likelihood of outages due to external transmission interruptions.

Though initial fuel cells will be designed to be fuelled with either natural gas or naphta (a widely available, low-cost refinery product), the cells require hydrogen. A chemical reformer that produces a hydrogen-rich gas allows the use of a variety of low-sulphur gaseous and liquid fuels. Advanced fuel cells will also be able to operate on gasified coal.

These planning flexibilities are not unique to fuel cells. Combined-cycle units and internal combustion engines each share some of these flexibilities. But combined with their other advantages, fuel cells are truly a unique energy resource.

2.3.5. Future Potential

The potential of fuel cells to break through the efficiency barriers now being faced by all conventional generation cannot be overemphasised!

As a developing technology, the phosphoric acid fuel cell should see significant improvements. More advanced fuel cell designs based on molten carbonate or solid oxides may be able to demonstrate quantum leaps in improvements, over the next 15 to 20 years.

The fuel cell's competitors, on the other hand, including gas turbines and internal combustion engines, are at a mature stage of development. Small incremental improvements are the most that can be expected from these technologies. And improvements in conventional technologies will come at the expense of higher operating temperatures and therefore greater nitrogen oxide air pollution.

2.4. Economic Benefits of Fuel Cell Systems

From the point of view of electricity producing companies, it was simply axiomatic that the generation and distribution of power should be regulated within large scale systems. These would be able to adjust to periodic or unusual power demands by centralized management and if needed, by transfer of power between the electricity networks – relying on transcontinental and sometimes decades – old international purchase contracts.

The flexibility and even the capability needed to provide more power to new and rapidly developing areas is limited, because it takes a lead time of at least 10 years to build conventional power plants. There is the possibility of using single-cycle combustion turbine plants in the 1–10 MW class, which can be installed within 2 or 3 years, but they are comparably inefficient and, if operated with hydrocarbon fuels, they produce higher levels of nitrogen oxides and are therefore environmentally less acceptable.

The use of fuel cells can make an important contribution to the problems of local and diversified power production. The first fuel cell power plants in the class of 1 to 10 MW can be installed within a year, and they can also operate at partial load (fitting the particular demand schedule) with high efficiency. From the standpoint of capital investment needs for industrial projects, this represents very considerable savings because the output of fuel cells can be added in accordance with the local requirements, even on a yearly basis. This also obviates the construction of expensive and sometimes objectionable transmission lines and high charges for short periods of peak power demands.

Such flexibility is not possible with conventional, capital-intensive, large base-load power plants. In the USA, the demand for electric power rose in the late 1990's, commensurate with industrial planning, but the present scheduled construction of new power plants by utility companies lags far behind the projected needs. There are several reasons, but high construction cost and uncertainties in geographical siting and future load distribution possibilities are the main causes for the delay in building new facilities. This is especially true for nuclear power plants, where new safety measures and regulations are to be imposed. In addition, older (coal) power plants will have to be closed due to their non-compliance with increasingly stringent laws for reducing pollution levels. Their relatively low efficiency and poor earnings rarely permit any refitting to modern standards.

An assessment of fuel cell benefits to utilities was published as early as 1986 in a report prepared under EPRI sponsorship by Temple, Baker and Loans [3]. However, the following years did not bring the introduction of fuel cell power plants as expected. The reasons are of a technical and financial nature.

W. Schaller [4] has made a list of requirements for the introduction of fuel cells and the criteria which the European utility companies will demand for their acceptance. The summary actually sounds trivial:

"Fuel cell power plants are granted all the frequently discussed advantages in efficiency, industrial, social and environmental benefits, but they must be available at an acceptable capital cost level per installed kW, pro-rated over the years of their life expectancy."

The total power plant cost will, of course, have to include the components needed to make it grid compatible. For PAFC systems, for example, this includes the reformer unit and the dc/ac converter. In a specific case for a 100 kW PAFC system, the fuel cell may represent only about one third of the cost. With all the proper credits applied to fuel cell plants, the total costs must be compatible with those of existing decentralized power plants, such as gas motors, diesel engines or small turbines.

The most sensitive factor is the service life expectancy of the fuel cell electrodes, or their permissible degradation slope with certain operating schedules: continuous or intermittent load profile, peak load conditions or nearly open circuit idling of the fuel cell. Knowing such data accurately – and having them recognized by the manufacturer and user alike – is important in order to be able to separate the cost of the conventional installation components (e.g. the heat exchanger system) from the fuel cell electrode assembly (stack, unit or module) replacement needs.

The capital investment and operating cost considerations are strongly influenced by the initially installed power level, the time of planning for later expansion to a higher output and for routine replacement of stack sections on a defined service schedule. This provides the basic number for all calculations. 40000 hours are usually claimed to be a reasonable, economically justifiable goal for PAFC systems, but it is definitely less at present and this is the main obstacle for the general introduction of fuel cells – except for special purposes which may serve as a very useful introductory application and as a help for further development. Some of these will be discussed in Section 2.5.

The tendency of manufacturers is therefore to make the electrode stack expend-able, and its production costs as low as possible, by using automated large-scale processes, low-cost materials (like carbon and plastic) and assuring that all noble metal catalysts are nearly completely recoverable by recycling. Simple calculations can be made by assuming that the expendable electrode stacks represent only about 1/3 of the real fuel cell value with the other components (pumps, fans, exchangers) represent-ing 2/3.

However, this fuel cell cost figure does not include the reformer and dc/ac converter. As mentioned previously, the total system cost (of a PAFC) installation is divided into 1/3 fuel cell, 1/3 reformer and 1/3 converter. The present converter technology will prob-ably fulfill the requirements of many years of plant life expectancy.

The reformer will require maintenance with respect to an exchange of the catalyst beds at specified intervals. The chemical industry has optimized reformers to have a high efficiency (65–75%), and the catalysts are now of relatively low cost. The total reformer efficiency, including the usable heat, is in the range of 85 to 93% and, from the point of view of the power companies, a reduction in the heat loss is an important factor. The lowering of insulation losses, maximum use of hot steam on one level, the use of low temperature effluents for warm water heating and the utilization of condensation heat etc., are all under consideration.

An important point for the use of reformers to produce hydrogen gas from natural gas or, in fact, any carbon-containing fuel (e.g. Biogas) is the possibility to separate carbon dioxide and possibly re-use or dispose of it. Several environmentally-benign feasibilities have been discussed. If a CO_2 tax is introduced in several countries (as proposed in Ger-many and elsewhere), this reforming capability would be of financial interest for the electric power companies. The higher the initial efficiency of conversion, the less CO_2 is emitted as pollutant per kW electric power produced. In this case, fuel cells can be credit-ed with another cost reduction factor.

Molten carbonate fuel cells (MCFC) do not need a separate reformer. At the present state of technology only rough estimates about their energy balances are available, but it is certain that the heat available especially from the internal reforming systems will be more useful. However, the life-time expectancy question still overrides all aspects of cost assessment. The same can be said for the solid oxide fuel cells (SOFC).

Cost estimates for the proton exchange membrane fuel cells (PEMFC) vary from US-$ 30000 per installed kW down to several 100 US-$/kW. The range is wide open for speculation.

Alkaline fuel cells (AFC) were considered for power plant applications before deve-lopment of the higher temperature systems. There are still possibilities for small remote area power sources using types of biogas in a pre-reformer, as suggested by O. Lind-stroem [5], but it is not likely that a power company would consider such a system. The alkaline system has a relatively short life time in the range of 2000 to 5000 hours and is therefore not suited for a large installation.

The suggestion of Kordesch to use alkaline fuel cells "only in the operating mode" like a gasoline engine or a motor generator (which also has a life time of 2000 to 5000 hours) may become important for the design of fuel cell operated vehicles, especially due to the low cost of alkaline fuel cells with circulating electrolyte (See Section 7.6. Technical Out look and Conclusion).

2.5. The Potential Fuel Cell Market, Production Volume and Estimated Costs

Nowadays, the fuel cell community is trying harder to break its isolation to attract the interests of the society and the business world. Commercialization has been a leading theme during recent fuel cell conferences. Many scenarios have been presented with more or less optimistic market projections.

The fuel cell community has also become aware of the competition, and the levels of permissible user costs. Some pessimistic conclusions have emerged from such analyses. Price tags have also been put on fuel cell benefits by "monetizing externalities", which has had a positive impact on the balance of comparisons for the fuel cell side.

Figure 2-4 is reproduced from the report of Penner (1985) as a typical example of forecasts. This projection was invalid a few years later.

Figure 2-5 illustrates how the predicted date for market entry is pushed ahead as time goes by. The graph shows the predicted year for market entry as a function of the year of the forecast. Present predictions evidently promise to make PAFC, MCFC and SOFC commercially viable in the utility sector before the turn of the century.

Forecasts of the Japanese fuel cell community are also very optimistic, and confirm the positive statements of the last paragraph (see Table 2-2). The development of fuel cell

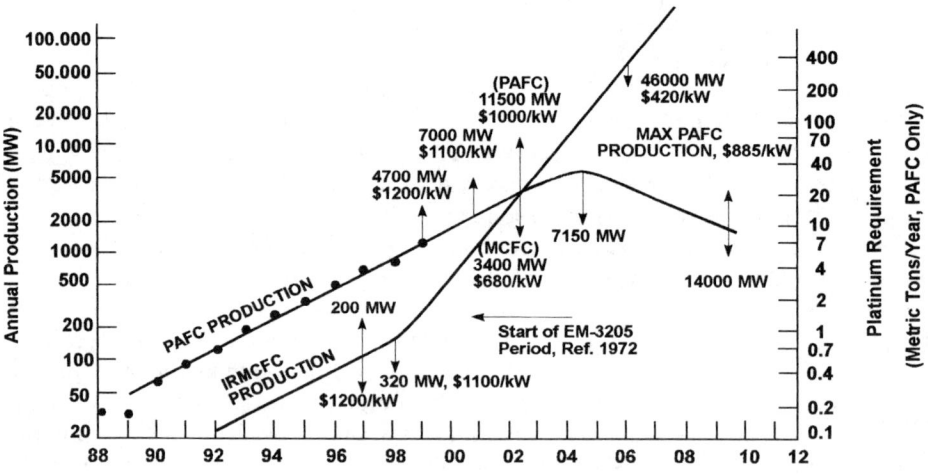

Figure 2-4. Example of Market Forecast [6]

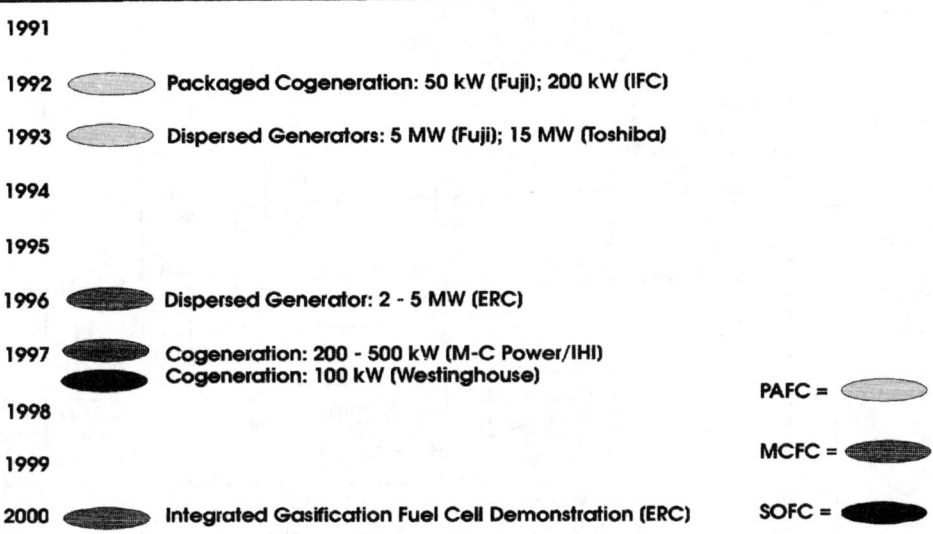

Figure 2-5. Forecast from the Fuel Cell Association [7]

Table 2-2. Target value of introduction of fuel cells [9]

	Year 2000	**Year 2010**
Commercial use	900 MW	2 800 MW
Industrial use	300 MW	2 400 MW
Utility use	1 050 MW	5 000 MW
Total	2 250 MW	10 700 MW

technology for utility usage (on-site and dispersed) has shown steady progress and gives hope for the future [8].

Viewed as a day-dream by some, the use of fuel cells as potential replacements for internal combustion engines in transportation is being actively considered and promoted. In 1987, the US Department of Energy initiated a program to develop PAFC-powered buses. This was followed in 1990 with a program to develop PEFC-powered automobiles [10]. Figure 2-6 shows a system diagram of a methanol-fueled, PEFC power source with a capacity of 60 kW. Fuel is reformed and converted analogously to the PAFC system, but an additional preferential oxidation step is needed to reduce the concentration of carbon monoxide to the low parts-per-million range. Unlike the 200 kW PAFC system, the PEFC system has to be pressurized and uses a turbo-compressor to pressurize the air supplied to the cathode and to recover the energy from the air leaving the cathode. This 60 kW PEFC

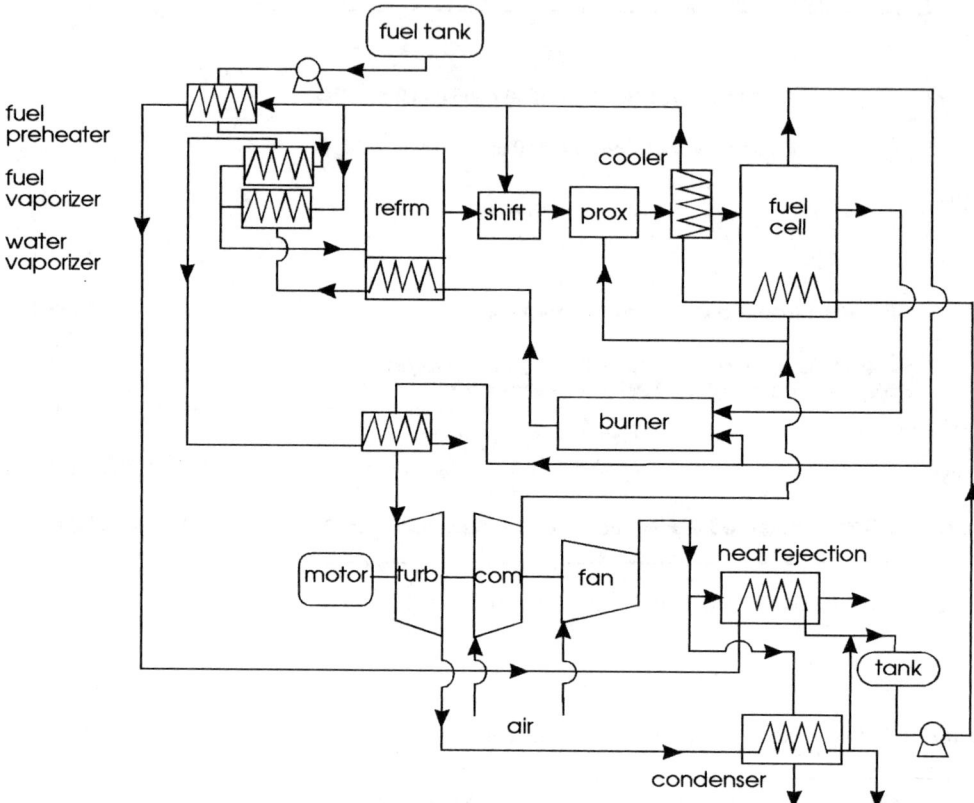

Figure 2-6. System Diagram of a Methanol-Fueled Polymer Electrolyte Fuel Cell Power Source for Transportation Appliations.

system is clearly more complex than the 200 kW stationary PAFC power plant, yet to be viable in transportation it would have to be manufactured for US-$ 50–100/kW.

A cost estimate by a major automobile manufacturer, based on the learning-curve experience in mass-producing automotive components, suggests that a PEFC system could be made for US-$ 90/kW if produced at a rate of one million units per year. Implicit in such a projection are technology improvements relative to the state-of-the-art PEFC systems. To put this projected cost in perspective, the relationship between the number of fuel cell systems manufactured and their cost is shown in Figure 2-7. The first 56 PAFC units from IFC/ONSI were sold for US-$ 2500/kW, although the actual costs were somewhat higher; the cost of the next series of 700 units with more advanced technology is expected to decrease to about US-$ 1500/kW [11].

Early SOFC systems are believed to have been sold for about US-$ 100000/kW, while prototype PEFC systems can now be obtained for approximately US-$ 25000/kW. On the log-log scale used in Figure 2-7, these data points correlate reasonably well with the US-$ 90/kW cost for one million fuel cell systems built per year. The slope of the line agrees with the generic approximation that production costs decrease by one order of magnitude as

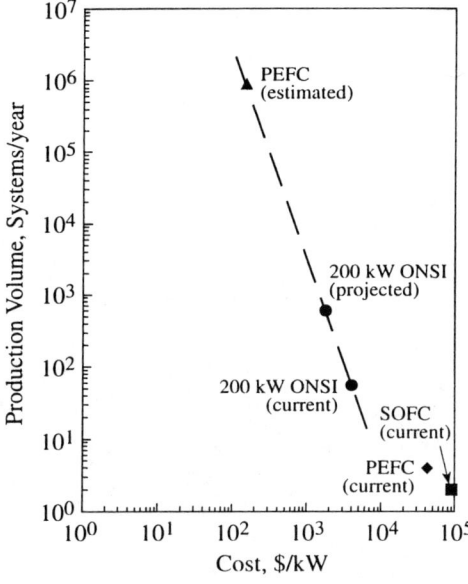

Figure 2-7. The Projected Relationship between the Number of Fuel Cell Systems Built per Year and their Manufacturing Cost

the production volume increases by two orders of magnitude. If these cost projections are borne out in the years to come, then fuel cell systems may indeed replace heat engines in many applications, and system complexity may well become a less critical issue than it is at present.

References for Chapter 2

[1] K. V. Kordesch, J. C. T. Oliveira, "Fuel Cells", Ullmann's Encyclopedia of Industrial Chemistry, Fifth Edition, VCH, Weinheim, Germany, Vol. A 12, pp. 55–83

[2] "Climate Change – The IPCC Response Strategies", Island Press, Washington, D.C., Covelo, California, 1991

[3] Anon. (Temple, Barker, and Loane, Inc.) "The Financial and Strategic Planning Benefits of Fuel Cell Power Plants", 2 Vols., EPRI EM-4511, Palo Alto, CA (1986)

[4] W. Schaller, "Die Interessenslage der Elektrizitätswirtschaft am Brennstoffzelleneinsatz", Symposium "Energieinnovation in der Elektrizitätswirtschaft und Elektrodindustrie", TU-Graz, Austria, 23. Jänner 1992

[5] O. Lindström, S. Schwartz, Y. Kiros, M. Ramanathan, A. Sampathrajan, "Small Scale AFC Stack Technology", 1994 – Fuel Cell Seminar, San-Diego (CA), pp. 323–324

[6] S. S. Penner (ed), "Assessment of research needs for advanced fuel cells", DOE/ER/30060-TL, November 1985

[7] Fuel Cell Association, PO BOX 66392, Washington DC 20035-6392, 1992

[8] G. Simader et al.; "Fuel Cell Developments in Japan", Final Report for the South Styrian Energy Cooperation, June 1993

[9] A. Fukutome, Journal of Power Sources **37** (1992) 53

[10] P. G. Patil, Journal of Power Sources **37** (1992) 171

[11] M. Krumpelt, R. Kumar, K. M. Myles, Journal of Power Sources, **49** (1994) 37

3. Basic Principles

3.1. Thermodynamics of Chemical Reactions [1, 2]

3.1.1. Free-Energy Change of a Chemical Reaction

Since it is more practical to carry out reactions at a constant temperature and pressure rather than at constant temperature and volume, the change in Gibbs free energy is more useful than the change in Helmholtz free energy.

The Gibbs free-energy change, ΔG, of the reaction

$$aA + bB \longrightarrow cC + dD \tag{3.1}$$

is given by the equation

$$\Delta G = c\mu_C + d\mu_D - d\mu_A - d\mu_B \tag{3.2}$$

where μ is the chemical potential ($\mu_i = \left(\partial G/\partial n_i\right)_{T,P,n_j}$; $j \neq i$) of the indicated species.

The free-energy change of a chemical reaction is a measure of the maximum net work obtainable from the reaction. It is equal to the enthalpy change of the reaction only if the entropy change, ΔS, is zero, as may be seen from the equation

$$\Delta G = \Delta H - T\,\Delta S. \tag{3.3}$$

In this connection, it is interesting to note that if in a chemical reaction the number of moles of gaseous products and reactants are equal, the entropy change of such a reaction is effectively zero. This is because the main contribution to the entropy change in a reaction is a change in translational entropies, and this is zero for a reaction involving no change in the number of molecules in the gas phase during the reaction. An example of a chemical reaction of this type is the oxidation of carbon to carbon dioxide:

$$C + O_2 \longrightarrow CO_2, \Delta G = -394.6 \text{ kJ/mol} \tag{3.4}$$

If the number of moles of gaseous products exceeds that of the gaseous reactants, the entropy change of the reaction is positive as a result of the increase in translational modes. Oxidation of carbon to carbon monoxide serves as a good example of this type:

$$2\,C + O_2 \longrightarrow 2\,CO, \Delta G = -137.3 \text{ kJ/mol} \tag{3.5}$$

For such a reaction, it follows from Eq. (3.3) that the free-energy change is more negative than the enthalpy change of the reaction.

More common are reactions in which the number of moles of gaseous reactants exceeds that of the products. In such cases, the entropy change is negative. The free-energy change is less negative than the enthalpy change for such a reaction. The gas-phase reaction between hydrogen and oxygen to produce water is of this type:

$$2\ H_2 + O_2 \longrightarrow 2\ H_2O,\ \Delta G = -237.3\ \text{kJ/mol} \tag{3.6}$$

(The free-energy and enthalpy changes for the various possible types of reactions carried out in fuel cells are presented in Table 3-1, see Section 3.6.3).

3.1.2. Standard Free-Energy Change of a Chemical Reaction

The chemical potential of any substance may be expressed by an equation of the form

$$\mu = \mu^0 + RT \ln a \tag{3.7}$$

where a is the activity of the substance and μ has the value μ^0 when a is unity. The standard free energy change ΔG_0 of the reaction (3.1) is then given by Eq. (3.2), with the chemical potentials of all species replaced by their standard chemical potentials:

$$\Delta G^0 = c\mu_{C^0} + \text{d}\,\mu_{D^0} - a\mu_{A^0} - b\mu_{B^0} \tag{3.8}$$

Substitution of Eq. (3.7) for each of the reactants and products, and Eq. (3.8) into Eq. (3.2) gives

$$\Delta G = \Delta G^0 + RT \ln \frac{a_C^c a_D^d}{a_A^a a_B^b} \tag{3.9}$$

For a process at constant temperature and pressure at equilibrium, the free-energy change is zero. With the free-energy change in Eq. 3.9 as zero, it follows that

$$\Delta G^0 = -RT \ln \frac{a_{C,e}^c a_{D,e}^d}{a_{A,e}^a a_{B,e}^b} = -RT \ln K. \tag{3.10}$$

The suffixes e in the activity terms indicate the values of the activities at equilibrium, and K is the equilibrium constant for the reaction.

The importance of a knowledge of ΔG^0 is that it allows ΔG to be calculated for any composition of a reaction mixture. Knowledge of ΔG indicates whether a reaction will occur or not. If ΔG is positive, a reaction cannot occur for the assumed composition of reactants and products. If ΔG is negative, a reaction can occur.

3.1.3. Relation between Free-Energy Change in a Cell Reaction and Cell Potential

The enthalpy change of the reaction may be written as

$$\Delta H = \Delta E + P\,\Delta V = Q - W + P\,\Delta V \tag{3.11}$$

where ΔE, Q and W are the internal energy of heat absorbed and work done by the system. If reaction (3.6) were carried out in a heat engine, then the only work done by the system would be the work of expansion, and Eq. (3.11) reduces to

$$\Delta H = Q. \tag{3.12}$$

The enthalpy change of the reaction is therefore equal to the heat evolved from the re-action, which is then the heat absorbed by the system.

If the same reaction is carried out electrochemically, W in Eq. (3.11) is not only the work of expansion of the gases produced, but is also the electrical work involved in transporting the charges around the circuit from the anode to the cathode – at which the potentials are $V_{rev,a}$ and $V_{rev,c}$, respectively. The maximum electrical work that can be done by the overall reaction (W') carried out in a cell, involving the transfer of n electrons is

$$W'_{el} = ne(V_{rev,c} - V_{rev,a})$$
(3.13)

for a hypothetical case, in which the internal resistance of the cell and the overpotential losses are negligible. To convert to molar quantities, it is necessary to multiply W'_{el} by N, the Avogadro number. As the product of electronic charge and Avogradro's number is the Faraday, it follows that

$$W_{el} = nF(V_{rev,c} - V_{rev,a}).$$
(3.14)

The only forms of work involved in the operation of the electrochemical cell are electrical work and work of expansion, therefore

$$W = W_{el} + P\,\Delta V.$$
(3.15)

If, in addition, the process is carried out reversibly,

$$Q = T\,\Delta S.$$
(3.16)

Using Eqs. (3.11) and (3.14) to (3.16), it follows that

$$\Delta H = T\,\Delta S - nF(V_{rev,c} - V_{rev,a}).$$
(3.17)

Comparison of Eq. (3.17) with Eq. (3.3), which holds for an isothermal process, results in:

$$\Delta G = -nF(V_{rev,c} - V_{rev,a})$$
(3.18)

Noting that

$$\left(V_{rev,c} - V_{rev,a}\right) = E,$$
(3.19)

Eq. (3.18) becomes

$$\Delta G = -nFE,$$
(3.20)

where E is the electromotive force of the cell.

If the reactants and products are all in their standard states, it follows that

$$\Delta G^0 = -nFE^0,$$
(3.21)

where E^0 is the standard electromotive force, commonly referred to as the standard reversible potential of the cell.

3.1.4. Temperature and Pressure Coefficients of Thermodynamic Reversible Cell Potentials

The variations of E with temperature at constant pressure can easily be obtained by using

$$\left(\frac{\partial \Delta G}{\partial T}\right)_P = -\Delta S \tag{3.22}$$

and Eqs. (3.20) and (3.21):

$$E = -\frac{\Delta H}{nF} + T\left(\frac{\partial E}{\partial T}\right)_P. \tag{3.23}$$

From Eqs. (3.10), (3.22), and (3.23), it can be seen that the second term in Eq. (3.23) is related to the entropy change of the cell reaction by the equation

$$nF\left(\frac{\partial E}{\partial T}\right)_P = \Delta S. \tag{3.24}$$

Equations (3.23) and (3.24) show that cell potentials at any temperature can be calculated theoretically, if data for entropy changes of the corresponding cell reactions are available.

Using

$$\Delta G_{P_2} = \Delta G_{P_1} + \Delta nRT \ln\frac{P_2}{P_1}$$

and Eq. (3.20), the thermodynamic reversible cell potential may be expressed as a function of pressure:

$$E_P = E_{P_0} - \Delta n\frac{RT}{nF}\ln\frac{P}{P_0}, \tag{3.25}$$

where E_P and E_{P_0} are the cell potentials at (total) pressures of P and P_0 respectively. Δn is the change in the number of gaseous molecules during the reaction. It is assumed that the gaseous reactants and products obey the ideal gas laws. The more general expression for the thermodynamic reversible cell potential as a function of pressure, as obtained from Eqs. (3.25) and $\left((\partial G/\partial P)_T = V\right)$, is

$$E_P = E_{P_0} - \frac{1}{nF}\int_{P_0}^{P} \Delta V\, dP, \tag{3.26}$$

where ΔV is the volume change of the reaction. According to this equation, it follows that the effect of pressure on the thermodynamic reversible cell potentials is small for reactions involving only liquids and solids. However, in the cases where gaseous reactants and products are involved, and if the volume change is significant (e.g., in the hydrogen-oxygen fuel cell), pressure effects must be taken into account. For example, increasing the total pressure from 1 to 10 atm in the H_2-O_2 fuel cell changes the cell potential by 45 mV.

3.2. Some Electrode Kinetic Aspects of Electrochemical Energy Conversion

3.2.1. Electrode Polarization (Overvoltage) – Nomenclature

The losses in galvanic elements under the conditions of delivering current are determined by the kinetics of the electrode reactions, by the physical structure (geometry) of the cell and by the type of electrolyte used. These losses have been called "polarization" since the early days of electrochemistry. This expression summarized a multitude of phenomena, which were related to the kinetics of the electrode reactions and to the state of the dynamic equilibrium. Specifically, they were influenced by the ionic mobility of the reaction partners (by the mass-transport situation in porous structures) and even responded to factors in the cell design – for example to the dependance of the cell resistance on the geometrical shape of the electrodes.

Instead of the expression "polarization", the more descriptive word "overvoltage" is often used. This expression can now be used for the current producing (discharge) process of a galvanic cell, or for the current uptake (charging, electrolyzing, plating) process in an electrochemical system. From a practical point of view, the "overvoltage" is the voltage difference which is measured between the "open circuit voltage" and the "terminal voltage" under the conditions of current flowing in either direction. The terminal voltage is therefore also termed "closed circuit voltage". On discharge, the terminal voltage is lower, on charge the terminal voltage is higher than the open circuit voltage. The "overvoltage" can simply be considered to be a measurable value for the losses appearing as a result of the current flow in either direction.

In what way the current flow was originated by the voltage differences and, vice versa, how the current flow in turn influenced these voltage differences are theoretical questions which were extensively discussed in the text books of physical chemistry and electrochemistry during the 1960s and 1970s [3].

3.2.2. Types of Overvoltage

Useful amounts of work (electrical energy) are obtained from a fuel cell only when a reasonably large current is drawn, but the cell potential will be decreased from its equilibrium potential because of irreversible losses. There are several sources that contribute to irreversible losses in a practical fuel cell. The losses are often called polarization, overpotential, or overvoltage (see Figure 3-1).

a) **Open Circuit Voltage** describes the voltage which can actually be measured at the terminals of an idling cell and is a commonly-used value for a cell (at a given time) under no-load conditions. The value is usually different from the often calculated – on the basis of thermodynamic data – "theoretical open circuit voltage", which is the same as the "electromotoric force" (EMF) of a cell and strictly determined under convention-defined standard conditions. There is also the expression "open circuit volt-

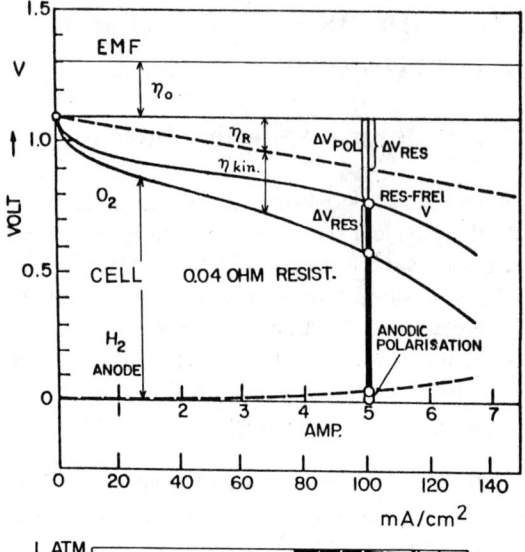

Figure 3-1. Polarization curves of a H_2-O_2 cell illustrating the components of the cell voltage [4].

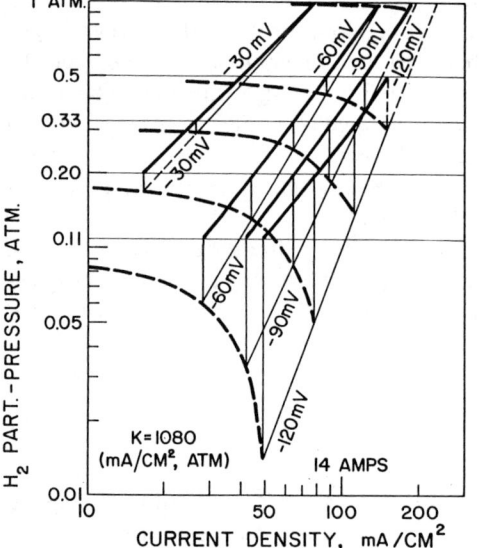

Figure 3-2. H_2 pressure drop within electrode: [5]

age difference". This describes the differences between similar cells under various conditions, and occasonally also the differences between the theoretical and actually measured voltages of the same cells.

b) **"Activation-Overvoltage"** is an expression for the voltage loss caused by the fact that the charge transport (charge transfer) in any material or process has a limited speed. The limitations which are considered here are those which typically can be influenced by applying an "activation catalyst". Temperature usually has a large influence on this type of overvoltage. Example: Catalysis of hydrogen peroxide decomposition.

c) **"Concentration-Overvoltage"** stands for voltage differences caused by diffusion processes (pressure gradients, changes in the usage rates of gases or liquids). The delay in reaching steady-state conditions, or the absence of equilibrium conditions as a result of the current flow using up or producing materials, are sources of concentration differences. Other parameters are, for example, the porosity of materials which influence the gas or liquid flow, or the permeability of membranes changing the ionic flow.

d) **"Reaction-Overvoltage"** is a term for the voltage difference appearing when an earlier or simultaneous (related) chemical reaction produces another compound which changes the operating conditions – for example, the water produced in a hydrogen-oxygen cell dilutes the electrolyte, which then causes an electrolyte concentration change at the electrode interface.

e) **"Transfer Overvoltage"** is unfortunately used vaguely in many connections and should be limited to the description of the changes in potential under load conditions as a fundamental interrelation of current and potential. Its theory is discussed in the section "Electrode characteristics".

f) **"Resistance-Overvoltage"** has no correlation with any chemical process at the electrodes and is simply the voltage drop across the resistive components of the cell. There is a differentiation to be noted here: the ohmic resistance of the electron-conductors (metals, carbon) and the resistance of the ionic conductors (electrolytes), which mathematically show the same essentially linear dependance (Ohm's law) but are based on ion-mobility characteristics. This is apparent when measurements are made with alternating currents at different frequencies or with d.c. pulse currents. The speed of charge transfer by electron flow compared with ionic flow is about 100 : 1. This difference is important for electrode reactions, which always occur on the interfaces between conductors and electrolytes. If the electrodes are porous, the supply of ions and the charge transfer is usually the limiting process. Fuel cell electrodes having a three-phase interphase (solid conductor, electrolyte, gaseous reactants) may also have a gas supply limitation determining the maximum current flow.

Resistance measurements can also be in error because the interfaces between the electrode surfaces and the electrolyte act as a capacitor (pseudo-capacitor). Such an "electrolytic capacitor" can store far larger amounts of electricity than a physical capacitor consisting of two plates and gas or vacuum as dielectricum. New "super-capacitors", which can be discharged in fractions of milli-seconds contain high surface-area materials (carbon, Ru-oxides) as electrodes and approach the storage capacity of batteries [6].

These two groups of overvoltages can be measured and separated by means of a circuit which loads the cell with a d.c. current and, at the same time, with a superimposed a.c. current of a defined frequency. More conveniently, a pulse current circuit can be used. This separates the "fast electronic" voltage differences which follow Ohm's law practically instantaneously, and measures the sum of all the "slow" polarization differences (of electrochemical origin) between the pulses. Such a measuring device, called the "Kordesch – Marko Bridge" was developed in 1960 and has found very wide use in battery and fuel cell technology [7, 8].

The circuit measures the voltage of a cell under load (the load being the average current of the pulses) but eliminates the voltage drop caused by the current flowing

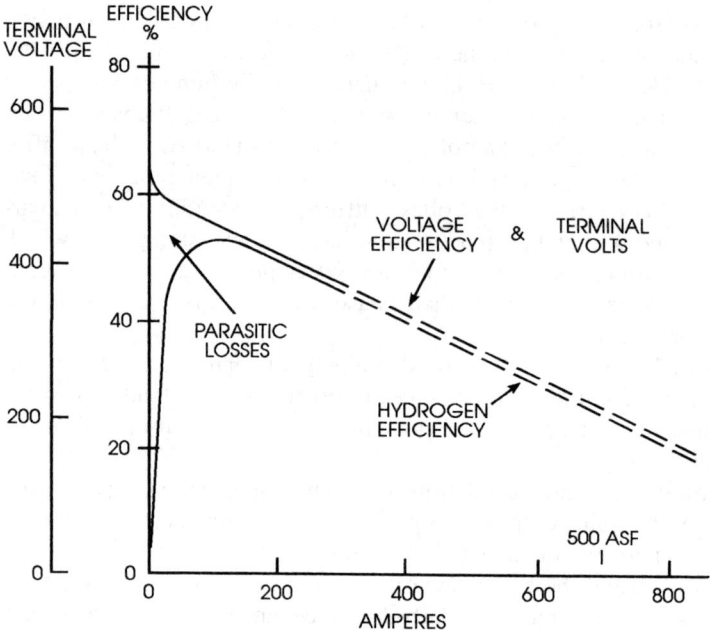

Figure 3-3. Fuel Cell power-plant terminal voltage and power-plant efficiency comparison at an electrolyte temperature of 150 °C [4]

through the rapidly responding "ohmic" resistances (solid electrode structure and the bulk electrolyte). The measured voltage is "resistance-free". The slow electrochemical processes cannot change significantly (only 1 to 2% in 1 millisecond) during the interruption period, and the "resistance-free voltage" is actually an "open circuit voltage under load conditions". The analytical value of such measurements has also been used extensively, also in sensor circuits for determination of the state of charge of batteries.

g) "Voltage Efficiency as Measurement of Fuel Consumption"
This consideration is often applied as a rough estimate for fuel cell efficiency. An example describing the performance of the General Motor's "Electrovan" fuel cell system is shown in Figure 3-3.

3.2.3. Electrode Characteristics

At current densities in the range between 0.1 to 1 mA/cm^2, resistance-polarization and ion concentration differences can be neglected for practical purposes. However, the changes in absolute or partial gas pressure affect the electrode potential. For a reversible hydrogen electrode supplied with a gas mixture, the potential is dependent on the partial pressure of H_2 in the fuel gas. Nernst's law is applicable.

The relationship between the current delivered by an electrode and its change in potential under that load is sometimes discussed as "transfer-polarization" (-overvoltage). However, theoretical treatment of the situation can be based on consideration of an equilibrium state, where current flow is a function of the voltage difference and, in turn the voltage difference determines the amount of current flow. If we assume that no other overvoltage type is present, which is the case if the current is small and therefore deviation from the equilibrium state is also small, then we can write the following equation for the current/voltage relation of a single electrode:

$$i = i_0 \left[\exp \frac{(1-\alpha) zF\eta_D}{RT} - \exp \frac{-\alpha F\eta_D}{RT} \right], \tag{3.27}$$

where i_0 is the exchange current density, α is the transfer factor and z is the number of electrons participating. For the case of a large overvoltage (practical case), the second part of the equation, which is the reaction in the opposite direction, can be neglected and the following equation results for the anodic current range i_+

$$i_+ = i_0 \exp \frac{\alpha zF\eta_D}{RT}. \tag{3.28}$$

The same relation is applicable for the cathodic current range i_-

$$i_- = i_0 \exp \frac{-(1-\alpha) zF\eta_D}{RT}. \tag{3.29}$$

If the two equations (3.28) and (3.29) are combined in a logarithmic form, and all constants are expressed by the values a and b, then the simple "Tafel equation" is obtained. It was originally determined experimentally and only much later confirmed theoretically.

$$\eta_D = a + b \log i. \tag{3.30}$$

This indicates, that the overvoltage needed to cause a certain current flow to or from an electrode is a measure of the thermodynamic irreversibility of the electrode reaction. The equation can be deduced by considering the discharge of hydrogen-ion in equilibrium with hydrogen gas (Erdey-Grzuz-Volmer-reaction) and also from the electrochemical theories developed by Butler-Volmer and Horiuti. If the overvoltage is negative, then electrons flow in the direction to the solution (de-electronation). If positive, then charges are accepted from the solution (electronation). Expressed in a different form,

$$i = A \exp(B\eta/RT), \tag{3.31}$$

the Tafel equation indicates how small potential differences can cause large changes in current flow. On the other hand, if the applied voltage is intentionally changed by even small amounts, reaction rates can change tremendously. The explanation can be found in the structure of the "Double Layer" existing at the interface of electrode and electrolyte. In a diagram, the slope of the function η versus log i is very helpful in finding the steps in a reaction sequence which are determining the speed of the overall reaction.

A special form of the Butler-Volmer-equation is applicable if the equilibrium potential change is small (overvoltage is less then 0.005– 0.01 V). The linear equation:

$$\eta = (RT/i_0 F) i = \text{proportional } i. \tag{3.32}$$

then applies. R, T and F are constants and therefore the slope of the line going through i_0 is the exchange current density, an important value for judging and comparing electrochemical reactions.

The third possibility, with the overvoltage considered to be zero, is applying the Nernst equation. It was believed for a long time that all reaction schemes could be explained thermodynamically and all overvoltage effects would obey Nernst's law. Research in the field of catalysts produced strong evidence of the value of electrokinetics, which includes concepts of transport mechanisms across interfaces and in electrolyte solutions.

3.3. The Electrode Mechanism of Fuel Cells

Although, there are several types of fuel cells, the past has shown that the H_2–O_2 fuel cells are of major significance. Especially the Alkaline Fuel Cell (AFC), which was used in space (space shuttle) and the Phosphoric Acid Fuel Cell (PAFC), which is mainly used in co-generation systems terrestrially, form the two best known systems. The electrode mechanism of these two types will be shown in more detail[1].

In an electrochemical cell making use of gaseous reactants, the anodic and cathodic reactants are fed into their respective chambers and an electrolyte layer is situated between the two electrodes. The half cell reaction at the anode yields electrons, which are transported through the external circuit and reach the cathode. These electrons are then transferred to the cathodic reactants. The circuit is completed by the transport of ions from one electrode to the other through the electrolyte. The electrode reactions of a hydrogen-oxygen fuel cell in acid or alkaline electrolyte, and the overall reactions, are as follows:

In acid electrolyte

Cathode:	$O_2 + 4\ H^+ + 4\ e^- \longrightarrow 2\ H_2O$	$E_0 = 1.229$ V
Anode:	$H_2 \longrightarrow 2\ H^+ + 2\ e^-$	$E_0 = 0.000$ V
Total:	$2\ H_2 + O_2 \longrightarrow 2\ H_2O$	$E_0 = 1.229$ V (3.33)

In alkaline electrolyte

Cathode:	$O_2 + 2\ H_2O + 4\ e^- \longrightarrow 4OH^-$	$E_0 = 0.401$ V
Anode:	$H_2 + 2\ OH^- \longrightarrow 2\ H_2O + 2\ e^-$	$E_0 = -0.828$ V
Total:	$2\ H_2 + O_2 \longrightarrow 2\ H_2O$	$E_0 = 1.229$ V (3.34)

The basic electrochemical cell consists of two electrodes separated by an electrolyte. The energy conversion reactions occur at the electrodes, with fuel oxidation at the anode and

---

1 PEFC: Breakthrough in Performance data came in 1987 (electrode reactions are shown in Section 4.3.)

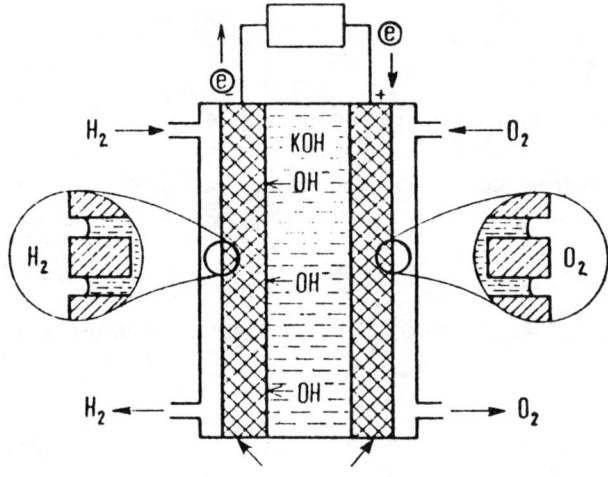

$$H_2 \rightarrow 2H^+ + 2\bar{e} \qquad\qquad 1/2\, O_2 + H_2O + 2\bar{e} \rightarrow 2OH^-$$

$$2H^+ + 2OH^- \rightarrow 2H_2O$$

Figure 3-4. Principle of the Alkaline Hydrogen Oxygen Cell

oxidant reduction at the cathode. The hydrogen oxygen fuel cell with alkaline electrolyte is schematically shown in Figure 3-4.

Hydrogen enters the pores of the anode and reaches the reaction zone, where gas, liquid electrolyte (aqueous potassium hydroxide solution) and the solid conducting structure meet. The hydrogen gas diffuses to the electrochemically-active site (e.g., a surface coated with a platinum catalyst), where it is adsorbed and dissolved by the electrolyte. It then dissociates and ionises to $2\, H^+$. The protons react with the hydroxyl ions of the electrolyte to form water, which dilutes the KOH electrolyte. Two electrons are made available to the electrical circuit. The hydroxyl ions used up are "replenished" from the cathode reaction, in which $1/2\, O_2$ reacts with water to produce $2\, OH^-$ ions, thereby taking up two electrons from the outer circuit. If the reaction is related to one oxygen molecule, the process produces a total of four electrons. In this case the following equations can be derived.

3.3.1. Anode Process

In comparison with the processes at the oxygen/air-cathode, it is considerably easier to interpret the oxidation of hydrogen molecules to hydrogen ions, because the potential at the hydrogen anode is in general constant. The overall reaction of the anodic hydrogen oxidation is in alkaline medium as follows:

$$1/2\, H_2 + OH^- \longrightarrow H_2O + e^- \qquad\qquad (3.35)$$

and in acid solution

$$1/2\ H_{2,ads} + H_2O \longrightarrow H_3O^+ + e^- \tag{3.36}$$

The result of numerous analyses divides the overall reaction into the following steps:

1. Transport of molecular hydrogen from the gas phase or from the electrolyte to the electrode and adsorption on the surface:

$$H_2 \longrightarrow H_{2,solv} \longrightarrow H_{2,ads} \tag{3.37}$$

2. Hydration and ionisation of the adsorbed hydrogen. The possibility of two different reaction mechanisms exists.
   a. Dissociation of the hydrogen molecule

$$H_{2,ads} \longrightarrow 2\,H_{ads}\ \text{(Tafel reaction)} \tag{3.38}$$

   with a subsequent Volmer reaction (ionisation and hydration on discreet points of the electrode surface):

$$H_{ads} + OH^- \longrightarrow H_2O + e^-\ \text{(alkaline)} \tag{3.39}$$

   or

$$H_{ads} + H_2O \longrightarrow H_3O^+ + e^-\ \text{(acid)} \tag{3.40}$$

   b. Hydration and ionisation proceed in one step (Heyrovsky-Volmer or Horuti-Volmer mechanism):

$$H_{2,ads} + OH^- \longrightarrow H_{ads} \cdot H_2O + e^- \longrightarrow H_{ads} + H_2O + e^- \tag{3.41}$$

   and

$$H_{2,ads} + H_2O \longrightarrow H_{ads} \cdot H_3O^+ + e^- \longrightarrow H_{ads} + H_3O^+ + e^- \tag{3.42}$$

   or

$$H_{ads} + OH^- \longrightarrow H_2O + e^- \tag{3.43}$$

   and

$$H_{ads} + H_2O \longrightarrow H_3O^+ + e^- \tag{3.44}$$

3. Desorption of products (H_3O^+, H_2O) and transport into the electrolyte (see Figure 3-5)

The validity of the mechanism depends in special cases on the surface structure of the electrodes and the current operating conditions. Special potentiostatic and potentiodynamic analyses on rotating and stationary ring-disc electrodes, as well as impulse methods are used for investigations. Evaluation of the determined cell potential/current density and charge curves allow conclusions to be drawn as to the possible electrode mechanism.

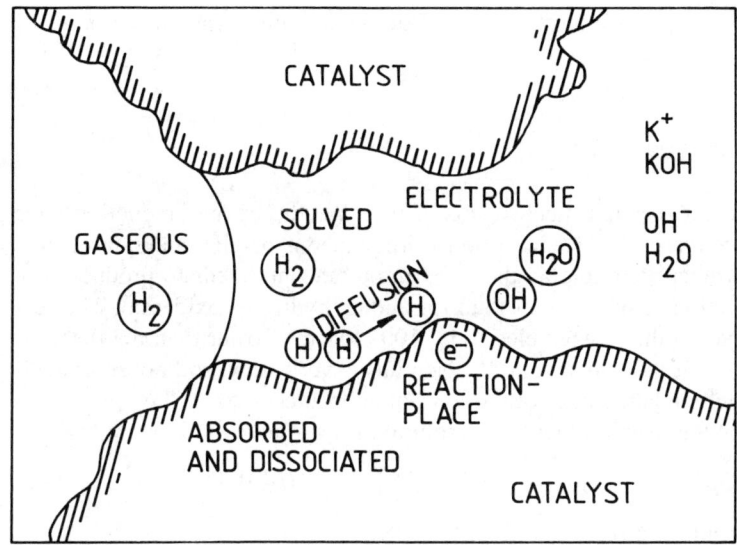

Figure 3-5. Reaction Process on the Three Phase Border of a H_2-Electrode

W. Vielstich stated that the metals can be divided into groups with regard to hydrogen oxidation. Metals such as Hg and Ag, on which hydrogen cannot be chemisorbed and therefore no oxidation reactions are possible belong to the first group. The second group includes the platinum metals, such as Pt, Ir, Rh and Pd, where the chemisorption and oxidation of hydrogen is less inhibited than in the first group, and transport processes limit the velocity of the reaction.

On investigation of the hydrogen electrode reaction, M. Kita proposed a periodic dependence of the exchange current density on the ordinal number of the metals.

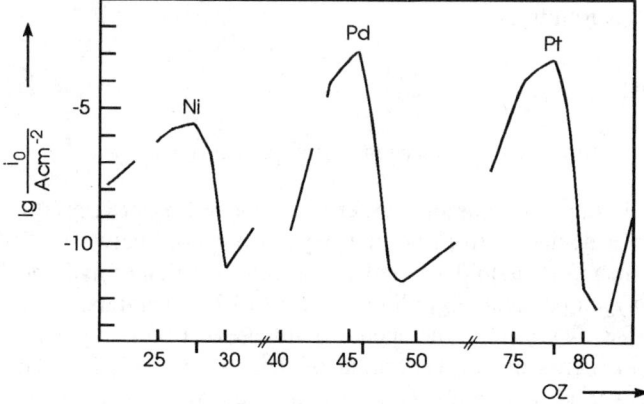

Figure 3-6. Periodic Dependence of the Exchange Current Density for hydrogen evolution on the work function of metals [9]

Transport phenomena form an important part of the kinetic processes in the pores of porous electrodes.

3.3.2. Cathode Process

Discussion of the different reaction mechanisms of the cathodic oxygen reduction have now continued for more than two decades, due to simultaneous parallel and consecutive reactions, forming hydrogen peroxide as the most important intermediate product. The number of possible reaction steps is larger than for anodic hydrogen oxidation. The reason for the low potential on the oxygen electrode (100–140 mV lower than the theoretical oxygen potential) is that the adjustment of potentials is very slow, and not reproducible, due to the very high bondage forces of the oxygen molecule.

The gross equation of the reaction in alkaline solution is as follows:

$$O_2 + 2\ H_2O + 4\ e^- \longrightarrow 4\ OH^- \qquad E_0 = 0.401\ V \qquad (3.45)$$

and the reaction in an acid solution:

$$O_2 + 4\ H^+ + 4\ e^- \longrightarrow 2\ H_2O \qquad E_0 = 1.229\ V \qquad (3.46)$$

At 25°C, a potential difference of 1.23 V to the hydrogen electrode would be expected. However only values between 1.05 and 1.15 V are obtained. This effect can be interpreted – after dissociation of the oxygen molecule – by the formation of hydrogen peroxide. The reduction:

$$O_2 + H_2O + 2e^- \longrightarrow HO_2^- + OH^- \qquad E_0 = -0.07\ V \qquad (3.47)$$

then follows. The standard potential quoted was corrected from -0.076 to -0.065 V, because detailed measurements on carbon electrodes showed that the potential for the reaction process according to the above-mentioned equation also depends on the oxygen partial pressure. Using the Nernst equation, the concentration dependency of the O_2/H_2O_2 electrode at 25 °C is as follows:

$$E = E_{0,H_2O_2} - 0,029 \log\left(a_{OH^-} a_{HO_2^-} \right) \Big/ \left(p_{O_2} a_{H_2O} \right) \qquad (3.48)$$

$E_{0,H_2O_2} = -0.07\ V$; a = activities; p_{O_2} = oxygen partial pressure.

The experimentally determined concentrations of the peroxyde ion were between 10^{-5} and 10^{-8} mol/l. This results in a deviation from normal oxygen potential between 150 and 240 mV. Active catalysts, such as Pt and Pd, can reduce the concentration of hydrogen peroxide to 10^{-11} mol/l. The oxygen potential according to Nernst (0.22 V) and the potential of the hydrogen electrode (0.83 V) result in the open cell voltage of 1.05 volt reached in practice. This increases with increasing concentration of the alkaline solution. The hydrogen peroxide will be converted and catalytically decomposed by consecutive reactions. The direct, four-electron mechanism only takes place on highly active catalysts at small current densities or at high temperatures.

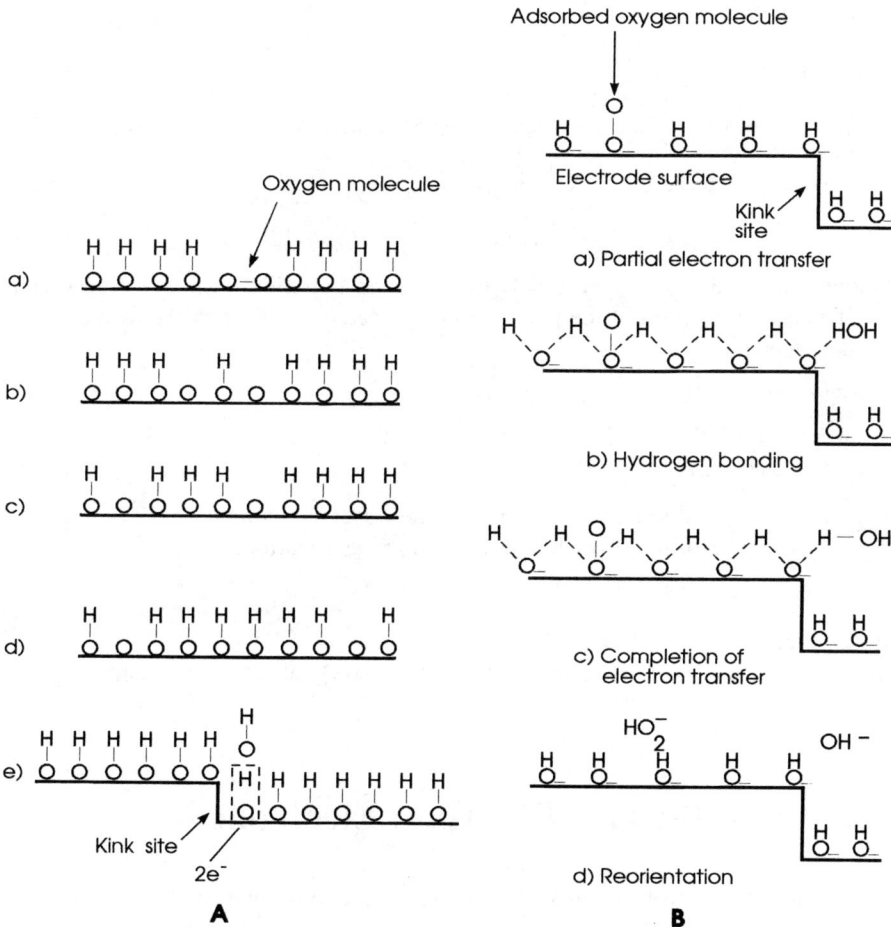

Figure 3-7. "Pseudo-splitting" models for oxygen reduction. (A) [10] (B) [11]

The following reaction mechanisms occurs for oxygen reduction in alkaline solutions.

$$O_2 + H_2O + 2e^- \longrightarrow HO_2^- + OH^- \qquad E_0 = -0.07 \text{ V} \qquad (3.49)$$

$$HO_2^- \longrightarrow OH^- + 1/2 \ O_2 \ \text{(decomposition reaction)} \qquad (3.50)$$

$$M-HO_2^- \longrightarrow OH^- + M-O \ \text{(intermediate reaction)}, \qquad (3.51)$$

whereby the following consecutive reaction has yet to be confirmed:

$$HO_2^- + H_2O + 2e^- \longrightarrow 3 \ OH^- \qquad E_0 = 0.867 \text{ V} \qquad (3.52)$$

In acid electrolytes:

$$O_2 + 2H^+ + 2e^- \longrightarrow H_2O_2 \qquad\qquad E_0 = 0.67 \text{ V} \qquad (3.53)$$

$$H_2O_2 \longrightarrow H_2O + 1/2\ O_2 \text{ (decomposition reaction)} \qquad (3.54)$$

$$M - H_2O_2 \longrightarrow H_2O + M - O \qquad (3.55)$$

$$H_2O_2 + 2\ H^+ + 2\ e^- \longrightarrow 2\ H_2O; \qquad\qquad E_0 = 1.77 \text{ V} \qquad (3.56)$$

From equations (3.49, 3.53) oxygen reacts to hydrogen peroxide with retention of the O–O-bond. The decomposition of the peroxide occurs catalytically with the formation of chemisorbed oxygen.

$$M - O + H_2O + 2e^- \longrightarrow M + 2\ OH^- \qquad (3.57)$$

or

$$M - O + 2H^+ + 2e^- \longrightarrow M + H_2O \qquad (3.58)$$

The energy of the M–O-bond is a measure for the observed voltage loss. The chemisorbed oxygen can be desorbed according to the following equation:

$$2M - O \longrightarrow 2\ M + O_2 \qquad (3.59)$$

and can once more react with H_2O or H^+.

A comprehensive summary of the description of the oxygen electrochemistry is given by K. Kinoshita [12].

3.4. Theory of the Gas Diffusion Electrode

The electrodes are the sites of the energy conversion reactions (fuel oxidation at the anode and oxidant reduction at the cathode). As reactants for fuel cells are usually gaseous, the most significant breakthrough in fuel cell technology has been the development of porous gas diffusion electrodes. Some fuel cells today, such as the monolithic solid oxide fuel cell, can provide power densities coming close to those produced in combustion engines.

The principle function of a porous gas diffusion electrode is to provide a large reaction zone area with a minimum of mass transport hindrance for the access of reactants and the removal of products. As electrode potential changes with the current density per unit area, a substantial increase in electrode area has an important effect on cell performance. This is the reason why, for example, smooth platinum electrodes can only be loaded at micro-amperes per square centimeter of surface, but porous electrodes can deliver amperes per cm^2 (geometrical) surface area.

In a porous gas-diffusion electrode, the contact zone where reactant, electrolyte and catalyst meet is called a three-phase zone. Large-area, three-phase zones can be achieved by the use of metal powders with a very high specific area ($100 \text{ m}^2/\text{g}$) or carbon (specific area $1000 \text{ m}^2/\text{g}$) for the manufacture of electrodes.

Porous electrodes can have surface areas which are orders of magnitudes larger than their geometrical ones. Fillers are sometimes mixed with the electrode components during manufacture and removed at the end of the manufacturing process, leaving voids that enhance the porosity of the structure. Depending on the basic structure (carbon or metal), the added binders and/or impregnation material used during their fabrication, the electrodes have a *hydrophobic* (carbon) or a *hydrophilic* (metal powder) surface.

Hydrophobic gas diffusion electrodes are made of fine carbon powder bonded with a plastic material, e.g. polytetrafluoroethylene (PTFE) (see Figure 3-8).

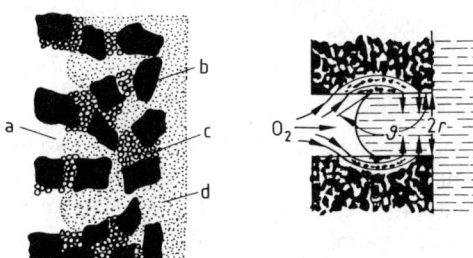

Figure 3-8. Section through a *Two-layer* Hydrophobic (PTFE-bonded) Gas Diffusion Carbon Electrode (δ = contact angle)
a) Gas; b) Carbon particle; c) Wet-proofing agent; d) Electrolyte

Carbon powders are light, have a large surface area and are suitable for the deposition of very active catalysts. PTFE bonded carbon electrodes are easy to manufacture on a large scale. To improve their conductivity, a wire screen as current collector is required. Without a metal collector, the conductivity of carbon is not sufficient for high currents (e.g. 200 mA/ cm^2) and only a bipolar flat-plate cell configuration is acceptable.

At least two layers can be distinguished in such PTFE bonded carbon electrodes: a highly hydrophobic, porous gas diffusion layer and an electrolyte-wettable thin layer. The electrochemical reactions take place at their interface, where the catalytically active material must also be present. The hydrophobicity of the diffusion layer prevents the electrolyte from penetrating deeper into the electrode structure, thereby keeping the pores free and facilitating gas access to the reaction sites.

Hydrophilic electrodes (see Figure 3-9) are made of sintered metal powders. The gas diffusion layer of these electrodes has larger pores than the reaction layer. Capillary forces retain the electrolyte in the small pores when an overpressure of the gases with respect to the electrolyte is applied. Porous metal electrodes are heavy, but have a high conductivity, which is a great advantage for monopolar plate electrodes where the current is collected from tabs at the edges of the electrodes. Raney metal structures provide a high catalytic activity at low temperature without using platinum.

The voltage of a single fuel cell is low (less then 1 V under load) and therefore many cells are packed together into multicell stacks or batteries, connected in series or parallel according to the voltage or current requirements. The stack construction also depends on the configuration of the system. A sequential stack design out of prefabricated units is, for example, necessary for the circulation of reactants and coolants in an *immobile* (matrix) electrolyte system and for the provision of electrolyte loops in *mobile* electrolyte systems.

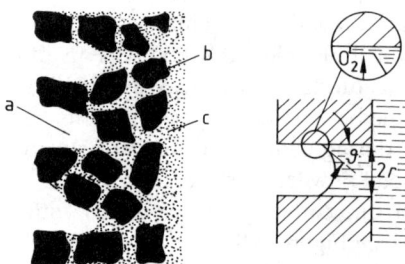

Figure 3-9. Section through a Hydrophilic Gas Diffusion Powdered Metal Electrode (ϑ = contact angle) a) Gas; b) Metal particle; c) Electrolyte

3.5. Electrocatalysts

Electrocatalysis may be defined as the acceleration of an electrode reaction by a substance which is not consumed in the overall reaction. The substance is generally the catalytically-active electrode surface (or catalyst), though, to a lesser extent, the influence of the electrolyte solution may also require consideration (example: catalysts act differently in acidic or alkaline solutions). Further, electrocatalysis can be classified as heterogenous catalysis because at least one step of the electrochemical reaction occurs at the electrode-solution interface. At a given electrode-solution potential difference, specifically the properties of the electrode surface affect the overall reaction rates.

In a multi-step electrochemical reaction in which several charge transfer steps must occur, the chemical processes are usually the ones which are kinetically slow, and therefore rate determining. Accordingly, as taught by the Butler-Volmer equation, the low exchange current density (which is an expression of the inhibited low speed reaction), i_0, must be compensated by a high overvoltage, η. The Butler-Volmer equation for $|\eta_\pm| > nF/RT$ is:

$$i_\pm = i_0 \left(a_{A_\pm}^a \big/ a_{A_\pm}^b \right)^{\alpha_\pm} \exp(\pm\beta_\lessgtr (nF/RT)\eta_\pm) \qquad (3.60)$$

where,

$i_\pm$ = current density as a measure of the reaction speed

i_0 = exchange current density (if small) as a measure for the inhibition of the electrochemical reaction .

$\beta_>, \beta_<$ = the anodic or cathodic charge transfer coefficients as a measure of the acceleration of the reaction by

η_+, η_- = anodic or cathodic overvoltage.

The chemical steps of an electrochemical reaction can be catalyzed by suitable heterogeneous catalysts, which are deposited on the outer surface of a compact (smooth) electrode or deposited on the inner surface of a porous electrode. There is no principal difference between "chemical" catalysis and "electrocatalysis". For specific cases, there are so-called "redox-catalysts", which open up a faster reaction path by changing the valency states of the chemical compounds involved. Such catalysts are RuO and mixed oxides of Co (e. g., perovskites and spinels). The group of heterogeneous catalysts (only accelerating parts of the reaction steps without a charge transfer) are represented by Pt and other Platinum metals or alloys, together with metals of the 8th group of the periodic system like Fe, Co and Ni in finely divided form (e.g. Raney-Ni). They catalyse the electrochemical reaction by accelerating splitting of the hydrogen molecule into H-atoms (and also the reverse reaction, combining H-atoms to H_2 molecules), which is then followed by ionisation.

$$H_2 \longrightarrow 2H^+ + 2e^- \tag{3.61}$$

The capability to catalyze these reactions can be correlated with the values for the adsorption enthalpy of hydrogen and the specific metals ("vulcano curves"). A medium adsorption enthalpy, as shown by Pt-metals, of approx. 230 kJ/mol is optimal.

Effective electrocatalysts can be found among the Group 8 metals and the semi-conductors. Their catalytic activity for hydrogen reactions correlates with the characteristic d-orbital features of these metals. The highest activity is shown by Pt, Pd and Ni. These metals also absorb hydrogen. Figure 3-10 shows the exchange current density for hydrogen electrodes consisting of such metals in relation to the molar free-reaction (standard) enthalpy of hydrogen adsorption consistent with the Tafel reaction. There are metals which adsorb hydrogen well and some which do not, although there are metals of poor catalytic activity in both groups. In connection with the cathodic oxygen reduction, catalysis of the decomposition of the hydrogen peroxide intermediate (to oxygen and water) is important. Heterogeneous metal oxide catalysts are also applied, here.

The action of heterogeneous catalysts is determined by the type of surface, the particle size and the crystal structure. Electronic interaction of the surface is responsible for the catalytic activity shown between the surface and the reaction partner. Active centers can be found at crystallographical faults or disturbances, structural edges with disturbed atomic or ionic bonds. The presence of acidic or alkaline groups can be of influence.

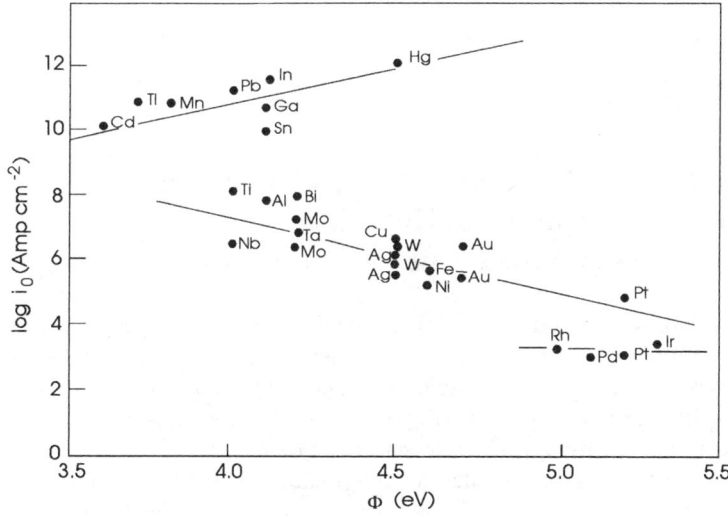

Figure 3-10. Dependence of exchange current density for hydrogen evolution on the work function of metals [13].

The main requirement for high catalytic activity is a large, accessible surface (porosity) or fine distribution of a powdered material. The reaction partners will be adsorbed at the catalyst surface and react at the interface. Gases may be absorbed in liquids wetting the surfaces. The reaction products are then desorbed. The acceleration is a result of new bonding possibilities at the surfaces, whereby it must be assumed that the original intramolecular bond structures in the adsorbed partner are weakened. Reactions which include the release or capture of electrons are catalyzed by substances which are themselves electron donors or acceptors (e. g. semi-conductors, certain metals).

Low-temperature fuel cells need catalysts for acceleration of the anodic and cathodic electrode reactions. In hydrogen-oxygen fuel cells with acidic electrolytes, platinum, palladium or tungsten carbide are commonly used for the hydrogen electrodes. The oxygen (air) electrodes usually contain either platinum, palladium, special carbons carrying spinels, phtalocyanines or porphyrines as catalysts. Platinum, palladium, nickel and nickelboride catalysts serve well for the oxygen electrodes in alkaline electrolytes. For cost-saving reasons, active carbon materials have these catalysts deposited on their large surface areas. Silver catalysts and spinels are also frequently incorporated [14].

Aging and depletion of a solid catalyst structure are caused by recrystallization processes, structural changes to a more ordered (less active) crystal type and certain phase or valency changes. Control of the deposition of impurities, which may act as "poisons" for the catalyst, is also important. Noble metal catalysts are especially sensitive to traces of carbon monoxide and sulfur compounds, which may often cover the surfaces irreversibly. Non-noble metal catalysts are usually less sensitive to these "poisons", one reason being that they are used in larger (not trace) quantities. Increased operating temperature are often beneficial for the life expectancy of the catalysts. A change from hydrogen operation to occasional oxygen operation can also "reactivate" deteriorated noble metal catalysts [15, 16].

3.6. Fuel Cell Efficiency

3.6.1. General Theoretical Efficiency of the Conversion of Heat Liberated from a Chemical Reaction into other Forms of Energy (Mechanical, Electrical)

At the present time, the common method of obtaining electrical energy is to convert the heat evolved during a chemical reaction to mechanical energy, which is then changed into electrical energy by means of a rotating generator. As a first step, the theoretical limitations of this method of energy conversion will be shown. The first stage is a chemical reaction (the combustion of a hydrocarbon to carbon dioxide), whereby a certain amount of heat is evolved because the heat content of the products is less than that of the reactants. This evolved heat (the enthalpy change of the reaction) causes a rise in temperature of the products and of any unconverted reactants. Since most of these are gases, they expand – a result of the increase in translational energy. In a heat engine, this

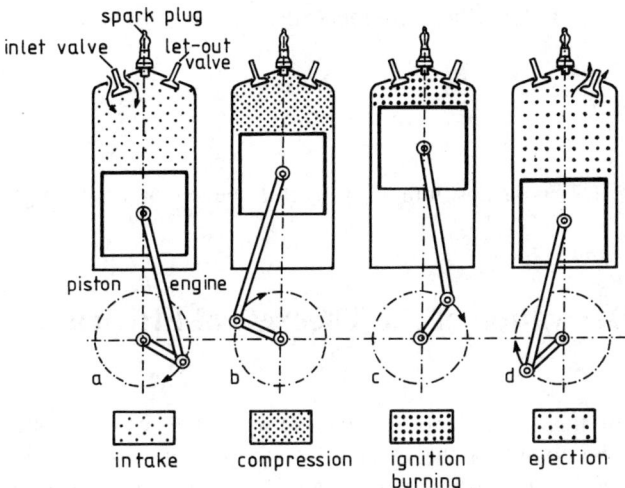

Figure 3-11. Internal Combustion Engine

expansion of gases produces mechanical work by causing the pistons to the engine move. The mechanical energy is then transferred to the wheels (see Figure 3-11).

The fact that the efficiency of this conversion is less than unity, even assuming an approach to an ideal system (by discounting the friction), can be seen from the following model. The essential act in the conversion of heat energy to work is the collision of gas molecules (which possess a high translational energy) with the piston. However, when two bodies collide, they do not transfer all their translational energy (cf. problems in elementary mechanics treating the collision of billiard balls). The energy that is not transferred to the piston in a heat engine appears as heat energy of the rejected gases. Hypothetically, the efficiency of a heat engine could be unity – but only if the kinetic energy of the rebounding gas molecules were zero. If this were so, the gases would leave the system at 0 K.

A simple derivation of the expression for the theoretical efficiency of a heat engine may be obtained in the following manner. The translational energy per mole of gas entering the cylinder is $3RT_1/2$, where T_1 is the source temperature; the energy per mole of gas leaving the cylinder after expansion is $3RT_2/2$, with T_2 the sink temperature. The amount of energy per mole of gas converted to useful work, W_u, during expansion of the piston per mole of gas is:

$$W_u = \frac{3R(T_1 - T_2)}{2}. \tag{3.62}$$

The energy input per mole of gas (W_i) is given by:

$$W_i = \frac{3RT_i}{2}. \tag{3.63}$$

Therefore, the theoretical efficiency, η, of a heat engine – the Carnot efficiency – is given by:

$$\eta = \frac{W_u}{W_i} = \frac{T_1 - T_2}{T_1}.$$

(3.64)

The efficiency can be unity only if T_2 is at absolute zero, or if T_1 becomes a very high value.

3.6.2. Thermodynamic Derivation of the Theoretical Efficiency of a Heat Engine

The thermodynamic derivation of the theoretical, i.e. maximum, efficiency of a heat engine is well-known and may be found in the standard textbooks. In essence, the derivation is made by assuming that the gas inside the cylinder first expands isothermally from a volume V_1 to V_2 at a temperature T_1, and then adiabatically from V_2 to V_3, during which process the gas cools from temperature T_1 to T_2. The gas returns again to its original state in two stages again, i.e., first isothermal contraction at T_2, follow-ed by adiabatic contraction (Figure 3-12). The theoretical efficiency for the entire process was first derived by Carnot and shown to satisfy Eq. (3.64). In general, it is difficult to have high source temperatures because of material problems. Furthermore, it is not practical to have T_2 much less than T_1 because of heat conduction bet-ween the high and low-temperature zones of the machine. From these practical considera-tions, the maximum Carnot efficiency was generally found to be between 40 and 50 percent. The actually observed efficiencies of practical heat engines are about half these values.

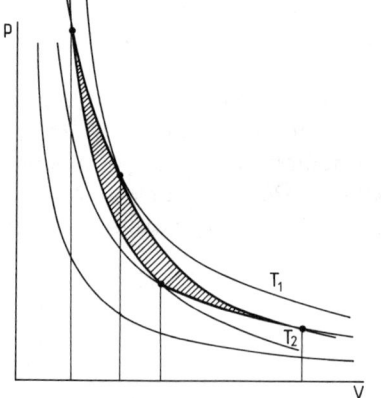
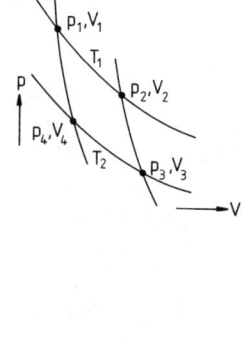

Figure 3-12. Intermediate Steps in a Carnot Cycle for Derivation of Theoretical Efficiency of a Heat Engine. Steps A to B and C to D are carried out isothermally; steps B to C and D to A are carried out adiabatically.

3.6.3. Thermodynamic Efficiency

For the case of an electrochemical energy converter working ideally, it has been shown that the free-energy change of the reaction may be totally converted to electrical energy. Thus, an electrochemical energy converter has a thermodynamic efficiency given by:

$$\eta_{th} = \frac{\Delta G}{\Delta H} = 1 - \frac{T\Delta S}{\Delta H} \qquad (3.65)$$

Table 3-1. Thermodynamic Data for some Candidate Fuel-Cell Reactions Under Standard Conditions at 25 °C.

Fuel	Reaction	n	$-\Delta H^0$ [kJ/mol]	$-\Delta G^0$ [kJ/mol]	E^0 rev. [V]	%
Hydrogen	$H_2 + 0.5\ O_2 \longrightarrow H_2O_{(l)}$	2	286.0	237.3	1.229	83.0
	$H_2 + Cl_2 \longrightarrow 2\ HCl_{(aq)}$	2	335.5	262.5	1.359	78.3
	$H_2 + Br_2 \longrightarrow 2\ HBr$	2	242.0	205.7	1.066	85.0
Methane	$CH_4 + 2\ O_2 \longrightarrow CO_2 + 2\ H_2O_{(l)}$	8	890.8	818.4	1.060	91.9
Propane	$C_3H_8 + 5\ O_2 \longrightarrow 3\ CO_2 + 4\ H_2O_{(l)}$	20	2221.1	2109.9	1.093	95.0
Decane	$C_{10}H_{22} + 15.5\ O_2 \longrightarrow 10\ CO_2 + 11\ H_2O_{(l)}$	66	6832.9	6590.5	1.102	96.5
Carbon monoxide	$CO + 1.5\ O_2 \longrightarrow CO_2$	2	283.1	257.2	1.066	90.9
Carbon	$C + 0.5\ O_2 \longrightarrow CO$	2	110.6	137.3	0.712	124.2
	$C + O_2 \longrightarrow CO_2$	4	393.7	394.6	1.020	100.2
Methanol	$CH_3OH + 1.5\ O_2 \longrightarrow CO_2 + 2\ H_2O_{(l)}$	6	726.6	702.5	1.214	96.7
Formalde- hyde	$CH_2O_{(g)} + O_2 \longrightarrow CO_2 + 2\ H_2O_{(l)}$	4	561.3	522.0	1.350	93.0
Formic- acid	$HCOOH + 0.5\ O_2 \longrightarrow CO_2 + H_2O_{(l)}$	2	270.3	285.5	1.480	105.6
Ammonia	$NH_3 + 0.75\ O_2 \longrightarrow 0.5\ N_2 + 1.5\ H_2O$	3	382.8	338.2	1.170	88.4
Hydrazine	$N_2H_4 + O_2 \longrightarrow N_2 + 2\ H_2O_{(l)}$	4	622.4	602.4	1.560	96.8

Table 3-1 gives the thermodynamic efficiencies for a number of fuel-cell reactions under standard conditions. It shows that the thermodynamic efficiency for nearly all the reactions in the table are greater than 90 per cent (cf. the thermodynamic efficiencies of heat engines, working ideally between practical operating temperatures of source and sink, which are mostly less than 40 per cent).

An interesting fact should be noted concerning the efficiencies of hydrocarbon-air fuel cells. At temperature below 100 °C, water, a product of the cell reactions, is in the liquid form. This causes the entropy changes in the corresponding reaction to be negative, and thus the thermodynamic efficiencies are less than 100 percent. However, at temperatures above 100 °C, water is liberated as one of the products. The entropy changes are therefore positive under these conditions (Table 3-1), corresponding to thermodynamic maximum efficiencies of over 100 per cent. At the present time, overpotential losses reduce the practical efficiencies of these fuel cells to values much less than the thermodynamic efficiencies. However, if sufficient advances in electrocatalysis could be made (by reduction of overpotentials), it might be possible to attain practical efficiencies of over 100 percent and close to the maximum thermodynamic efficiencies. This is particularly relevant for those situations in which the power density is low, so that the current density and overpotential can also be low.

Effect of Temperature on Thermodynamic Efficiency

If the entropy change of a reaction is negative, it follows from Eq. (3.65) that (recalling the negative sign of ΔH) the thermodynamic efficiency would decrease with an increase of temperature. Therefore, a hydrogen-oxygen fuel cell has thermodynamic efficiencies of 0.83 at 25 °C and 0.78 at 100 °C. An important difference between this result and the corresponding one for a thermal engine must be made here. As the temperature of the hot stage of the thermal energy converter increases, the difference between the temperatures of the heat source and the heat sink also increases. Consequently the corresponding Carnot efficiency increases with temperature. However, caution is needed when making any theoretical forecast on the variation of efficiency of a thermal engine with temperature. It cannot simply be assumed that the temperature of the heat sink (T_2) is always at ambient temperature and is independent of the temperature of the heat source. Increasing the temperature of the heat source would automatically cause some increase in temperature of the heat sink. Thus, the increase in thermodynamic efficiency of thermal engines with the temperature of the heat source in these practical cases would be less than if it is assumed that the heat-sink temperature is constant and equal to the temperature of the surroundings. In Figure 3-13, the variation of thermodynamic maximum efficiency of a hydrogen-oxygen fuel cell and of heat engines as a function of temperature is shown.

The contrasting behaviour of thermal engines and fuel cells with regard to changes in thermodynamic efficiency as a function of the temperature reduces some of the advantages of fuel cells at high temperatures. At higher temperatures, however, the need for expensive electrocatalysts is diminished because the temperature itself increases the reaction rate, and therefore makes the overpotential necessary for providing a given current density less than that at lower temperatures.

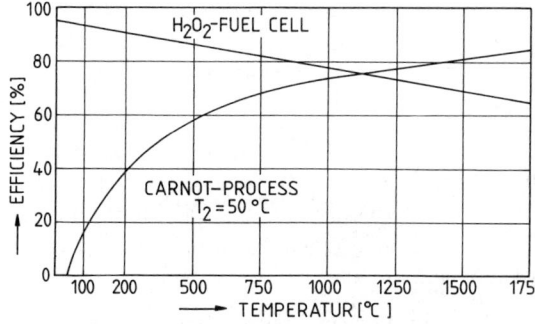

Figure 3-13. Comparison between the Carnot and the Thermodynamic Efficiency of a Fuel Cell

The entropy change for the oxidation of carbon to carbon monoxide is positive. For this case, η_{th} according to Eq. (3.65) exceeds unity and increases with temperature. For the complete oxidation of carbon to carbon dioxide, ΔS is approximately equal to zero and η_{th} is nearly unity with hardly any temperature variation. The entropy change in the overall reaction for various hydrocarbon-oxygen fuel cells operating at temperatures of over 100 °C is also larger than zero.

3.6.4. Electrochemical Efficiency

A measure of the quality of a fuel cell – apart from operational life, weight, costs and other parameters – is the electrochemical efficiency and the electrochemical kinetics of the electrodes. An electrochemical cell can only, at best, supply ΔG. If cells of different technical designs, in which the same reaction with the same enthalpy occurs, are compared, differences in the electrochemical efficiencies, η_{el}, will be found. The electrochemical efficiency is also termed voltage efficiency, and is defined as:

$$\eta_{el} = -nF E_K / \Delta G = E_K / E_0 \tag{3.66}$$

Electrochemical efficiencies observed in H_2–O_2 fuel cells are as high as 0.9 at low current densities and decrease only slowly with increasing current drawn from the cell until a limiting value is reached.

3.6.5. Practical Efficiency

It is useful to state the practical efficiency, η_P, under load conditions, as this takes account of the influence of some irreversible processes.

$$\eta_P = E_K / \left(E_0 - T \left(dE_0 / dT \right) \right) = -nFE_K / \Delta H \tag{3.67}$$

where E_K is the operating potential and E_0 the standard electrode potential.

The reasons why the potential E_K is lower than the standard electrode potential E_0 are to be found in the polarisation effects and in the voltage drop due to the ohmic resistance.

3.6.6. Faradaic Efficiency

The faradaic efficiency is defined as:

$$\eta_f = I/I_m \tag{3.68}$$

I is the observed current from the fuel cell. I_m is the theoretically expected current on the basis of the amount of reactants consumed, assuming that the overall reaction in the fuel cell proceeds to completion. Faradaic efficiency is analogous to current efficiency in conventional electrochemical cells. In most fuel cells, η_f may be less than 1 due to:

(1) parallel electrochemical reactions yielding fewer electrons per mole of reactant consumed,
(2) chemical reactions of reactants catalysed by the electrodes, and
(3) a direct chemical reaction of the two electrode reactants.

3.6.7. Total Efficiency

The total efficiency, η, of a fuel cell system includes (excluding parasitic currents and self-discharging) pumping of the fuel and the oxidation material, and also the necessary energy for heating, cooling, electrolyte pipes, compression and auxiliary installations. Exact data depend heavily on the system design and can only be obtained experimentally (see Figure 3-14).

Fuel cell systems are often used as co-generation systems, and total efficiency often includes the thermal and electrical efficiency. Because the electrical energy is an energy form of a higher quality than the thermal energy, it is very important to make an exergy analysis which also includes application of the second law of thermodynamics (see Section 6).

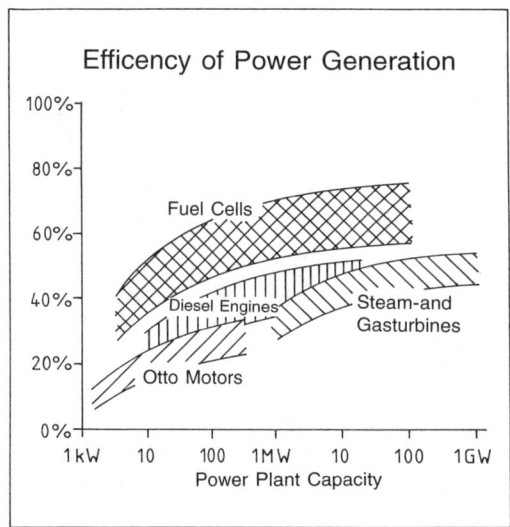

Figure 3-14. Efficiencies of Different Technologies as a Function of Scale

3.7. Monopolar and Bipolar Electrode Constructions

The two basic designs use *monopolar* or *bipolar* construction (see Figure 3-15). For current collection purposes, *monopolar electrodes* have tabs and are connected to one another at their edges. The edge or frame collection of the current requires good electrode conductivity. This is no problem with porous metal electrodes, but with carbon electrodes metal screens, expanded metal or metal felt structures are required. Current distribution at high current densities is an important factor in all monopolar cells. Non-uniformities are transferred throughout the stack being equalised by the resistance of the electrolyte layers. In general, an electrode size up to 400 cm^2 can be handled with frame or edge-current collection. Beyond that size, uneven current distribution, due to the voltage drop on the tapering ohmic resistance, is too high to be acceptable. On the other hand, monopolar cells can be connected individually in the stack, depending on the voltage-current demands. A single cell can also be disconnected in the case of failure without seriously disturbing performance.

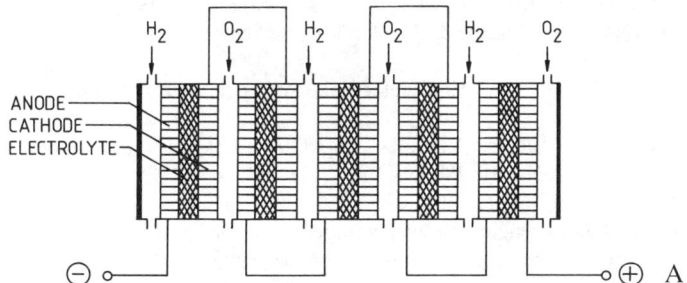

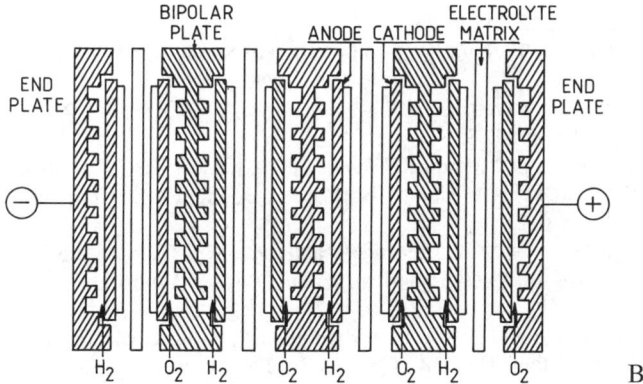

Figure 3-15. Types of Cell Construction
A) Monopolar
B) Bipolar

Bipolar current collection is a logical consequence of the stacking of many cells in series to obtain a high stack voltage. This method of current flow often results in interfacial contact problems, which must be solved by pressure devices acting on the stack. Bipolar cells are not limited in size. In this configuration, the current flows perpendicular to the electrode surface and current collection is realised over the whole area of the electrodes. Electrode materials with low conductivity, such as carbon can be used. A disadvantage of this concept is that a single cell failure leads to malfunctioning of the whole stack. However, the bipolar configuration has undisputed advantages over the monopolar concept for cell areas > 400 cm^2.

References for Chapter 3

[1] J. O'M. Bockris, S. Srinivasan, "Fuel Cells: Their Electrochemistry", McGraw-Hill Book Company, New York, 1969, Chapter 3 "Thermodynamic Aspects of Electrochemical Energy Conversion", pp. 144–176

[2] M. Eisenberg "Thermodynamics of Electrochemical Fuel Cells" Chapter 2 in Will Mitchell, "Fuel Cells", A Series of Monographs – Volume 1, Academic Press, New York and London, 1963

[3] K. Kordesch "Brennstoffbatterien", Innovative Energietechnik, Springer Verlag, Wien–New York, 1984, pp. 24–46

[4] Karl V. Kordesch, "Power Sources for Electric Vehicles", in J O'M Bockris, B. E. Conway, "Modern Aspects of Electrochemistry", No. 10, Plenum Press, New York, 1975, p. 376

[5] K. V. Kordesch, Electrochemica Acta 16 (1971) 597

[6] B. E. Conway, "Similarities and Differencies Between Supercapacitors and Batteries for Electrical Energy Storage", Lecture, University of Ottawa, February 1995

[7] K. Kordesch, A. Marko, J. Electrochem. Soc. **107** (1960) 480–483

[8] K. V. Kordesch, J. Electrochem. Soc., **119** (1972) 1053–55

[9] M. Kita, J. Electrochem. Soc. **113** (1966) 1095

[10] U. R. Evans, Electrochimica Acta **14** (1969) 197

[11] J. R. Goldstein, A.C.C. Tseug, Nature **222** (1969) 869

[12] K. Kinoshita, "Electrochemical Oxygen Technology", John-Wiley & Sons, Inc., New York, 1992, Chapter 2, pp. 19–112

[13] J. O'M Bockris, S. Srinivasan, "A Brief Outline of Electrochatalysis", 19th Annual Power Sources Conference, 18–20 May 1965, pp. 4–8

[14] E. Yeager, Oxygen Cathodes, Proceedings, "Renewable Fuels and Advanced Power Sources for Transportation" Workshop at SERAI, Los Alamos Natl. Lab., Boulder, Colorado, June 1982.

[15] Catalytic Materials: Relationship between Structure and Reactivity, ACS-State of the Art Symposium, San Francisco, June 13-16, 1983, American Chem. Soc., Symposium Series 248, Washington, DC, 1984.

[16] L. I. Krishtalik, "Hydrogen Overvoltage and Adsorption Phenomena", Advances in Electrochemistry and Chemical Engineering, Paul Delahay, Charles W. Tobias, eds., Vol. 7, John Wiley & Sons, Inc., New York 1970.

4. Fuel Cell Systems

4.1. Introduction

Fuel cell systems can be classified according to the working temperature: high, medium and low (ambient) temperature systems, or referring to the pressure of operation: high, medium and low (atmospheric) pressure systems. They may be further distinguished by the fuels and/or the oxidants they use: (1) gaseous reactants (such as hydrogen, ammonia, air and oxygen), (2) liquid fuels (alcohols, hydrazine, hydrocarbons) or (3) solid fuels (e.g., coal, hydrides). Strictly speaking zinc-air or aluminum-air systems, which have a replenishable metal supply (in powder, rod- or sheet form) and a continuous removal process for the produced oxides, are fuel cell systems as well. However, it is unconventional to call them so.

For practical reasons fuel cell systems are simply distinguished by the type of electrolyte used and the following names and abbreviations are now frequently used in publications: Alkaline fuel cells (AFC), phosphoric acid fuel cells (PAFC), molten carbonate fuel cells (MCFC), solid oxide fuel cells (SOFC) and proton exchange membrane fuel cells (PEMFC).

Subsequent classifications are discussed with the systems. The electrolyte of an alkaline or acidic fuel cell can be for instance in liquid form (as solution), then it is called a *mobile electrolyte system*. If the electrolyte is soaked up in a porous material such as asbestos, it is called an *immobile or matrix system*.

The principle division into direct and indirect fuel cells has its basis at the electrode level. For instance, hydrogen gas is ionized at the hydrogen electrode and participates directly in the electric power generating reaction. In indirect cells, chemicals e.g., ammonia, alcohol, cyclohexane, methane or aviation turbine fuels are cracked or steam reformed to hydrogen containing mixtures of gases. This is usually done outside of the fuel cell with the aid of catalysts at high temperatures. Only the hydrogen in the gas mixture reacts electrochemically at the electrode.

Catalytic fuel decomposition may also occur at the electrode surface as in the case of hydrazine which is added to the cell electrolyte. The electrode uses the hydrogen produced on-site. An example for an indirect *liquid oxidant* is a solution of hydrogen peroxide, catalytically decomposed to oxygen and water (either in a separate reactor or at the electrode itself).

H_2/O_2-Fuel Cell Types:

Alkaline Fuel Cell: The electrolyte in this fuel cell is concentrated (35–50 wt.-%) KOH for low temperature applications. The Alkaline Apollo Fuel Cell worked at 250 °C (KOH: 85 wt.-%). Oxygen reduction kinetics are more rapid in alkaline electrolytes than

Table 4-1. Applications for technical H_2/O_2-fuel cell systems

Fuel Cell System	Temperature Range	Efficiency (Cell)	Electrolyte	Application Area
Alkaline Fuel cell (AFC)	60–90 °C	50–60%	35–50% KOH	Space Applications/ Traction Applications
Polymer Electrolyte Fuel Cell (PEFC)	50–80 °C	50–60%	Polymer Membrane (Nafion/Dow)	Traction Applications/ Space Applications
Phosphoric Acid Fuel cell (PAFC)	160–220 °C	55%	Concentrated Phosphoric Acid	Predominantly Dispersed Power Applications (50–500 kW, 1 MW, 5 MW, 11 MW)
Molten Carbonate Fuel Cell (MCFC)	620–660 °C	60–65%	Molten Carbonate Melts (Li_2CO_3/ Na_2CO_3)	Power Generation
Solid Oxide Fuel Cell (SOFC)	800–1000 °C	55–65%	Yttrium-stabilized Zirkondioxide (ZrO_2/Y_2O_3)	Power Generation

in acid electrolytes, and the use of non-noble metal electrocatalysts in AFCs is feasible. A major disadvantage of AFCs is that alkaline electrolytes (i.e., NaOH, KOH) do not reject CO_2. The consequence of this property is that AFCs are currently restricted to specialized applications where pure H_2 and O_2 are utilized.

Polymer Electrolyte Fuel Cell: The electrolyte in this fuel cell is an ion exchange membrane (fluorinated sulfonic acid polymer or other similar polymers) which is a good proton conductor. The only liquid in this fuel cell is water, thus corrosion problems are minimal. Water management in the membrane is critical for efficient performance; the fuel cell must operate under conditions where the byproduct water does not evaporate faster than it is produced because the membrane must be hydrated. Because of the limitation on the operating temperature imposed by the polymer and problems with water balance, usually less than 100 °C, a H_2-rich gas with little or no CO is used and higher catalysts loadings (Pt in most cases) than those used in PAFCs are required in the anode and cathode.

Phosphoric Acid Fuel Cell: Concentrated phosphoric acid is used for the electrolyte in this fuel cell, which operates at 160–220 °C. At lower temperatures, phosphoric acid is a poor ionic conductor and CO poisoning of the Pt electrocatalyst in the anode becomes more severe. The relative stability of concentrated phosphoric acid is high compared to other common acids, consequently the PAFC is capable of operating at the high end of the acid temperature range (100–220 °C). In addition, the use of concentrated acid (~100%) minimizes the water vapor pressure so water management in the cell is not difficult. The matrix universally used to retain the acid is silicon carbide, and the electrocatalyst in both the anode and cathode is Pt.

Molten Carbonate Fuel Cell: The electrolyte in this fuel cell is usually a combination of alkali (Na, K, Li) carbonates, which is retained in a ceramic matrix of $LiAlO_2$. The fuel cell operates at 600–700 °C, where the alkali carbonates form a highly conductive molten salt, with carbonate ions providing ionic conduction. At the high operating temperatures in MCFCs, Ni (anode) and nickel oxide (cathode) are adequate to promote reaction and noble metals are not required.

Solid Oxide Fuel Cell: The electrolyte in this fuel cell is a solid, nonporous metal oxide, usually Y_2O_3-stabilized ZrO_2. The cell operates at 650–1000 °C where ionic conduction by oxygen ions takes place. Typically, the anode is $Co\text{-}ZrO_2$ or $Ni\text{-}ZrO_2$ cermet, and the cathode is Sr-doped $LaMnO_3$.

Table 4-2. Typical Electrochemical Reactions in Fuel Cells

Fuel Cell	Anode Reaction	Cathode Reaction
Alkaline Fuel Cells	$H_2 + 2(OH)^- \longrightarrow 2\,H_2O + 2e^-$	$\tfrac{1}{2}O_2 + H_2O + 2e^- \longrightarrow 2(OH)^-$
Polymer Electrolyte Fuel Cells	$H_2 \longrightarrow 2\,H^+ + 2e^-$	$\tfrac{1}{2}O_2 + 2\,H^+ + 2e^- \longrightarrow H_2O$
Phosphoric Acid Fuel Cells	$H_2 \longrightarrow 2\,H^+ + 2e^-$	$\tfrac{1}{2}O_2 + 2\,H^+ + 2e^- \longrightarrow H_2O$
Molten Carbonate Fuel Cells	$H_2 + CO_3^{2-} \longrightarrow H_2O + CO_2 + 2e^-$	$\tfrac{1}{2}O_2 + CO_2 + 2e^- \longrightarrow CO_3^{2-}$
Solid Oxide Fuel Cells	$H_2 + O^{2-} \longrightarrow H_2O + 2e^-$	$\tfrac{1}{2}O_2 + 2e^- \longrightarrow O^{2-}$

The overall electrochemical reactions corresponding to the individual electrode reactions listed in Table 4-2 are given in Table 4-3, along with the appropriate form of the Nernst equation. The Nernst equation provides a relationship between the standard potential (E^0) for the cell reaction and the equilibrium potential (E) at various temperatures and partial pressures (activities) of reactants and products. According to the Nernst equation, the equilibrium cell potential at a given temperature can be increased by operating at higher reactant pressures and improvements in fuel cell performance have, in fact, been observed at higher pressures (often contra life-time).

Table 4-3. Corresponding Nernst Equations

Cell Reactions (total)	Nernst Equation
$H_2 + \tfrac{1}{2}O_2 \longrightarrow H_2O$	$E = E^0 + (RT/2F)\ln\left(P_{H_2}/P_{H_2O}\right) + (RT/2F)\ln\left(P_{O_2}^{1/2}\right)$
$H_2 + \tfrac{1}{2}O_2 + CO_2(c) \longrightarrow H_2O + CO_2(a)$	$E = E^0 + (RT/2F)\ln\left[P_{H_2}/P_{H_2O}\left(P_{CO_2}\right)\right]_a + (RT/2F)\ln\left[P_{O_2}^{1/2}\left(P_{CO_2}\right)\right]_c$

(a) – anode	P – gas pressure
(c) – cathode	R – universal gas constant
E – equilibrium potential	T – Temperature
E^0 – standard potential	

Chapters (4.2.–4.6.) cover mainly these types of H_2/O_2-Fuel Cell Systems. Direct type fuel cell systems using fuels such as CH_3OH or its derivatives are noted as potentially transportable power sources, as a liquid fuel is best transported and converted into energy from the liquid state. Methanol is the favoured fuel. The direct methanol/air fuel cell (DMFC) uses a sulfuric acid electrolyte because of the need to reject the CO_2 that is produced during the electrooxidation of methanol. It is also imperative to employ a catalyst that is platinum-metal based. Nevertheless, a major scientific problem is the catalyst poisoning caused by residues of an aldehyde, carboxylic acid and other intermediates that are produced during the electrooxidation of methanol (Chapter 4.7.). Direct fuel cells operating on glucose and oxygen extracted from the blood have received increased interest for medical user in the last years. It is believed that this type of fuel cell is a very attractive power source to drive artificial hearts and should be mentioned therefore for reasons of completeness.

In the last few years increasing interest was given to metal air cells. These systems have a very favourable energy density which is achieved through not requiring to incorporate positive active components within the cell. To a large extent, the impetus for research and development in this field has arisen from possible electric vehicle applications where energy density is a critical parameter. These considerations have long intrigued electrochemists, and many attempts have been made over the years to develop practical systems. Zinc has received by far the most attention because it is a metal which can be efficiently electroplated from aqueous solution. All these aspects are discussed in Section 4.8. Hybrid Cells.

4.2. Alkaline Fuel Cells (AFC)

4.2.1. General Principle

Alkaline fuel cells use an aqueous solution of potassium hydroxide as electrolyte. The concentration varies from 30–45 wt.-%, depending on the system. The O_2 reduction in alkaline electrolytes is more favorable than in acid electrolytes [1]. Due to the alkaline nature of the electrolyte no acidic impurities, e.g., carbon dioxide, are allowed in either reactants.

In the alkaline medium carbon dioxide forms carbonates which would block electrolyte pathways and electrode pores (see Figure 4-1).

$$CO_2 + 2\,OH^- \longrightarrow CO_3^{2-} + H_2O \qquad (4\text{-}1)$$

This is one of the detrimental effects for the poor long term stability of AFCs beside the long-term oxidation of carbon electrodes in alkaline electrolytes (degradation of the electrode structure). Removal of the 0.03% carbon dioxide from the air can be accomplished by chemical absorption in a tower filled with, e.g., soda lime. An industrial temperature swing absorption desorption process, using amines, is very efficient, but presently not available for small units. In practice, alkaline systems use high purity hy-

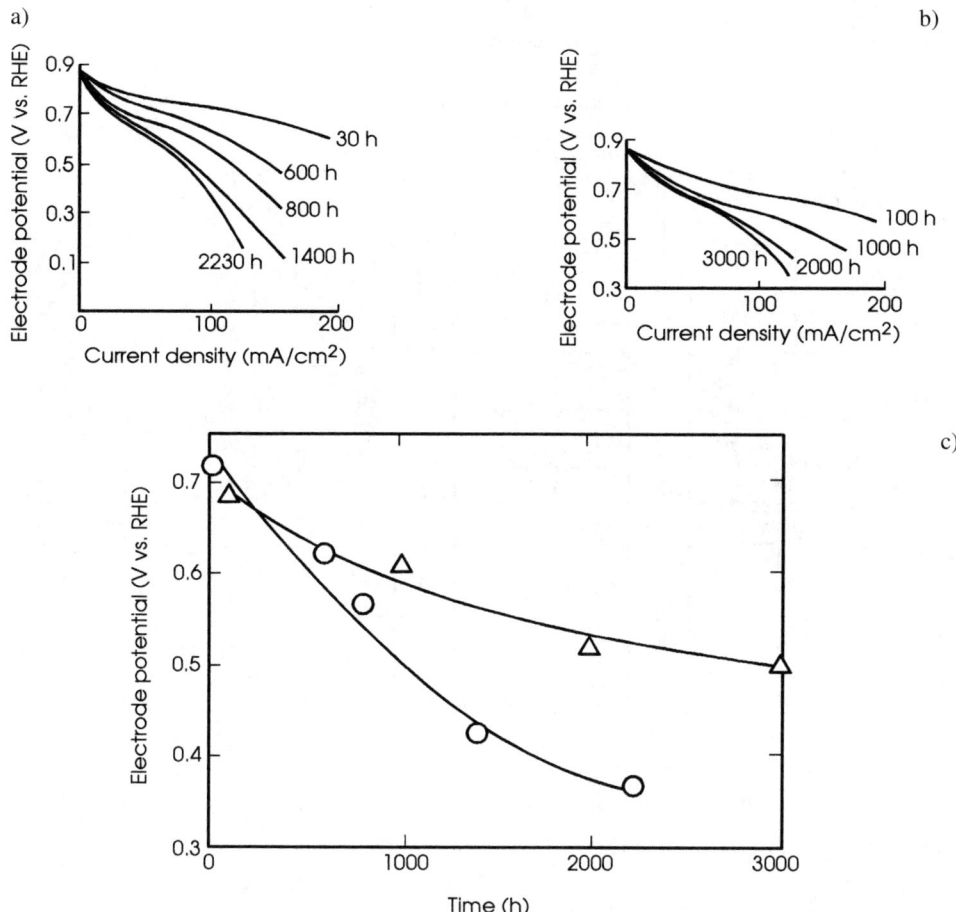

Figure 4-1. Influence of Carbon Dioxide on the Air Performance of Supported Pt Cathodes (0.2 mg/cm² Pt Supported on Carbon Black) in 6 N KOH at 50 C. a) CO_2 Containing Air, b) CO_2 Free Air. Current-Voltage Curves Were Measured Periodically for Electrodes Continuously Operated at 32 mA/cm², c) Variation in Electrode Potential as a Function of Time at 100 mA/cm². O, CO_2-Containing Air, Δ, CO_2-Free Air [2].

drogen from electrolysis or ammonia plants [3]. New investigations reject this detrimental effect for Ni and Ag fuel cell electrodes [4, 5].

Alkaline systems operate well at room temperature, yield the highest voltage (at comparable current densities) of all fuel cell systems, and cell and electrodes can be built from low-cost carbon and plastics. Because of a good compatibility with many construction materials AFCs can achieve a long operating life (15000 h have been demonstrated). The choice of catalysts available for alkaline cells is greater than for acidic cells, which are presently limited to platinum group metals and tungsten carbides.

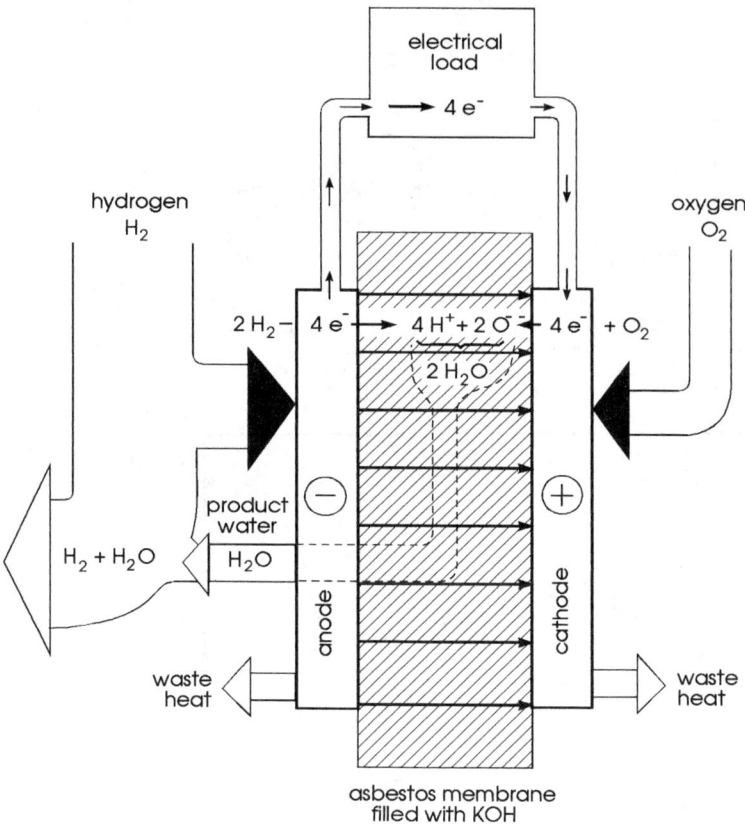

Figure 4-2. Principle of the Alkaline Fuel Cell with Static Electrolyte

Contrary to matrix cells, fuel cells with a *circulating electrolyte* establish a reliable barrier against reactant gas leakage from the electrodes and can use the electrolyte as a cleansing medium. Accumulated impurities and carbonates can easily be removed. The main advantage of the circulating electrolyte is its use as a cooling liquid and as a water removal vehicle.

Some systems use the circulating electrolyte to carry the reaction water to cooling subsystems and separate water vaporizing units (regenerators) outside the stacks. In such systems, the reactants hydrogen and oxygen can be supplied dead-ended, care must only be taken that inerts building up with time are removed by a continuous or periodic purging of the gases.

Another way of water removal is by evaporation through the electrodes into excess reactant gas streams. Vaporization cooling is an additional bonus. In alkaline fuel cells (contrary to acidic cells) the reaction water is produced at the hydrogen electrode (the anode). Therefore, the electrolyte concentration inside the anode structure is lower than in the bulk electrolyte. This increases the water vapor pressure and thus its evaporation rate. The excess hydrogen stream is obtained by circulating and cooling the hydrogen by

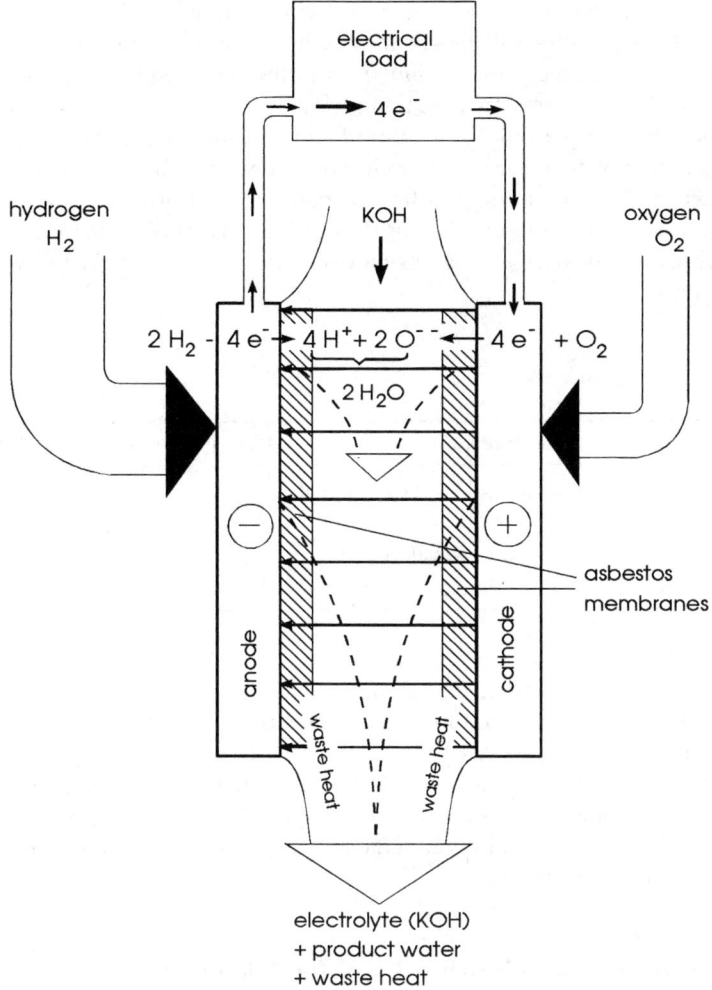

Figure 4-3. Principle of the Alkaline Fuel Cell with Circulating Electrolyte

means of a gas pump or, by mounting a gasjet into the gas loop between the condenser and the gas port leading into the stack. In the latter case the gas circulation rate becomes automatically proportional to the gas consumption. Oxygen circulation is not as effective as hydrogen circulation, because the electrolyte concentrates in the cathode and its water vapor pressure is lower than that of the bulk electrolyte. A schematic view of an Alkaline Fuel Cell with circulating electrolyte is demonstrated in Figure 4-3.

In air-operated fuel cells, excess air must be blown continuously through the cathodes to remove nitrogen from the air manifolds which would otherwise block them. The increasing necessity for carbon dioxide removal is the price which must be paid for this simple way of water and nitrogen removal without any condensers.

The disadvantage of *circulating electrolyte systems* in which the electrolyte is fed through a common manifold into a multi-cell stack is the occurrence of parasitic currents, through the ionically conductive liquid electrolyte junctions of the series-connected cells and parallel electrolyte flow: these currents must be minimized.

Regarding the R & D activities in the field of alkaline fuel cells, the international scientific community has been mainly changed its research topics. The main interest is now in the field of PEFC, MCFC and SOFC. In the last few years research efforts in Alkaline Fuel Cells were mainly carried out by some European Research teams[1]. Following these changes in fuel cell activities, the next chapter gives an overview of Alkaline Fuel Cell Systems developed in the past.

Table 4-4. European AFC Research Activities

Country	Organization	Markets	Materials	Stack	System
Belgium	Elenco, N.V.	Space, Defense, Traction		O	
Finland	Helsinki University		O		
Germany	Siemens	Utility, Naval, Space	O	O	O
Sweden	Royal Institute of Technology	Traction, on-site	O		

The AFCs for remote applications (space, undersea, military) are not strongly constrained by cost. On the other hand, the consumer and industrial markets require the development of low-cost components to successfully compete with alternative technologies. Much of the recent interest in AFCs for mobile and stationary terrestrial applications has addressed the development of low-cost cell components (carbon-based porous electrodes) [6, 7]. In this regard, the following chapter focuses on the different application of AFCs mainly as a historic survey.

4.2.2. Early Developments of Alkaline Fuel Cell Systems

Bacon Fuel Cell Systems

The 5 kW fuel cell built by BACON in the mid 1950s was the first operational multi-kilowatt fuel cell (see Figure 4-4). It consisted of porous nickel anodes and lithiated porous nickel oxide cathodes separated by a circulating 30% aqueous potassium hydroxide solution.

Operating temperature was 200 °C and a pressure of up to 5 MPa was used to prevent the electrolyte from boiling. The three phase zone was maintained by differential gas pressure, because no wet-proofing agent, stable at that temperature was available for porous metal electrodes at that time. Reactants were pure hydrogen and oxygen.

---

1 ASTRIS, Inc., Missisauga, ON, (Canada) is also active in the field of AFC

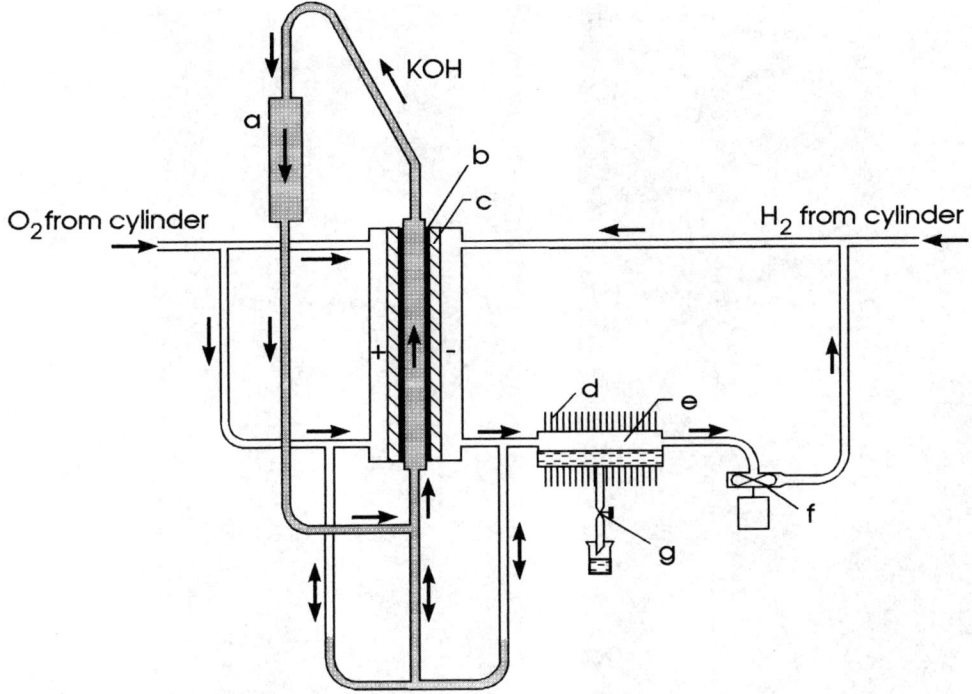

Figure 4-4. The Bacon Fuel Cell System
a) Electrolyte circulation; b) Porous nickel (max. 16 μm diameter pores), c) Porous nickel (ca. 30 μm diameter pores), d) Cooling fins; e) Condenser; f) Hydrogen circulator; g) Condensate release valve

The Patterson-Moos Research Division of Leesona Corp. developed the monopolar Bacon cell into a bipolar version (Hydrox-cell) and Pratt & Whitney Aircraft Division modified the Bacon cell such that it became the basis for the development of the NASA Apollo AFC.

Apollo Alkaline Fuel Cell

Pratt & Whitney Aircraft ultimately developed the alkaline fuel cell system model PC3A-2 for the NASA Apollo space program from 1960 to 1965 [8].

The Apollo cells used in principle the structure of BACON'S double porosity sintered nickel electrodes, but activated them additionally with platinum metal catalysts to boost performance in spite of lowering the operating pressure to ca. 0.33 MPa. Because of the lower pressure a much more concentrated electrolyte must be used to prevent boiling. The electrolyte (85% KOH) was solid at room temperature, it liquefied on heating over 100°C, and required a complex starting procedure. Operating temperature was between 200–230 °C, which was detrimental to the life of the electrolyte circulating pumps.

The electrodes were circular with a diameter of 200 mm and were ca. 2.5 mm thick. Each single cell was individually packed between nickel sheets. 31 of these cells were stacked together and electrically connected in series.

Figure 4-5. Apollo Fuel Cell Power-Plant [9]

Water balance was maintained by recirculation of hydrogen passing through a water condenser and a gas liquid separator. The cell stack was built into a container where circulating nitrogen promoted the removal of waste heat. Three stacks were connected in parallel. A PC3A-2 powerplant is shown in Figure 4-5.

92 of these fuel cells have been delivered, 54 have been utilized which includes nine flights to the moon, three Skylab- and also the Apollo-Soyuz mission. The nominal rating of the earlier stacks was 1.5 kW with overload capability of 2.3 kW; the weight was 109 kg, 10750 h of operation were logged altogether.

Space Shuttle Fuel Cell System

Continued development of the Apollo system by NASA and United Technologies Corporation (UTC) produced the NASA space shuttle fuel cell system. Each of the manned space shuttles is equipped with three fuel cell powerplant modules located under the payload bay in the forward portion of the mid-fuselage of the vehicle. Each fuel cell module contains a stack of 32 cells (465 cm^2 active area). The development program

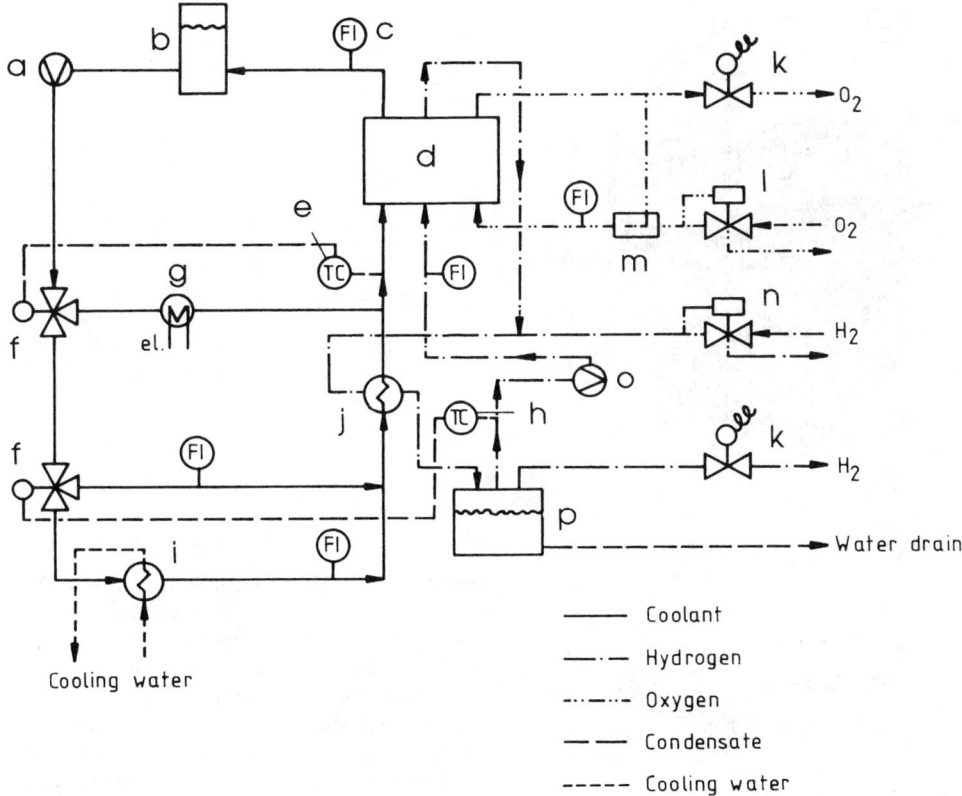

Figure 4-6. The NASA Space Shuttle Fuel Cell
a) coolant pump; b) Accumulator; c) Flowmeter; d) Module; e) Coolant inlet and temperature control valve; f) Heat exchanger bypass valve; g) Start-up heater; h) Temperature controller; i) Heat exchange; j) Condenser; k) Solenoid vent; I) Oxygen regulator; m) Venturi recycle; n) Hydrogen regulator; o) Hydrogen pump; p) Water tank.

started in January 1974 and the first development powerplant tests began in October 1975. It was qualified for manned space flight in June 1979 and utilized in the first orbital flight in April 1981. The fuel cell system provides the electrical power on board the space shuttle vehicle (orbiter) and, on a seven day mission, consumes 750 kg of hydrogen and oxygen, producing potable water for consumption by the crew and for cooling purposes. Each of the three power plant modules has a maximum power rating of 12 kW (27.5 V and 436 A) with an emergency overload capability of 16 kW The system is smaller than the Apollo modules, weighs about 18 kg less and delivers 8 times their power.

The stack has a bipolar configuration. The *bipolar plates* are light-weight, silver-plated magnesium foils which also aid in the heat transfer. The electrolyte is 35–45% aqueous potassium hydroxide totally immobilized by a reconstituted asbestos separator. The *anode* consists of PTFE bonded carbon carrying 10 mg/cm^2 platinum palladium (ratio

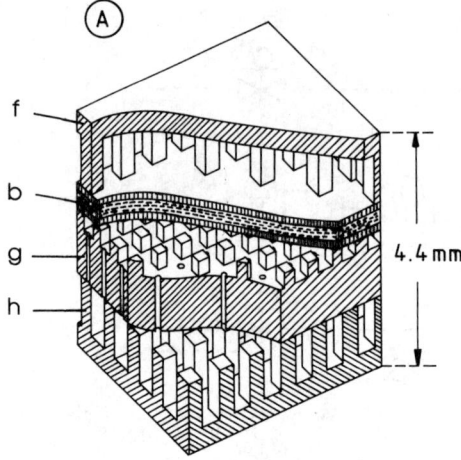

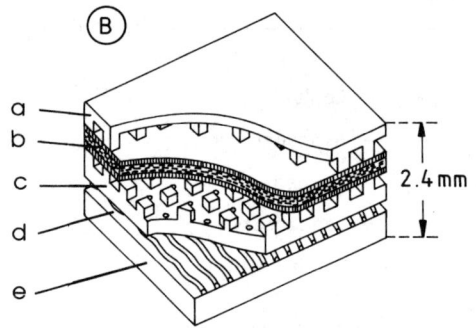

Figure 4-7. Comparison of Initial A) and Advanced B) Space Shuttle Fuel Cell Assembly
a) Oxygen flow field (plastic); b) Cell (screen electrode/asbestos matrix/screen electrode); c) Electrolyte reservoir plate (plastic); d) hydrogen flow field (plastic); e) Water removal assembly (plastic asbestos teflon); f) Oxygen flow field (gold-plated magnesium); g) Electrolyte reservoir plate (nickel); h) Hydrogen flow field (gold-plated magnesium)

80:20) as catalyst, pressed on a silver-plated nickel screen. The *cathode* catalyst consists of gold (90 wt.-%), and platinum (10 wt.-%) and is deposited in an amount of 20 mg/cm^2 on a gold-plated nickel screen. It is claimed that gold works as the main catalyst and platinum as the sintering inhibitor. The cell operates at 92 °C and 0.40–0.44 MPa.

The water balance is maintained by water vapor transfer into the hydrogen gas. The hydrogen is circulated through a condenser into a centrifugal separating device, which collects the suspended droplets of liquid water (gravity free environment !) from the hydrogen gas and pumps the water into a storage tank (bladder).

The gas spaces behind the oxygen electrodes are essentially dead-ended, just purging is necessary. The heat generated in the stack is transferred to heat-conducting foils which form cooling chambers, located between each 2nd (or 4th) cell of the stack. An isoelectric liquid is specified as heat-exchanging circulating fluid.

Each cell contains an electrolyte reservoir plate to compensate for the concentration changes of the electrolyte during different load profile operations. The electrolyte reservoir plate consists of a porous sintered nickel sheet in contact with the anode, with holes pierced to allow passage of hydrogen. The electrolyte reservoir plates represented 50% of the total stack weight in the earlier models. In subsequent models, porous graphite plates were used. A schematic of the fuel cell system and its components is presented in Figure 4-6.

These fuel cells have demonstrated life-spans of 2000 h with refurbishments between missions. Single cell performance is 0.86 V at 470 mA/cm^2 (12 kW). The specific power of the complete powerplant is 275 W/kg.

Cell design has been improved in further developments and considerable specific weight gains have been achieved. The main improvements include: (1) Thinner platinum-on-carbon hydrogen anode, (2) butyl-bonded potassium titanate matrix, (3) light-weight graphite and metallized porous plastic electrolyte reservoir plates, (4) gold-plated perforated nickel foils (photoetched) as electrode substrates, (5) polyphenylene sulfide edge-framing material for cells, (6) electroformed nickel-foil replacing the gold-plated magnesium plates. The changes in construction can be seen by comparing the design of the new light-weight cell with the former cell (see Figure 4-7).

Further development of the NASA space shuttle fuel cells was carried out at UTC in contract with International Fuel Cells (IFC). The performance regarding power:weight and energy:volume ratios supersedes that of any other galvanic system, with the possible exception of future PEFC systems. A lifetime of 15000 h was demonstrated in smaller units.

Allis Chalmers Manufacturing Fuel Cell

This fuel cell developed in the 1960s consisted of catalyst-coated porous sintered nickel plaque electrodes mounted in a bipolar assembly. Both electrodes, anode and cathode, contained platinum palladium catalysts. The electrolyte was immobilized, absorbed in a sheet of microporous asbestos. Nickel-plated magnesium plates were used as the dividing members to separate the cells. In the original version water was removed by circulation of hydrogen with a water condenser in the stream. This system was improved during the years 1962–1967. The essential change was the new design of the water removal system which ensured that it reacted quicker to follow load changes (the matrix had a slowing down effect on the water equilibrium). The new concept was called the static water-vapor control method for removing water (see Figure 4-8) and became a model for many asbestos-matrix fuel cell constructions [10], e.g., in cells for many NASA contracts. Fuel cells of 1.5 kW were built and tested. Operation temperature was maintained between 50–65°C by conduction along the bipolar assembly with forced air cooling of protruding cell edges.

The first large "civilian" vehicle equipped with a fuel cell system was demonstrated by ALLIS CHALMERS in the late 1950s. It was a farm tractor propelled by a 15 kW hydrogen-oxygen fuel cell power plant.

Union Carbide Fuel Cell System

Union Carbide (UCC), produced porous carbon plate electrodes (3 mm thick) in neoprene and ABS (acrylonitrile butadiene styrene) frames for fuel cells in the early 1960s. A prototype of the bipolar construction was used by the US navy. In 1961 UCC

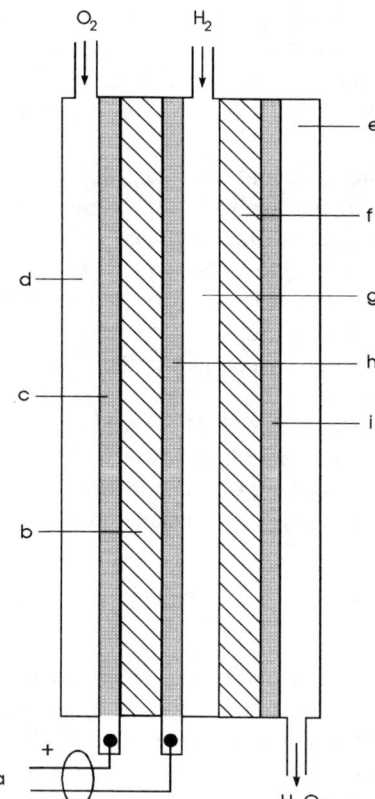

Figure 4-8. Allis Chalmers Static Water Vapor Control
a) Electrical power; b) Electrolyte; c) Porous oxygen electrode; d) Oxygen cavity; e) Moisture removal cavity; f) Moisture removal membrane; g) Hydrogen cavity; h) Porous hydrogen electrode; i) Porous support plaque

developed flexible, thin electrodes for light-weight fuel cell systems. The UCC electrodes contained PTFE powder and platinum catalyst coated carbon applied in several layers of increasing hydrophobicity to a porous nickel foil by means of wet-spraying. Subsequently the electrodes were rolled or pressed and sintered [11]. They became known as "fixed zone" electrodes and a cross-section through such an electrode is shown in Figure 4-9.

The porous nickel foil served as a support as well as a current collector. Due to the good conductivity of nickel, current collection along the edges of the 200 cm²-area electrodes presented no difficulties.

All UCC systems used *circulating* potassium hydroxide electrolyte for concentration control and heat exchange. The reaction water was removed by transferring the water vapor through the porous electrodes (anodes and cathodes) into the two excess gas streams circulated by jetpumps. The electrolyte temperature of 65–75 °C assured a self-regulation of the operating conditions as a function of the load profile up to current densities of 100 mA/cm². The materials used at that time (1970) did not allow operating temperatures of 80–85°C, at which the hydrogen loop alone would have sufficed.

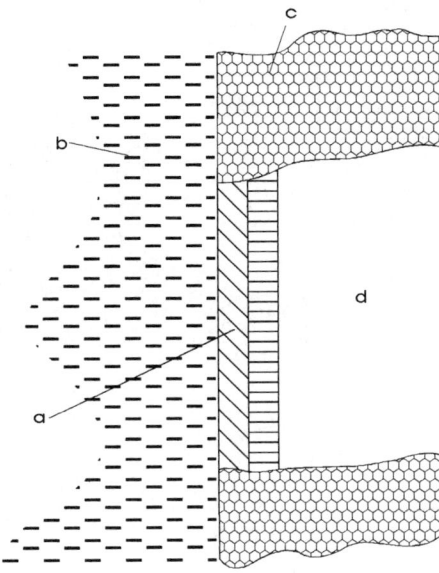

Figure 4-9. Union Carbide Fixed Zone Electrode
a) Pores filled with solution; b) KOH electrolyte;
c) Pores semiwet; d) Filled with dry gas

Eloflux System

The Varta Eloflux system as described in 1965 by WENDTLAND and WINSEL operates either as a fuel or an electrolysis cell. It includes a multiplicity of porous hydrogen and oxygen electrodes positioned alternately with porous electronically non-conductive diaphragms. The electrodes, as well as the diaphragms contain an interconnecting system of narrow pores in which the electrolyte is dispersed. The pressurized gases, hydrogen and oxygen, are dispersed in the coarse pores system of the double porous electrodes. In improved cells, for better support of the mass transport processes the electrolyte and the gases are moved independent of each other in cross flow fashion [12].

The principle of operation is as follows (Figure 4-10 A) [13]. The fuel cell (a) contains one fuel electrode (b) and two oxygen electrodes (c). Two diaphragms (d) are between them. Two additional diaphragms (e) are on the outer faces of the two oxygen electrodes which are supported by two porous support plates (f). The potassium hydroxide electrolyte is pumped from the right (g) to the left chamber (h) by means of a pump (i) via the connection pipe (j). The gases are fed through the pipes (k, hydrogen) (1, oxygen) under pressure into the big pores of the electrodes. The gases do not enter the small pores of the electrodes which are filled with electrolyte. The liquid retaining force of the small capillaries in the electrodes and in the diaphragms (asbestos) is larger than the gas pressure and a steady state is achieved, creating a large interfacial reaction zone. The diaphragm perfection is critical. Any gas cross leakage to the other electrode would cause a malfunction of the cell.

Figure 4-10 B shows an improved electrolyte circulation loop with central input and end collection of potassium hydroxide and also a method for the reconcentration of the electrolyte in the device which could be a dialytic reconcentrator or a water vaporizer.

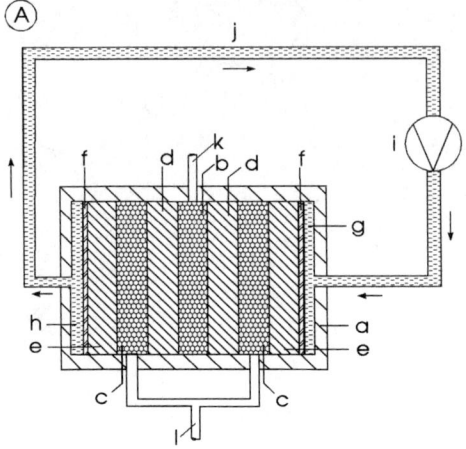

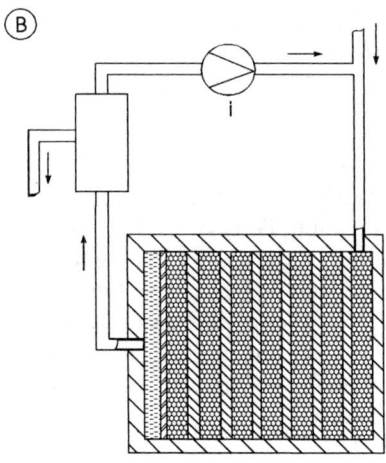

Figure 4-10. Eloflux Fuel Cell Concepts a) Fuel cell; b) Fuel electrode; c) Oxygen electrode; d) Diaphragm; e) Additional diaphragm; f) Support plate; g) Right chamber; h) Left chamber; i) Pump; j) Connection pipe; k) Hydrogen pipe; I) Oxygen pipe

Water and inert gas removal make use of the water vapor transport feature of the gas which circulates through a condenser in a portion of the stack, venting the accumulated inert gases from a remote end cell of the stack. The last cell is also used as a measuring cell, actuating the venting relay.

Elenco Fuel Cell System

Elenco (Electrochemische Energie Conversie) was founded by a consortium of the Belgian Atomic Energy Co. (AEC), the Dutch State Mines (DSM) and the Belgian Company Bekaert in the early 1970s.

The Elenco fuel cell system has a circulating 6.6 n aqueous KOH electrolyte, which is also used as coolant [14]. Operating temperatures are between 65–70 °C. In a monopolar

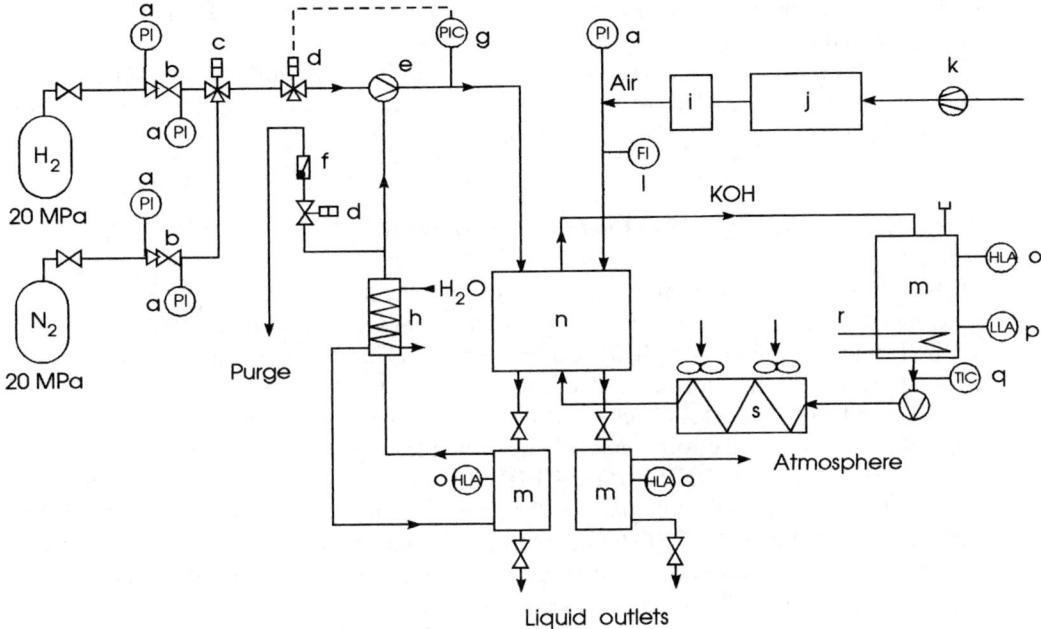

Figure 4-11. Elenco Fuel Cell System Flow Chart
a) Pressure indicator; b) Pressure regulator; c) 3/2 Way-electro valve; d) 2/2 Way-electro valve; e) Ejector; f) Non-return valve; g) Pressure indicator-controller; h) Condenser; i) Filter; j) CO_2 scrubber; k) Blower; I) Flow indicator; m) Tank, n) Fuel cell; o) High level alarm; p) Low level alarm; q) Temperature indicator-controller; r) Heater; s) Heat exchanger

cell assembly edge current collection is employed. In air cells water balance is maintained by evaporation of water through the cathodes alone. It can also be achieved by evaporation through the anodes alone, or both gas loops may be used at high current densities. Reactant gases are fed at ambient pressure. Operation with air, oxygen enriched air, or pure oxygen is possible.

The anodes and cathodes are multilayer gas diffusion electrodes. They consist of a PTFE powder-carbon-catalyst mixture which is rolled extensively to cross link the PTFE (no sintering at higher temperature is needed!). The layers are then pressed into a supporting nickel mesh. The gas side of the electrodes is covered by a porous hydrophobic (PTFE) layer. Electrode thickness is only 0.4 mm, electrode size is 17×17 cm. In 1989 the company had a theoretical production capacity of 250000 electrodes per year. The electrodes are mounted in injection-molded frames and are stacked up by a vibration welding method into modules. Present standard modules contain 24 cells, bigger units are assembled by combining different numbers of modules. Current densities for commercial cells reached 100 mA/cm^2 at 0.7V and 65°C, with minimal noble metal catalyst loadings of 0.15 mg platinum per square centimeter. Systems in the 1.5 kW, 15 kW and 50 kW range have been investigated. They were, e.g., intended to substitute small motor generators (1.5 kW), or to power electric vehicles (15 kW), and a special battery was used as electric power source in a trailer for geological activities (50 kW). The systems are represented by the flow chart shown in Figure 4-11.

A 20-module hydrogen-air battery has been tested in a van in a hybrid configuration with lead-acid batteries. The vehicle fuel cell modules are rated at 10 kW (see also Chapter 7.5.1).

European Space Agency (ESA) Fuel Cell Power Plant

1984 to 1992, ESA has been funding the development of the fuel cell power source for manned space applications. Following a technology selection performed by Dornier (G) as Prime Contractor for HERMES Program, the immobile KOH low temperature fuel cell technology has been selected. Two European technologies, a baseline with Siemens (G) and a fuel cell stack back-up with Elenco (B) have been developed.

In addition and on the similar working principle, two feasibility studies with TST (G) / IFC (USA) and DORNIER (G) / UEIP (Russia) have been made to assess the possibility of using these technologies for the first missions of HERMES. The HERMES Fuel Cell Power Plant (HFCP) should supply electrical power to the Power Subsystem and drinkable water produced by the reactants conversion to the crew members via the Liquid Management System [15, 16] (see Figure 4-12).

The low temperature (< 150 °C) and low pressure (< 5 bar) aqueous alkaline electrolyte (KOH) immobile fuel cell technology was selected in December 1988. The following table describes the main requirements for the development of this fuel cell category and the main results and development steps achieved before the program was discontinued.

Table 4-5. Requirements for the ESA fuel cell power plant

Power, min / mean / max / overload	1 kW/2 kW/6.5 kW/10 kW for 1 sec.
Voltage, min / max	75 V/115 V
Efficiency at 1 kW/4 kW/6.5 kW	60%/40%/55%
Mission Efficiency (acc. to load profile)	60%
Gas Quality, H_2/O_2	4.0/4.0
Mass/Volume	100 kg/156 dm^3
Lifetime (Storage Time/Number of Restarts)	4000 h/8900 h/200
Waste Heat, max.	5.3 kW
Product Water Quality, min/max	6.5 pH/8.5 pH
Leakage, max.	1 mg/h H_2 – 100 mg/h O_2 – 10 mg/h H_2O – 1 µg/h KOH
Start-Up	External Power Supply for 10 sec max
Switch-Off	Max. 10 sec to be safe

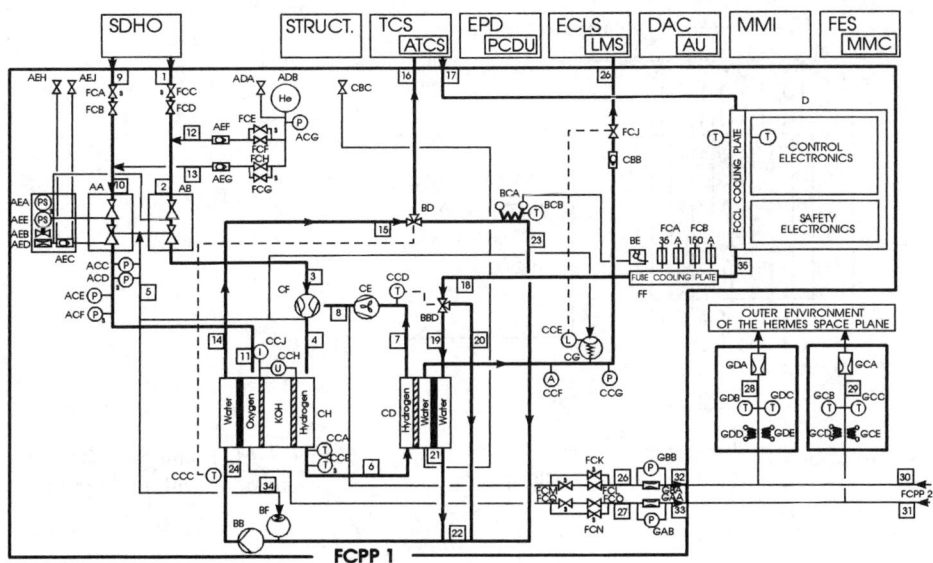

Figure 4-12. HFCP Function Diagram [17]

Equipments:

H-1JAB	A...	FCGM	Fuel Cell Gas Management Equipment
	B...	FCCO	Fuel Cell Cooling Equipment
	C...	FCSWS	Fuel Cell Stack with Water Separation Equipment
	D...	FCCL	Fuel Cell Controller Equipment
	E...	FCH	Fuel Cell Harness
	G...	FCSA	Fuel Cell Structure and Auxilliaries
	S		Sensors or Valves for Safety Electronics

Sensor ID:	A	Analysis
	F	Flow
	I	Current
	L	Level
	P	Pressure
	PS	Press. Switch
	T	Temperature
	U	Voltage

4.2.3. Further Developments

Alkaline Falling Film Fuel Cell (FFFC) by Hoechst

This type of fuel cell was being developed by Hoechst AG as a result from their research and development program for alkaline electrolysis. The increasing cost of electricity has forced Hoechst to develop Chlor-alkali electrolysis cells with a very low energy demand. The results of this development were membrane cells with an area from 1 to 3 m^2, marketed by the subsidiary UHDE GmbH.

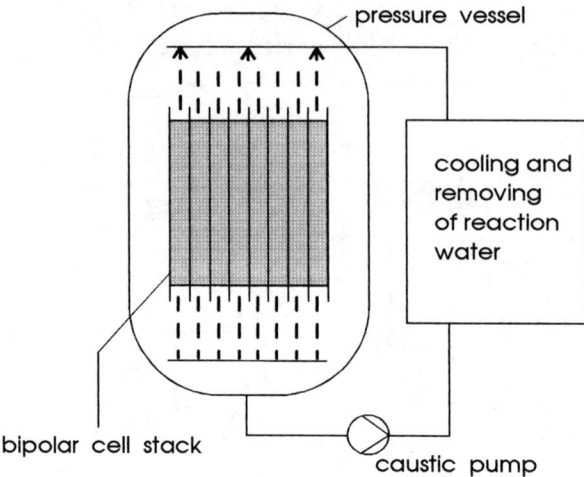

pressure vessel

cooling and
removing
of reaction
water

bipolar cell stack

caustic pump

Figure 4-13. Technical concept for
a power station with falling film

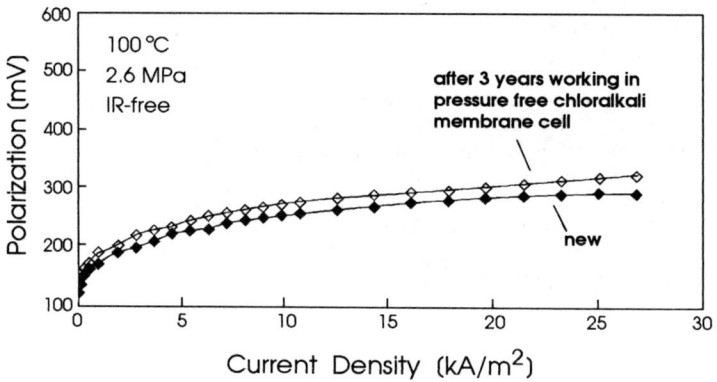

Figure 4-14. Degradation of Silflon®, operated three years in a chlor alkali membrane cell and
tested at high pressure [18]

Hoechst received their best results by its Silflon® electrodes. They contained silver and
Hostaflon®. Figure 4-14 depicts the degradation of such standard cathodes, which had
been operated for three years in a chlor-alkali membrane cell at 3 kA/m² in 80 °C NaOH.
The measured degradation over this period is ~20 mV at 2.6 MPa.

Bipolar AFC Concept by the Technical University of Graz

At the Technical University of Graz (Austria) a program aimed at the development of
low-cost PTFE bonded carbon fuel cells to be operated with air and hydrogen was started
in 1985. The goal was the design and testing of a bipolar all-carbon system (which can be

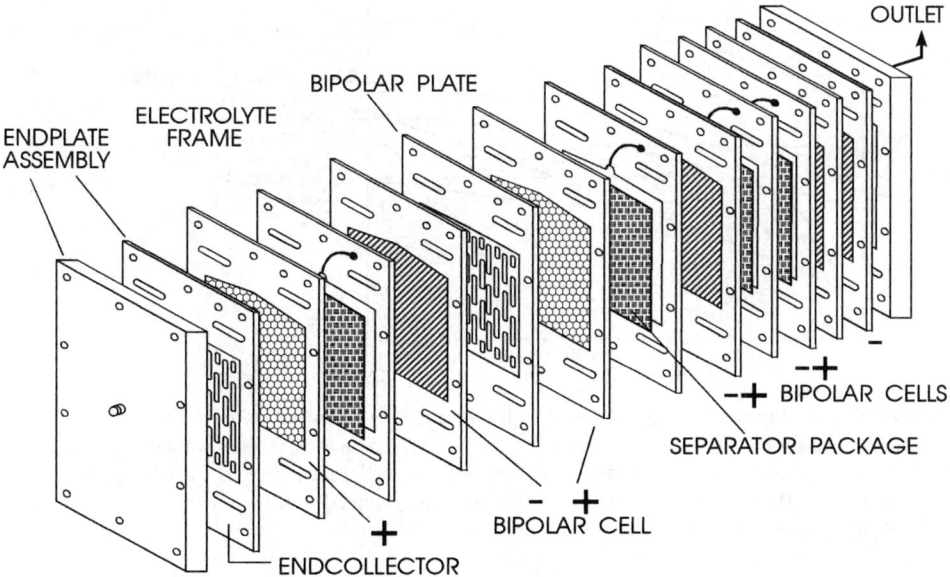

Figure 4-15. Bipolar alkaline fuel cell battery concept developed at the Technical University Graz

used with alkaline and acidic electrolytes). The system and components should be mass-producible, a minimum of noble metal catalyst should be used. Extruded carbon-polypropylene plates are used as bipolar plates and manifolds. The applications are seen in the field of fuel cells for remote locations, for the replacement of rechargeable batteries and for the propulsion of electric vehicles. Figure 4–15 shows an experimental bipolar fuel cell stack [19].

Small Scale AFC Stack Technology

India and Sweden are involved in joint development of biofuels power plants with alkaline fuel cell generators. Such village power plants with outputs of 500 kW may operate standing alone or connected to a typically distributed generation network. The ongoing second three year phase shall culminate in a process development demonstration unit comprising primary gas generation, hydrogen generation, gas clean up, cold combustion and DC/AC conversion [20–23].

O. Lindström published a very critical assessment of fuel cell technology, which indicates that the alkaline fuel cell technology has been underevaluated over the last years. He indicates in his works that most of the problems regarding the PEM Systems were minimized in contrast to the AFC systems [24]. In contrast to the fuel cell community O. Lindström is convinced that the CO_2 can be removed cost effective from the hydrogen feed (which is practiced in every NH_3 plant).

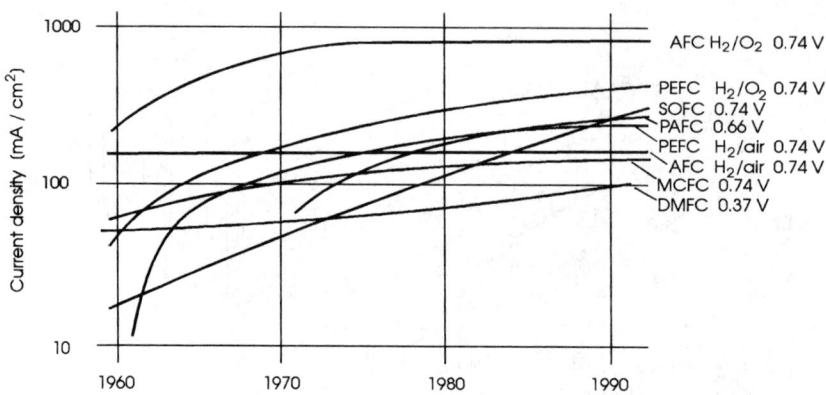

Figure 4-16. Development Curves for Fuel Cell Options "best verified state of art" [25] (A standardized approach was used for this figure; current densities at a reference cell voltage of 0.74 V have been used, under standard conditions (50% efficiency related to the higher heating value, HHV). For PAFCs, the reference voltage has to be reduced to 0.66 V corresponding to 45% efficiency because of the low operating voltage of the PAFC.)

4.3. Polymer-Electrolyte Fuel Cells (PEFCs)[1]

4.3.1. Introduction

Application of the polymer electrolyte fuel cell (PEFC) as a primary power source in electric vehicles has received increasing attention in the last few years. This attention has been fueled by a combination of significant technical advances and by the initiation of projects for the demonstration of a complete, PEFC-based power system in a bus or in a passenger car. Whether such vehicles should carry methanol fuel (to be reformed on board) or hydrogen fuel (in pressurized, liquefied, or hydride form), the PEFC cathode will operate on air in all of these terrestrial transportation systems.

The PEFC uses a polymer membrane as its electrolyte. This membrane is an electronic insulator, but an excellent conductor of hydrogen ions. The materials used to date consist of a fluorocarbon polymer backbone, similar to Teflon, to which sulfonic acid groups are attached. The acid molecules are fixed to the polymer and cannot "leak" out, but the protons on these acid groups are free to migrate through the membrane. The advantages of PEFCs against AFCs are:

– there is no free corrosive liquid in the cell,
– it is simple to fabricate the cell,
– they are able to withstand large pressure differentials
– material corrosion problems are minimal, and
– they have demonstrated long life.

---

1 The authors decided to use "PE instead of "PEM", "IEM" and "SPE".

The disadvantages of PEFCs are:

- the fluorinated polymer electrolyte is traditionally expensive and cell costs are high,
- water-management in the membrane is critical for efficient operation,
- CO tolerance is poor,
- long-term, high performance with low catalyst loadings in the electrodes needs to be demonstrated [26], and
- difficulty in thermally integrating with a reformer.

The electrode reactions in the PEFC are analogous to those in the PAFC. Hydrogen from the fuel gas stream is consumed at the anode, yielding electrons to the anode and producing hydrogen ions which enter the electrolyte. At the cathode, oxygen combines with electrons from the cathode and hydrogen ions from the electrolyte to produce water.

$$\textbf{anode: } 2\,H_2 \longrightarrow 4\,H^+ + 4e^- \tag{4-2}$$

$$\textbf{cathode: } O_2 + 4\,H^+ + 4e^- \longrightarrow 2\,H_2O \tag{4-3}$$

The water does not dissolve in the electrolyte and is, instead, expelled from the back of the cathode into the oxidant gas stream. As the PEFC operates at about 80 °C, the water is produced as liquid and is carried out of the fuel cell by excess oxidant flow (see Chapter 4.3.3).

4.3.2. History of Development

Historically, the first major application of fuel cell systems was in the Gemini space flights, and for this purpose the solid-polymer electrolyte fuel cell system of the General Electric Company was chosen. The requirement of fuel cell systems for space applications are the high power and energy densities with respect to weight and volume, high efficiency, few moving parts, minimum noise and vibration, and reliability. The General Electric solid-polymer electrolyte fuel cell system, with a power output of 1 kW, was used as an auxiliary power source in the space vehicles, and the pure water – the product of the fuel cell reaction – was used for drinking purposes by the astronauts [27].

Though the fuel cells performed quite well for space missions of 1–2 weeks duration, there were some problems with this technology, e.g. the power densities attained were still not high enough (< 50 mW/cm^2); the polystyrene sulfonate ion-exchange membrane was not stable under the electrochemical environments in the cell and the platinum loading was quite high. For these reasons, the alkaline fuel cell system was chosen for the later Apollo program and the space shuttle flights.

The General Electric Company pursued the development of solid polymer electrolyte fuel cells after their use in the Gemini space flights. The major breakthrough was the identification of the perfluorinated, sulfonic acid polymer, Nafion (produced by DuPont), as the electrolyte. The main advantages of Nafion over polystyrene sulfonic-acid membranes are:

- higher acidity because of the presence of fluorocarbon rather than hydrocarbon groups (the fluorine atom is electrophilic), and
- higher stability of the C-F compared with the C-H bond in the electrochemical environment.

However, progress in this technology was slow due to the "drying out" of the membrane during operation of the systems.

The major problem encountered with the General Electric fuel cell was keeping the membrane wet under operating conditions. Proper humidification of the reactant gases is necessary. The General Electric Company solved the water management problem by:

– internal humidification of the reactant gases in a separate chamber; and
– differential pressurization, i.e. using a higher pressure on the cathode side (10 atm with air) than on the anode side (2 atm with hydrogen).

The noble metal loading in the General Electric /Hamilton Standards-United Technology Corporation (GE/HS-UTC) fuel cell was high (4 mg/cm² Pt on the anode as well as on the cathode) because unsupported platinum particles were used as electrocatalyst. Good performances were obtained with hydrogen/oxygen as reactants at 105 °C with a pressure of 2 atm on the anode side and 10 atm on the cathode side. Thus, a cell potential of 0.825 V at 300 mA/cm² was obtained. These cells generated current densities of 1 A/cm² at a cell potential of 0.5 V. The slope of the cell potential versus current density plot in the linear region was 0.3 ohm cm².

The performance of the GE/HS-UTC was not satisfactory with air as the cathode reactant. In spite of the high pressure, the cells reached a limiting current density at only about 300 mA/cm² and 105 °C.

The Canadian Department of National Defense and the National Research Council took interest in the SPE fuel cell in 1983 as a potential power source for military appli-

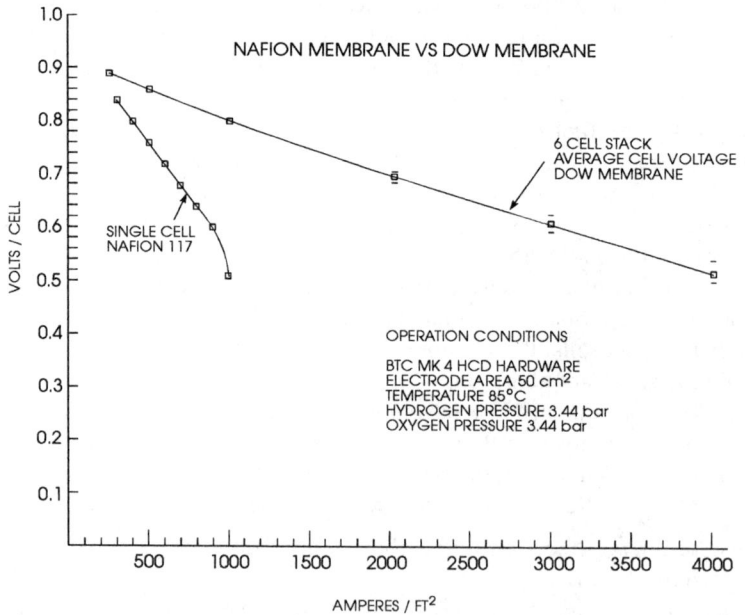

Figure 4-17. Ballard's 1987 Fuel Cell Performance with Improved Dow Electrolyte [28, 29]

cations. It was realized that the fuel cell would have to be modified to make it more affordable for this purpose. Ballard Power Systems first began work with the SPE cell in 1984 under a contract from DND. Their first objective was to design a cell which could run on air rather than pure oxygen. Another objective for reducing the cost was to replace the niobium plates in the NASA cell with a less expensive graphite version. A longer-term goal was to develop a cell that would operate on reformed hydrocarbon fuels and to develop a selective oxidation process to oxidize CO (harmful to the cell) in the reformed gas.

In 1987, Ballard made a breakthrough in PEFC performance using a new, sulfonated fluorocarbon polymer membrane from Dow Chemical, similar to the earlier Nafion membrane developed by DuPont, but capable of producing four times the current at the same operating voltages that the Nafion membrane produced. When placed in the improved Ballard hardware, the new Dow membrane produced 4000 A/ft^2 (4.3 A/cm^2) at 0.5 V/cell on hydrogen and oxygen at 50 psig (3.5 bar). This represents an area power density of 2000 W/ft^2 (2.15 W/cm^2). On air at 50 psig (3.5 bar) 1000 A/ft^2 (1.1 A/cm^2).

4.3.3. Operating Principle of PEFCs

4.3.3.1. General

The cross-section of a single polymer-electrolyte fuel cell is shown schematically in Figure 4-18. The cell typically consists of graphite bipolar plates, which are pressed against the membrane-electrode assembly. These plates have a manifold of grooves to distribute the reactant gases to the electrodes. They are also sufficiently electrically conductive to pass the generated current to the adjacent cell. Los Alamos National Laboratory (LANL) also experimented with bipolar plates made of conductive plastic and of several different metals [30].

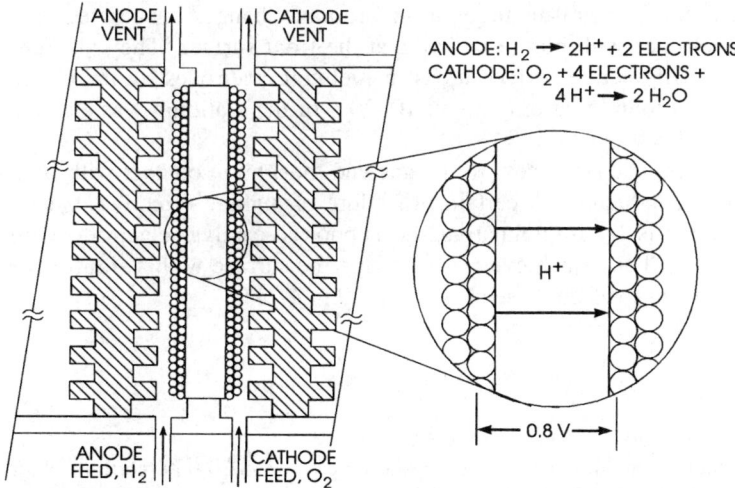

Figure 4-18. Schematic Cross Section of a Polymer Electrolyte Fuel Cell

The membrane-electrode assembly is the electrochemical heart of the system. On the anode, or hydrogen side, hydrogen gas is catalytically disassociated according to the reaction:

$$\textbf{anode: } 2\,H_2 \longrightarrow 4\,H^+ + 4e^- \tag{4-4}$$

The hydrogen ions pass through the polymer electrolyte to the cathode, or oxygen side, of the cell. There they are combined catalytically with oxygen and electrons from the adjacent cell to form water according to the reaction:

$$\textbf{cathode: } O_2 + 4\,H^+ + 4e^- \longrightarrow 2\,H_2O \tag{4-5}$$

The membrane functions twofold, it acts as the electrolyte which provides ionic communication between the anode and the cathode and it serves as a separator for the two reactant gases. Optimized proton and water transport properties of the membrane and proper water management are crucial for efficient fuel cell operation. Dehydration of the membrane reduces proton conductivity and excess of water can lead to flooding of the electrodes. Both conditions result in poor cell performance.

4.3.3.2. Electrodes

The *electrodes* are typical gas-diffusion electrodes[2]. The backing is a porous carbon cloth with a hydrophobic coating. In order for the electrochemical reactions to take place at a useful efficiency, they must be catalyzed [31–33]. To date, platinum has proven to be the best catalyst for both the hydrogen oxidation and the oxygen-reduction reactions. The problem of high platinum loading (typically 4 mg of platinum per square centimeter)[3] was solved by the use of supported platinum catalysts similar to those used in liquid-electrolyte fuel cells. These consist of 2 to 5 nm diameter Pt particles on the surface of fine carbon particles[4]. This greatly increases the effective surface area of the platinum. Ticianelli and co-workers used several methods to incorporate changes in the electrode structure and to have electrodes with platinum layers at the front surface. They also determined the electrochemically-active surface areas in fuel cell electrodes by the cyclic voltammetric technique, and found that only about 10–20% of the platinum was electrochemically active in the fuel cell reaction [34–36].

To function, the catalyst must have access to the gas and must be in contact with both the electrical and protonic conductors. A certain procedure of making effective contact with the protonic conductor is by impregnating the supported-catalyst electrode with protonic conducting material. This is achieved by covering the surface with a solution of solubilized membrane material (Nafion®) (see Figures 4-19 to 4-21).

---

2 E-TEK, Inc., 6 Mercer Road, Natick (MA) 01760, USA
3 With such a platinum loading, an automobile would require up to US-$ 10 000 worth of platinum. Surface area is between 140 and 25 m^2/g
4 Surface area is between 140 and 25 m^2/g

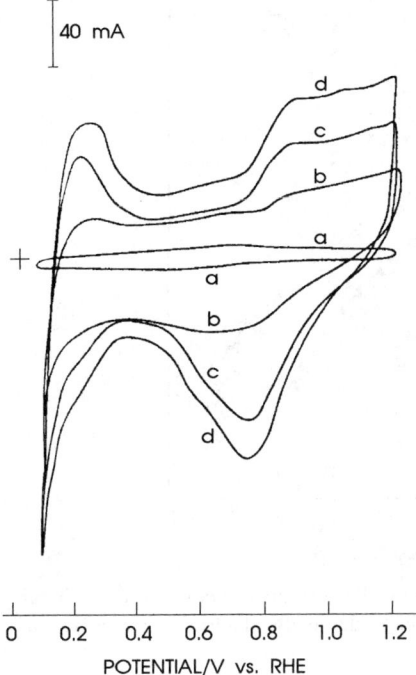

Figure 4-19. Cyclic Voltammograms at 50 °C on Prototech Electrodes: (a) uncatalyzed; (b) 10 wt.% Pt/C; c) 20 wt.% Pt/C plus 50 nm sputtered film of Pt (v = 100 mV/s). Note the increase of electro-chemically-active surface are measured by hydrogen or oxygen adsorption/desorption change with increased localization of Pt near front surface [37].

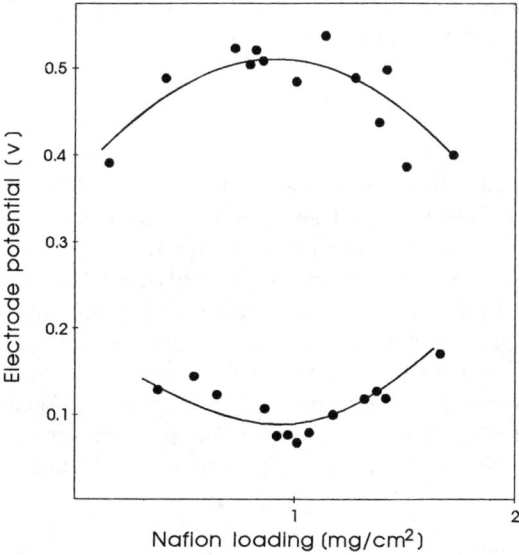

Figure 4-20. Effect of Nafion® Loading on Anodic and Cathodic Voltage of 200 mA/ cm² [38]

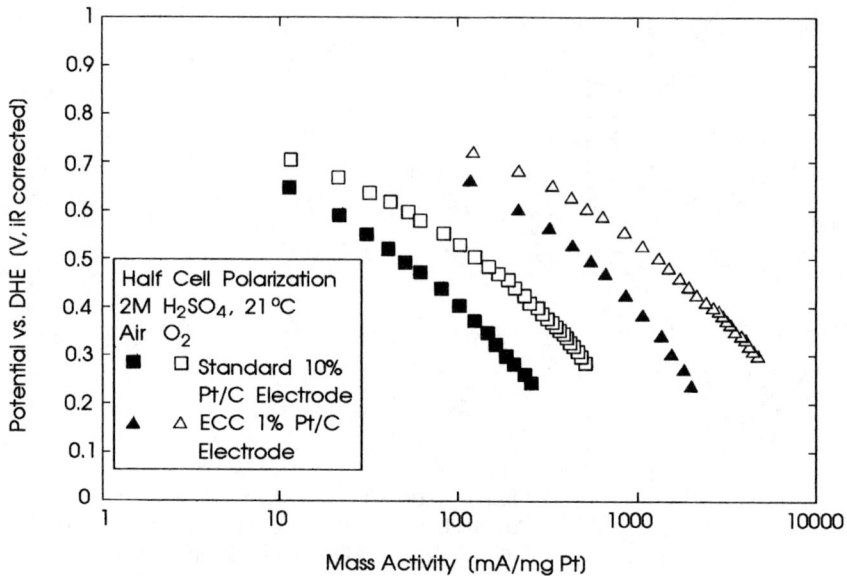

Figure 4-21. Comparison of O_2 Reduction Polarization Performance between PSIT High-Utilization PT/C Electrode and Standard 10% PT/C electrode. Both electrodes were Nafion® impregnated using the PSIT technique [41].

New electrochemical canalization techniques for producing gas-diffusion electrodes for PEFC applications, in which platinum catalyst particles are in regions of the electrode that are in ionic contact with the proton-exchange membrane and in electronic contact with the carbon support, propose 0.05 mg Pt/cm^2 [39, 40].

Attempts to reduce the high platinum content in the electrodes (4 mg/cm²) were carried out by various research groups, but mainly by LANL [42–46].

4.3.3.3. Electrolytes-Membranes

The proton-exchange membrane fuel cell (also known as the solid polymer fuel cell) uses one of several ion-exchange polymers as its electrolyte. These polymers are electronic insulators, but excellent conductors of hydrogen ions. The early membranes (U.S. Gemini space program) tested in PEFCs include the hydrocarbon-type polymers such as crosslinked, polystyrene-divinylbenzene sulfonic acids and sulfonated phenolformaldehyde. Hydrocarbon-type polymers are unstable because of C-H bond cleavage, particularly at the α-H sites where the functional groups are attached. When these polystyrenes were replaced with fluorine-substituted polystyrenes (e.g. polytrifluorostyrene sulfonic acid), the life of the PEFCs was extended by a factor of four to five. The research started in the early 1960s with the development of Nafion®5. This material, which is still used to

---

5 Registered trademark of E.I. DuPont de Nemours

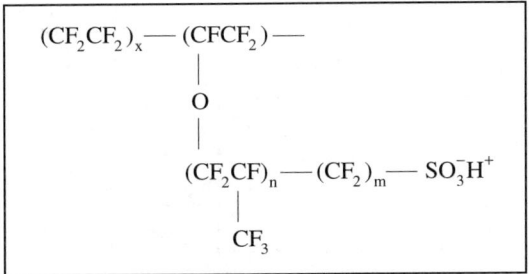

Membrane	Equiv. weight g/mol SO_3^-	Thickness in dry state, μm	Water Content %	Conductivity $\Omega^{-1}\,cm^{-1}$
Dow	800	125	54	0.114
Aciplex®-S	1000	120	43	0.108
Nafion®-115	1100	100	34	0.059

Figure 4-22. Structure of perfluorocarbon ion exchange polymers

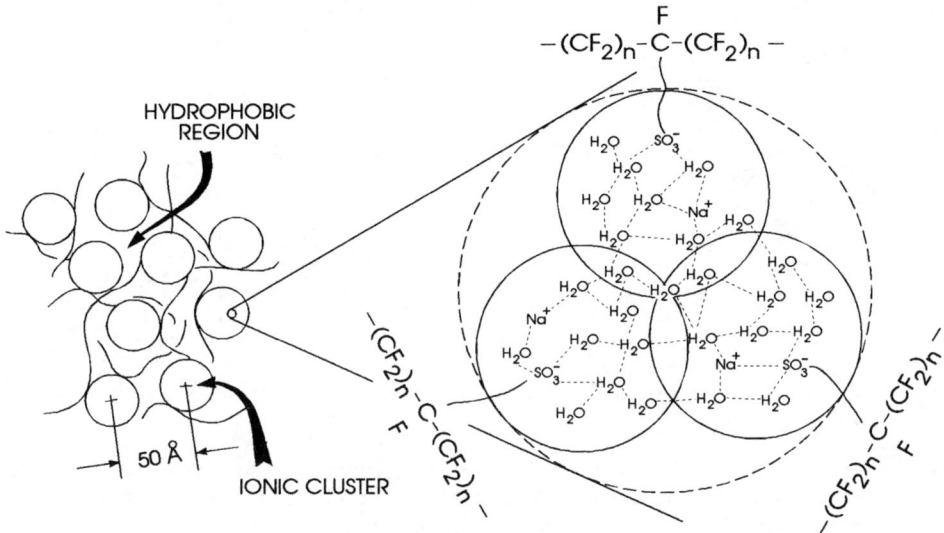

Figure 4-23. Schematic Representation of Ion Clustering in Nafion® [47, 48]

date, consists of a fluorocarbon polymer backbone, similar to Teflon, to which sulfonic acid groups have been chemically bonded. The acid molecules are fixed to the polymer and cannot be leached out, but the protons on these acid groups are free to migrate through the electrolyte (see Figures 4-22 and 4-23).

The electrolyte is used in the form of a very thin membrane, typically 50 to 175 μm thick, and can be handled easily and safely. The material is a relatively dilute acid with

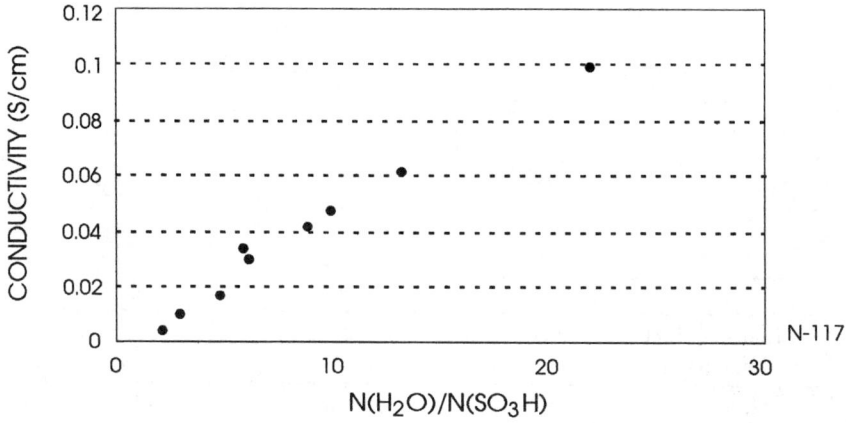

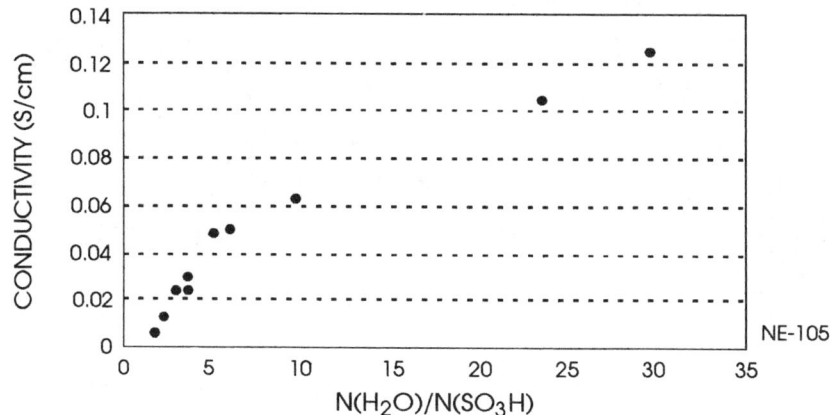

Figure 4-24. Conductivity of Nafion® Membranes as a Function of Water Content at 30 °C [49]

about the same conductivity as 1 M sulfuric acid. It behaves as a typical aqueous acid from the viewpoint of conductivity, so that its use is limited to temperatures below the boiling point of water.

The latest version from DuPont is Nafion®-105 (see Figure 4-24), which has similar ionic properties to materials from the Dow Chemical Company (for which n = 0 and m = 2) and the materials known as Aciplex-S® from the Asahi Chemical Industry Company, in which n = 0 – 2, and m = 2 – 5. The materials are made using a slightly different chemistry and have slightly different properties, with equivalent weights[6] of 1100 800

---

6 Defined as the weight of polymer that will neutralize one equivalent of base, and is inversely proportional to the ion-exchange capacity.

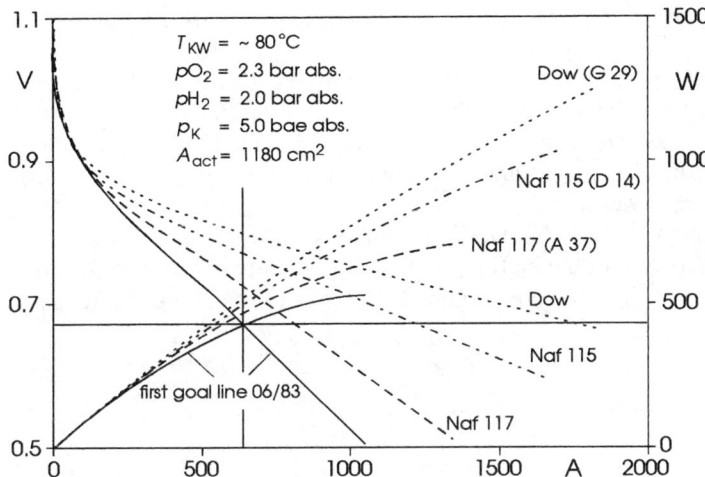

Figure 4-25. Potential Performance Increases with the Use of Different Nafion® and Dow Membranes [50, 51] (Courtesy of Siemens, Erlangen)

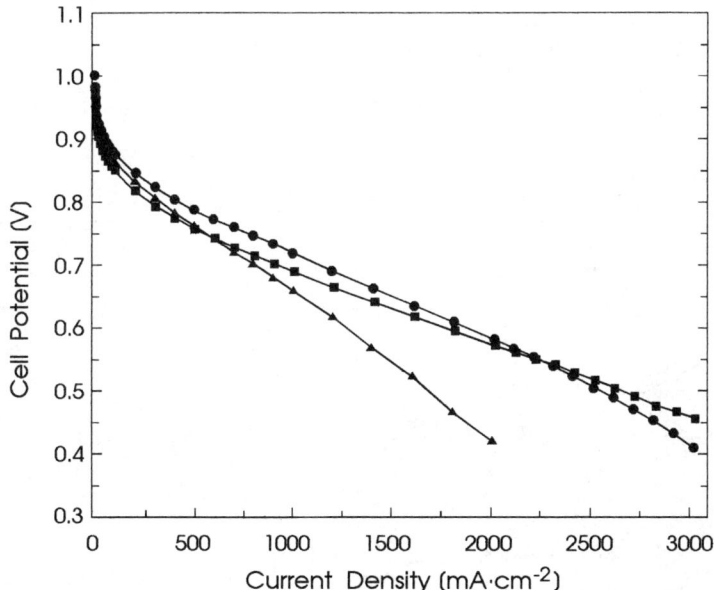

Figure 4-26. Effect of Proton Conducting Membrane on PEFC Performance – H_2/O_2 Reactants (E-TEC electrodes – 20% Pt/C, 0,4 mg cm^{-2}), 95 °C, 5 atm; Aciplex-S® 1004 (●); Dow (■); Nafion®-115 (▲) [52]

and 1000 respectively. Figure 4-25 shows performance data achieved with different Nafion membranes.

However, the open-circuit potential of the thinner membranes (Nafion®-105) show a lower open circuit Voltage (50–100 mV!) than the thicker ones (Nafion®-117) because of cross-over of the hydrogen gas. The cross-over effect is also reflected by a lower potential in the cell with the 50 µm thick membrane compared to the cell with a 175 or 125 µm thick membranes at low current densities.

The PEFCs with Aciplex-S® and the Dow membranes show considerably better performances than the PEMFC with the Nafion® membrane, mainly because of the considerably lower R values for these two cells. Figure 4-26 also shows that the PEFC with

a)

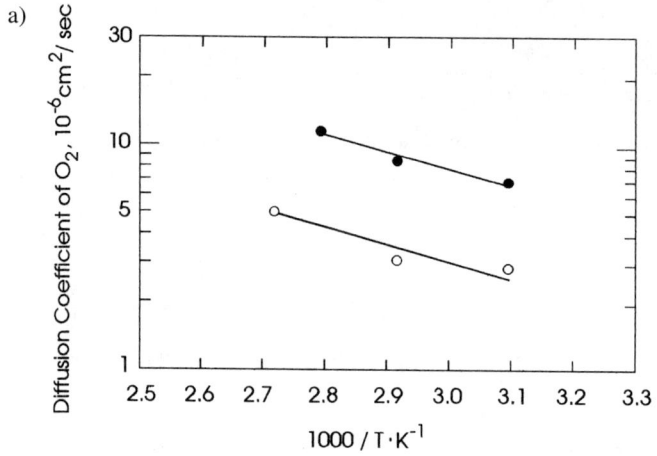

b)

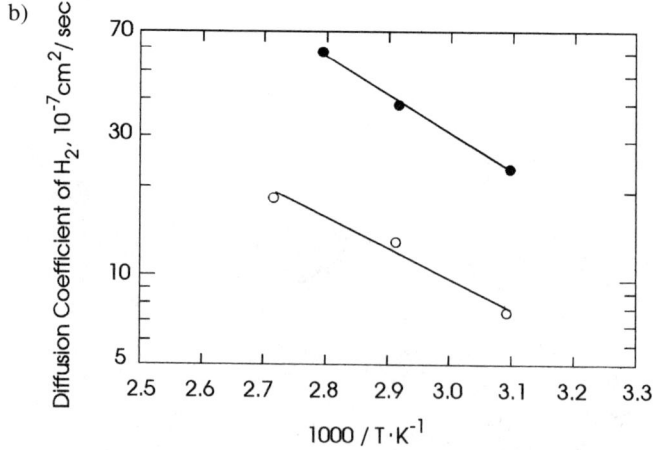

Figure 4-27. Arrhenius plots for diffusion coefficients of H_2 and O_2 in PEMFCs with a) Aciplex®-S (µ) and b) Nafion®-117 membrane (λ) [55]

the Dow membrane exhibits the pseudo-linear behavior even at current densities higher than 2 A/cm^2, but above this current density the PEFC with the Aciplex-S$^®$ membrane shows some mass-transport limitations. Comparisons between Nafion$^®$117 and Aciplex-S$^®$ in which the transport parameters of hydrogen and oxygen were investigated are reported in references [53, 54]. Arrhenius plots for diffusion coefficients are shown in Figure 4-27.

Further Developments

Ballard Advanced Materials (BAM) (Canada) reported results from their project "Low-Cost Membrane for Commercial PEFC". Parameters were optimized to produce a cheap membrane for a target price of Can-\$ 110–150/m^2. Details of the monomer and polymer synthesis have not been published yet. The high performance of a BAM3G membrane in a MK 4 cell in comparison to a Nafion$^®$117 and a Dow membrane is displayed in Figure 4-28.

The Fuel Cell Group at the Paul Scherrer Institute (PSI) (Switzerland) utilizes the radiation-grafting method for membrane preparation [58, 59]. Highly crosslinked membranes of the system FEP (perfluoroethylene-perfluoropropylene copolymer)-styrene/

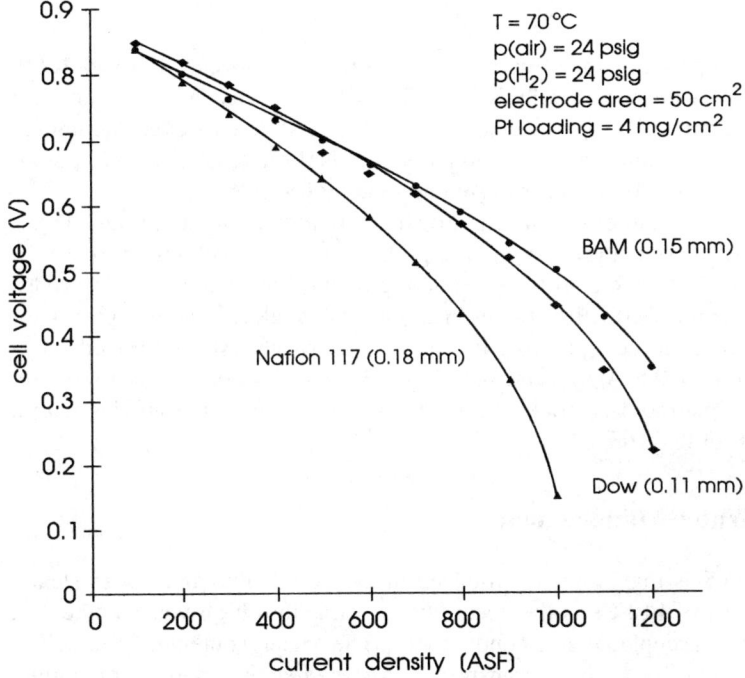

Figure 4-28. Performance Data for Ballard's BAM3G in Comparison to Nafion$^®$117 and Dow-Membrane [56, 57]

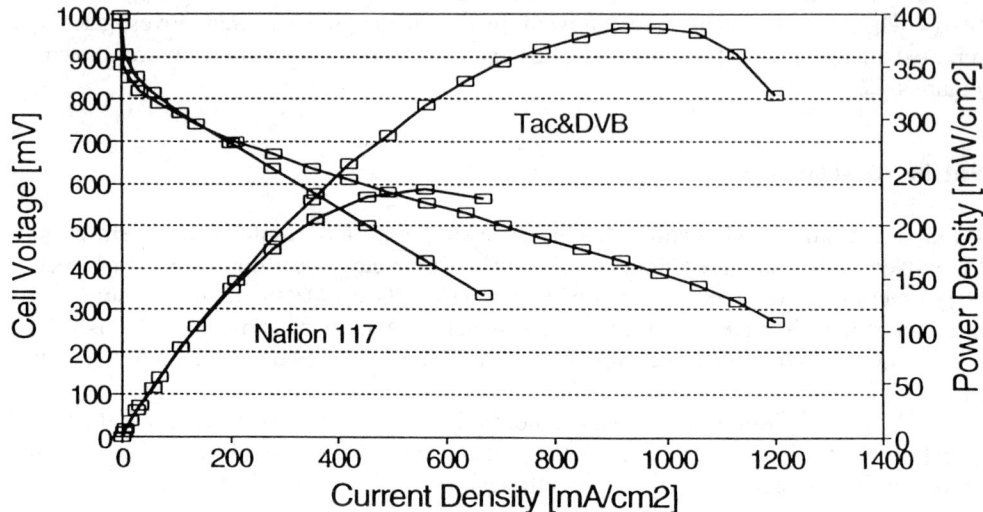

Figure 4-29. Performance Data for TAC&DVB and Nafion®117 after 400 h at 60 °C (ambient air pressure [62]

divinyl benzene were prepared. Typical equivalent weights are between 800 and 1000. Membranes with triallylcyanurate in addition to divinylbenzene as crosslinkers were prepared by the same method. The performance of a DVB-TAC crosslinked membrane in a fuel cell is displayed in Figure 4-29 in comparison to Nafion®117. A power output of close to 400 mW/cm^2 was published for this membrane by PSI [60, 61].

In addition to the development efforts at Ballard and PSI, alternative membrane development is also being carried out at the Loughborough University of Technology (UK).

The secret of using PEM-film electrolytes lies in the method of bonding of electrodes to the surface, to give the most effective, three-phase electrolyte electrocatalyst gas-phase boundary, thereby making the best possible use of the electrocatalyst. Optimized interfaces may achieve 0.6 V at 0.5 A/cm^2 and 0.7 V at 0.3 A/cm^2 at atmospheric pressure using carbon-supported platinum cathodes with 0.1 mg/cm^2 loading, and anodes with a loading of 0.05 mg/cm^2 or less [63, 64].

4.3.3.4. Heat and Water Management

One distinction of PEFCs is that water is produced in the liquid state and not as steam. One of the critical issues of PEFCs is the necessity to maintain a high water-content in the membrane to obtain acceptable ion conductivity. The water content in the cell is determined by the balance of water or its transport during the reactive mode of operation. Water-transport processes are a function of the current and the characteristics of both the membrane and the electrodes. Influencing the water transport are the water drag through the cell, back diffusion from the cathode, and the diffusion of any water in the membrane and the electrodes themselves. A detailed description of the many proposed mechanisms of water transport is beyond the scope of this publication.

Furthermore, the discussion of experimental values referring to water transport still continues. Starting with 3.5–4 H_2O molecules per H^+ in 1977 [65], Koch et al. reports no water transport at current densities up to 500 mA/cm^2 in a cell with Prototech electrodes and Dow membrane at 50 °C and atmospheric pressure [66]. On the other hand, water transport of 0.67 H_2O molecules per equivalent of H^+ was derived at 500 mA/cm^2 from the results of Springer et al. in a cell with a Nafion$^®$117 membrane at 80 °C [67]. In 1991 Springer et al. reported a value of 0.2 mol H_2O/H^+ in a cell at 80 °C and 1 atm (Nafion$^®$117) [68, 69]. A one-dimensional, steady-state model of complete PEFC was used in this experiment.

Several contributions have been made to the discussion of modeling and experimental diagnostics in Polymer-Electrolyte Fuel Cells [70–80]. Perhaps the most important result of such attempts should be a set of diagnostic experiments which, with the aid of the model, would enable an effective identification of the origin and magnitude of losses in the operating cell.

Most of the product water from a PEM fuel cell is traditionally removed from the cathode, where it is produced, by excess flow of oxidant gas. Typically, the oxidant flow represents twice the stoichiometric amount of oxygen required in the fuel cell. A novel method of water management has been recently developed at Ballard. With appropriate stack design, liquid water accumulated in the cathode can be drawn across the membrane to the anode by a concentration gradient, and removed in the fuel stream. This method of

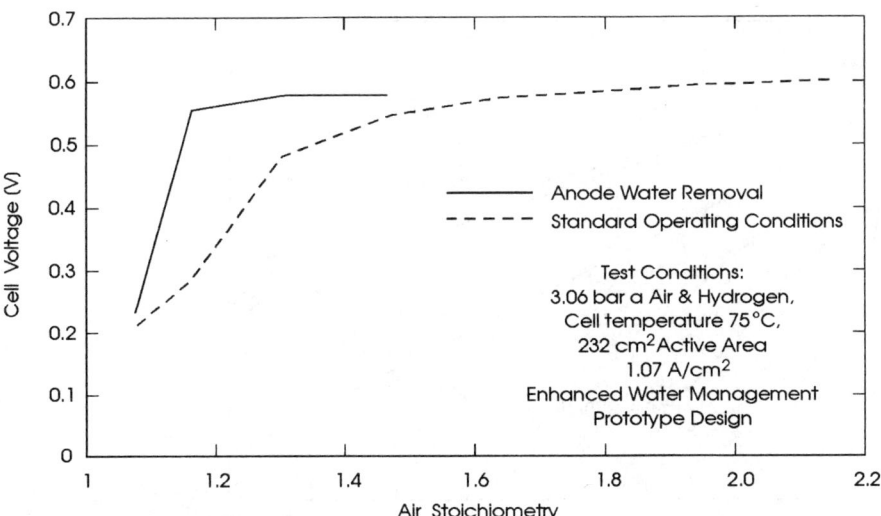

Figure 4-30. Effect of Anode Water Removal at Low Air Stoichiometry on Cell Performance at 1.08 A/cm^2 (232 cm^2 active area, Dow XUS-13204.10 membrane, $P_{H2} = P_{Air} = 3$ bar, $T_{cell} = 80$ °C) [81].

water management, "anode water removal", can significantly reduce parasitic loads associated with the oxidant side of the fuel cell, and results in a totally new approach to fuel cell system design. Using this approach, Ballard had been able to run fuel cells at air stoichiometries close to 1.0 with no significant loss in performance (see Figure 4-30).

Another approach for water management in PEFCs is the direct supply of liquid water through wicks (porous polyester threads) fitted directly to the membrane. The main advantage of this kind of humidification is its easy management [82]. A. Cisar et al. reported the development of an internally humidified membrane whereby the water is fed directly to the entire area of the membrane [83].

4.3.4. Performance of Polymer-Electrolyte Fuel Cells (PEFCs)

A summary of the performance levels achieved with PEFCs since the mid-1960s is presented in Figure 4-31. Due to changes in the operating conditions (involving pressure, temperature, reactant gases etc.) a wide range of performance levels has been obtained. The performance of the PEFC in the US Gemini Space Program was 37 mA/cm^2 at 0.78 V in a 32-cell stack that typically operated at 50 °C and 2 atm [84]. Current technology yields performance levels which are vastly superior. Recent results from the Los Alamos National Laboratory show that a performance of 0.5 V at 3.0 A/cm^2 ($T_{cell}/T_{humidifiers}$ = 80/105 °C – anode/cathode pressure = 3/5 atm) can be obtained in PEFCs for a directly-catalyzed, developmental Dow membrane with catalyst loadings of 0.13 mg Pt/cm^2/ electrode at 80 °C [85].

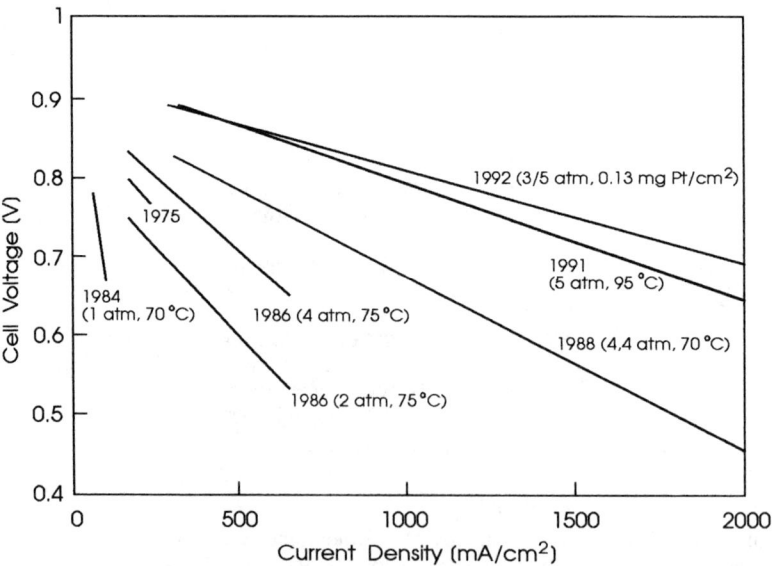

Figure 4-31. Performance Changes in PEFCs [86–89]

Performance analyses of the cell voltage (V) – current density (i) relationship for PEFC electrodes and systems were made by Srinivasan [90], RMC [91] and DSTO [92]. Attempts for dynamic modeling of PEFC are emphasised of research in the University of Ottawa [93].

4.3.4.1. Influence of Temperature

Both the temperature and pressure of operation of the cell have a significant influence on its performance. The effect of temperature on performance is illustrated in Figure 4-32 for the temperatures 50 to 95 °C at a pressure of 5 atm. The slope of the linear region of the cell potential versus current density plot decreases, which is indicative of a lowering of the internal resistance of the cell. This decrease is predominantly due to the decrease in ohmic resistance of the electrolyte. The mass transport limitations, caused by diffusion of reactants through the PEM assembly to the active Pt sites, the movement of protons from the anode to the cathode and the removal of product water are also reduced with the increase in temperature.

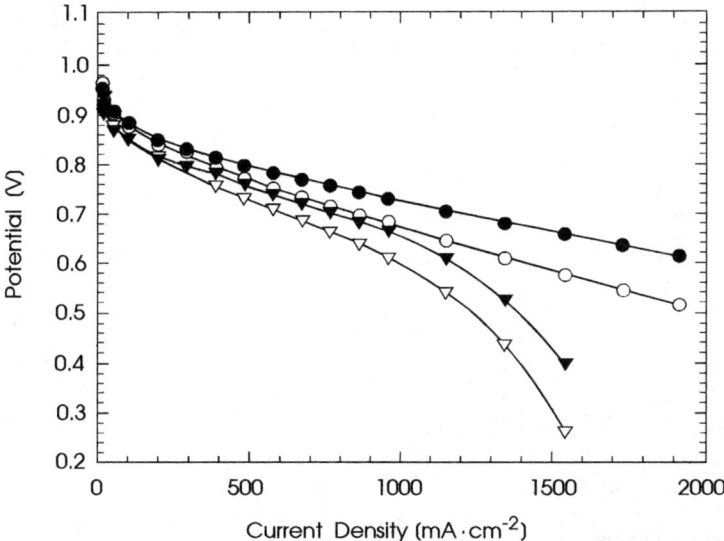

Figure 4-32. Effect of Temperature and Reactant Gas on the Performance of a PEM Fuel Cell, Pt loading of electrodes 0.45 mg/cm², Dow membrane, 5 atm. (●) 95 °C oxygen (○) 50 °C oxygen, (▼) 95 °C air, (▽) 50 °C air.

4.3.4.2. Influence of Cathodic Reactant Composition and Pressure

For terrestrial applications, the source of oxygen is air, whereas for extraterrestrial (space, underwater) applications pure oxygen is used. Experiments using both these gases in single cells operating at pressures from 1 to 5 atm show the significant influ-

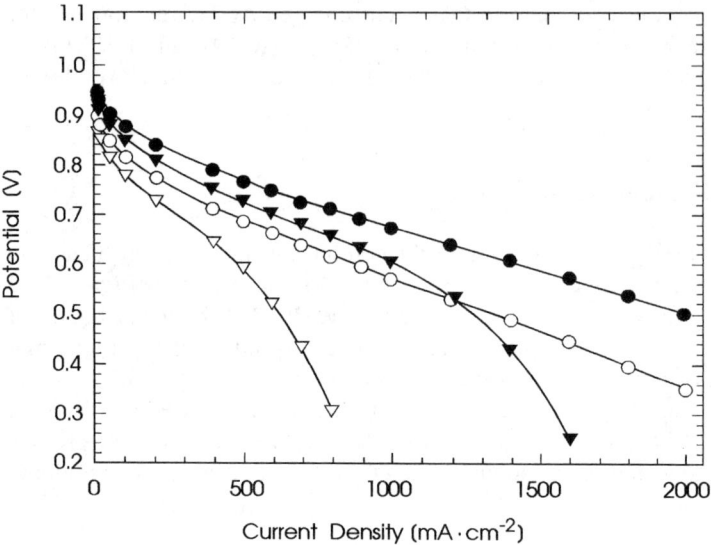

Figure 4-33. Effect of Pressure and Reactant Gas on the Performance of a PEM Fuel Cell; Pt loading of electrodes 0.45 mg /cm^2, Dow membrane, 50 °C, (●) 5 atm, O_2, (○) 1 atm, O_2, (▼) 5 atm, air, (▽) 1 atm, air.

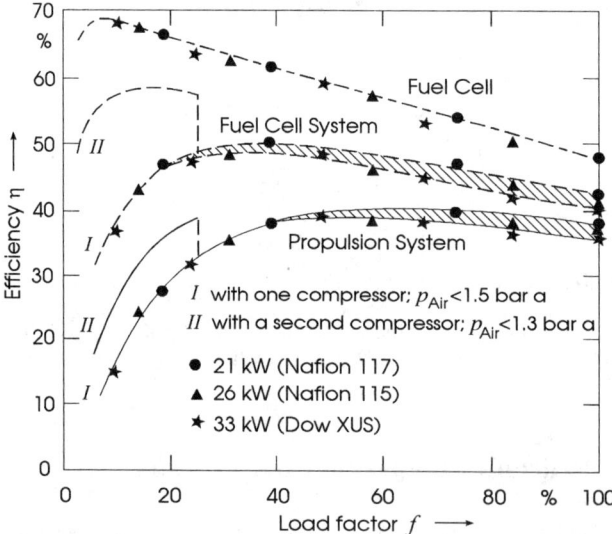

Figure 4-34. Performance Losses in a PEM Fuel Cell with Varying Load Factor [94] (Courtesy of Siemens, Erlangen)

ences of mass transport phenomena in the cell. Figure 4-33 demonstrates the effect of pressure at 50 °C on the cell potential versus current density plot using H_2/O_2 and H_2/air as reactants.

Two effects are apparent on changing over from oxygen to air as the cathodic reactant:

–   the slope of the linear region of the V-i plot is 50% higher, and
–   deviation from the linear region is visible at lower current densities.

In the higher-temperature fuel cells, (PAFC, MCFC, SOFC), the oxygen gain in potential is constant at all current densities in the linear region, and thus the slopes of the V-i plot are the same with H_2/O_2 and H_2/air as reactants. The dependence of the slope of the V-i plot on the oxygen partial pressure is indicative of mass-transport limitations, which also cause a linear variation of V with i at lower current densities. It is generally accepted that there is a 'nitrogen barrier layer effect' which causes this mass transport limitation. The second effect, i.e. deviation from the linear region of the V versus i plot at lower current densities with air, is clearly due to the effect of oxygen partial pressure. Figure 4-34 shows the effect of parasitic losses for air compression on performance efficiency.

4.3.4.3.   Influence of CO in the Fuel Gas

One of the problems with methanol-water reformers is that the reaction is often not complete, leaving small concentrations (approximately 1%) of carbon monoxide in the fuel stream. This carbon monoxide is an extremely effective poison for the fuel cell catalyst. Even a few parts per million CO produce a substantial degradation in fuel cell performance, particularly at high current densities [95, 96] (see Figure 4-35).

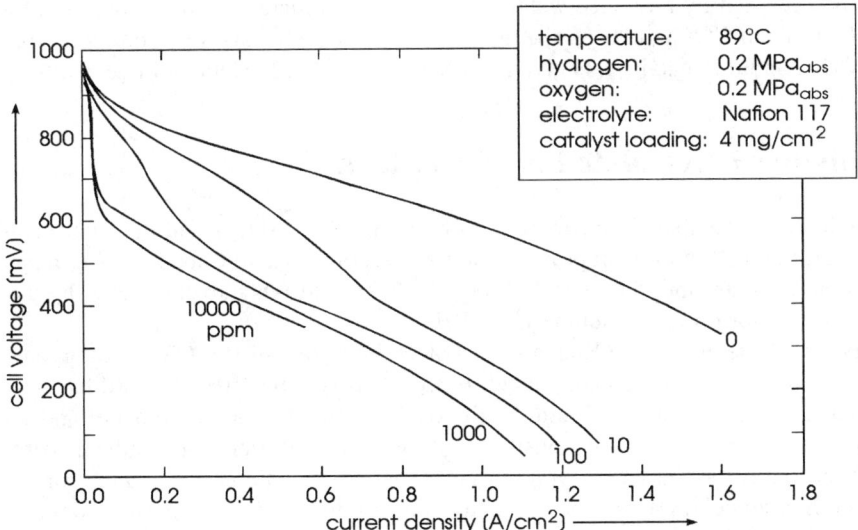

Figure 4-35. Influence of CO on the V-I characteristics of a PEM Single Cell [97]

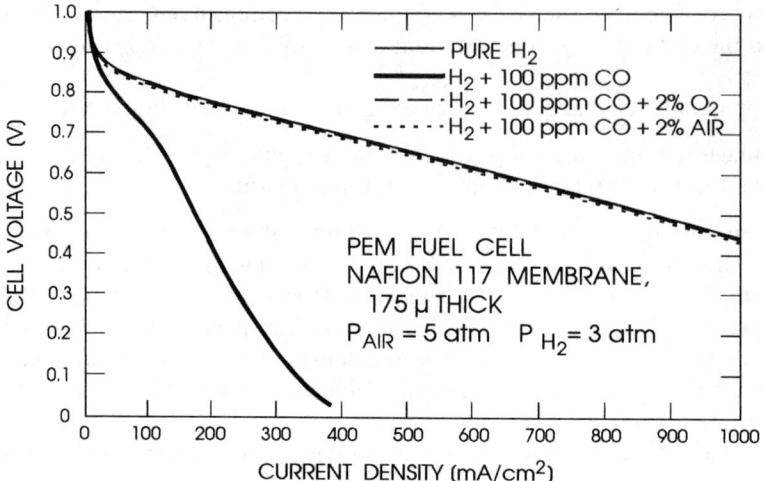

Figure 4-36. Residual Catalyst Poisoning by Trace Amounts of CO With and Without Injecting ~2% Air into the Fuel Feed Stream [99, 100]

One approach to solve this problem is to reduce the carbon monoxide content of the reformate. This has been achieved by selectively oxidizing the CO to CO_2. To achieve this, the reformate passes through a small reactor containing a platinum catalyst. The injection of a small amount of oxygen or air into this reactor causes a significant reduction of the CO content with relatively little hydrogen consumption [98] (see Figure 4-36).

The removal of CO in the reformate by partial oxidation to below shift-equilibrium levels may be thwarted by a reverse water-gas shift of reformate, i.e. the formation of a few ppm of CO from CO_2 and H_2 on the anode catalyst, which causes poisoning. A discussion of CO-tolerant Pt catalysts is given in Chapter 4.7. Direct Methanol Fuel Cells.

4.3.5. Polymer Electrolyte Fuel Cell Stacks

Few companies have published information on building PEFC stacks and their performances, although fuel cell work is in progress in many laboratories in Canada, USA, Japan and other countries around the world. Siemens (Germany) is developing a hydrogen/oxygen system for powering submarines [101].

Ballard Power Systems Inc. in Canada has advanced the state of PEFC technology and has demonstrated substantial increases in cell power density [102–108]. Ballard has also demonstrated fuel cell operation with air as the oxidant. In addition, a methanol and air PEFC system was demonstrated. It consists of a methanol reformer, an oxidizer to remove CO from the reformed gas stream, an air compressor and control systems for the reformer and the fuel cell system. The company is also involved in a program to demonstrate the operation of a transit bus using a 105 kW PEFC system [109]. Activities in stationary PEM applications were carried out in cooperation with the Dow Chemical

Figure 4-37. Ballard's 30 kW on-site power plant (Courtesy of Ballard Power Systems)

Company [110]. The construction and operation of a 30 kW hydrogen-fueled power plant and a 10 kW natural-gas fueled power plant marked the completion of the first phase of their utility demonstration program [111] (see Figure 4-37).

A typical PEM fuel cell stack is shown in Figure 4-38. This stack produces about 5 kW of power when supplied with hydrogen and air at 3 atm (30 psig). The stack is 25 cm×25 cm×46 cm in size and weighs 45 kg. Ballard dramatically improved the stack power density from 5 kW (1991) to 10 kW (1993) and 20 kW (1994) within three years. All these fuel cell stacks have about the same weight and volume. Ballard's goal in 1995 is 25 kW from this weight and volume, the threshold for powering an automobile.

At the International GROVE-Fuel Cell Conference in London, September 19–21, 1995, Ballard Power Systems, Inc. announced in a press release that this goal was already exceeded. Their fuel cell can now deliver 700 W per kg and 1000 W per litre.

A number of companies in the USA are developing PEFC stacks and the Dow Chemical Company and DuPont are also developing membranes. General Motors Corporation, in collaboration with Ballard Power Systems, Dow Chemical and LANL, is involved in a program sponsored by the US Department of Energy for the development of PEFC stacks for electric-vehicle applications.

Figure 4-38. Ballard MK-5 fuel cell stack (Courtesy of Ballard Power Systems)

The target of all of these automotive programs is the development of fuel cell stacks with higher power density, in the range of 500 to 1000 W/kg, which are manufacturable in quantity at low cost.

4.3.6. Regenerative Fuel Cell Systems

The PEFC would be useful in many terrestrial and space applications. However, the fuel cell would serve better if it were regenerative, i.e., if it could perform both as a fuel cell and an electrolyzer. This versatility is desirable because a solar array could conveniently provide power to electrolyze water during the sunlit portion of a space-vehicle or space-station orbit and the same unit could then function as a fuel cell to generate power from hydrogen and oxygen during the dark portion of the orbit.

In the design of a regenerative fuel cell, two approaches can be adopted:

– separate fuel cell and electrolysis units, or
– only one unit functions as a fuel cell and an electrolyzer. Such a fuel cell is known as a 'unitized regenerative fuel cell'. In the present state of technology, a PEFC can only provide power and cannot function as an electrolyzer. In fact, if by chance the unit is polarized to split water, the fuel cell fails. A schematic of such a unitized, regenerative fuel cell system is shown in Figure 4-39.

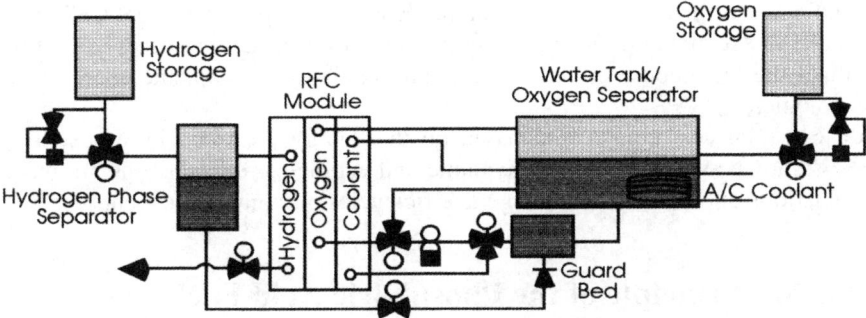

Figure 4-39. Schematic of a Unitized Regenerative Fuel Cell [112]

Considerations for the development of this fuel cell system have been published by various authors [113–116]. Investigations to improve the bifunctional electrodes in the various system designs are still necessary, and focus mainly on activity and stability of the oxygen electrode and on cell and electrode design. Systems using hydrogen – halogen concepts for regenerative fuel cells have been also discussed in the past. The hydrogen – bromine system is particularly favourable with respect to the electrochemical reaction kinetics [117–119].

4.4. Phosphoric Acid Fuel Cells (PAFC)

4.4.1. Introduction

The only fuel cell close to commercialization to date (not counting the alkaline fuel cell for spacecrafts) is the phosphoric acid cell. Acid Fuel Cells with phosphoric acid as electrolyte use relatively clean, reformed fuels (Natural Gas, LPG, light distillates, etc.) or cleaned-up coal gas from a gasifier. The two applications envisioned for the initial marketing efforts are:

1) dispersed power plants, in sizes of 5 ~ 20 MW AC, mainly using reformed hydrocarbons (Natural Gas), (Section 5) and
2) on-site cogeneration plants of about 50–1000 kW AC, intended for the supply of electricity from reformed Natural gas to commercial premises, apartments, or utility buildings, with Fuel Cell waste energy used for water and space heating, absorption, air-conditioning cycles, etc. (Section 6).

These are the only major market segments that have so far been identified for cost-effective technological use of current PAFCs in mature volume production. In each sector, penetration will be determined by the usual market-place considerations: final product cost compared with costs for competing technologies. Nevertheless the community and the legislator have to decide to give support for the introduction of sustainable energy conversion technologies with such important features like those apparent in Fuel Cell technology. This means excellent efficiencies, conservation of fossil fuel resources, reduction of acid rain and the greenhouse effect, and the introduction of sustainable energy conversion technologies.

In chapter 6 a total cost comparison based on the energy produced by competitive technologies is shown. Moreover thermodynamic and ecological comparisons are made based on evaluations of energetic and exergetic efficiencies of competitive technologies.

4.4.2. Reaction Principle of the Phosphoric Acid Fuel Cell

The slow kinetics of oxygen reduction is one factor that limits the performance of fuel cells utilizing oxygen electrodes [120] (see Table 4-6).

Table 4-6. Activation Polarization[a] for Oxygen Cathodes in Various Fuel Cells

Fuel Cell	Anode	Cathode	Temperature [T]	Polarization [V]
AFC	Pt	Pt	65	0.1[b]–0.25 (0.35)
PEMFC	Pt	Pt	100	0.25
PAFC	Pt	Pt	~190	0.4
MCFC	Ni	NiO	650	<0.1

a 200 mA/cm^2 in air
b Pure oxygen at cathode instead of air.

Oxygen reduction kinetics is more rapid in alkaline electrolytes than in acid electrolytes, consequently, the use of noble metal electrocatalysts is necessary for PAFCs. Examples of polarization curves for oxygen reduction (0.1 MPa O_2) in gas-diffusion electrodes in acid and alkaline electrolytes are presented in Figure 4-40.

Beside of these disadvantages phosphoric acid offers many other advantages as a fuel cell electrolyte, including (1) excellent thermal, chemical and electrochemical stability, (2) relatively low volatility at temperatures above 150 °C Other inorganic acids such as $HClO_4$, H_2SO_4, HF, and HCl were evaluated in laboratory-scale fuel cells. Compared to H_3PO_4, these acids have lower thermochemical stability and higher vapor pressure, and therefore are not acceptable at high temperatures (~200 °C). For these reasons, H_3PO_4 is the only inorganic acid that is being developed for use in fuel cells by utility and industrial companies. A discussion of the so-called "Superacids" is beyond the scope of this publication [122–124], they did not progress beyond the research state.

In Figure 4-41 the principle of the PAFC's is shown. The phosphoric acid with the function of the electrolyte is fixed in a porous layer between the electrodes.

The advantage of this type of fuel cells is their relatively simple construction, mainly based on the use of carbon, PTFE and SiC which all can be processed by methods which had been available at the very beginning of fuel cell development and could be adjusted in relatively short time to the demands of the fabrication of PAFC fuel cell components.

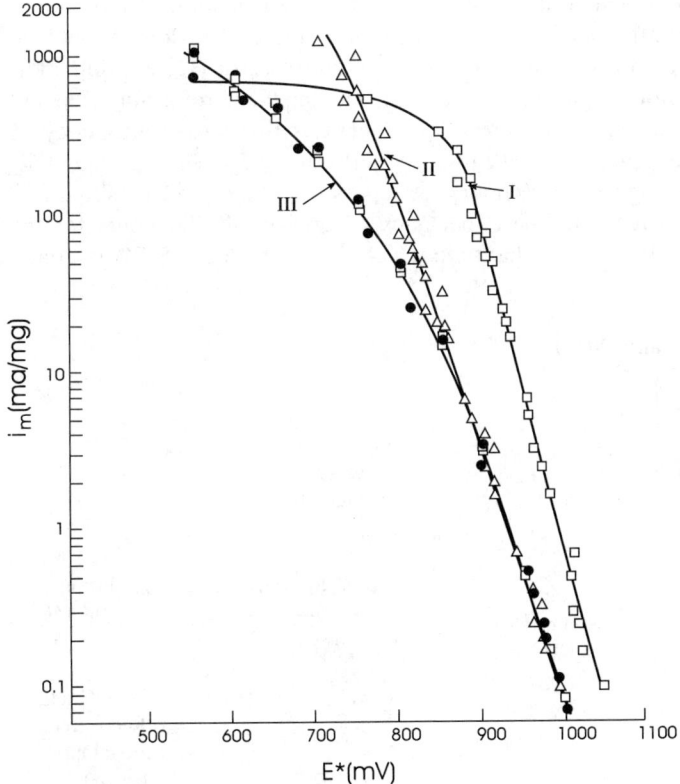

Figure 4-40. Polarization Curves of Oxygen Reduction on PTFE-Bonded Pt-Black [121]
(1) 30% KOH, 70 °C; (2) 20% H_2SO_4; (3) 50% H_3PO_4, 70 °C, (4) 85% H_3PO_4, 120 °C

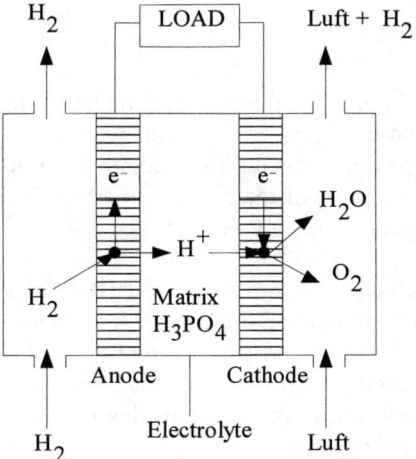

Figure 4-41. Reaction schematics of a PAFC

Furthermore, this type of cells, which is suitable for electricity generation from natural gas, has a high total energy efficiency, although it requires the conversion of methane into a type of synthesis gas comprising approx. 20% CO_2 and 80% of H_2 (omitting additional water vapor) in an additional gas processing unit. This methane reforming step and the consecutive shift reaction (exothermic reaction) increases the overall efficiency of PAFC power plants to more than 40% (LHV) with a cell efficiency of more than 50%. Life time experiments extending to 24.000 hours are reported, but the producers like IFC claim that their units have a real lifetime expectancy of approx. 40000 hours for the stack, including an efficiency loss during that time which does not exceed 5–7% overall.

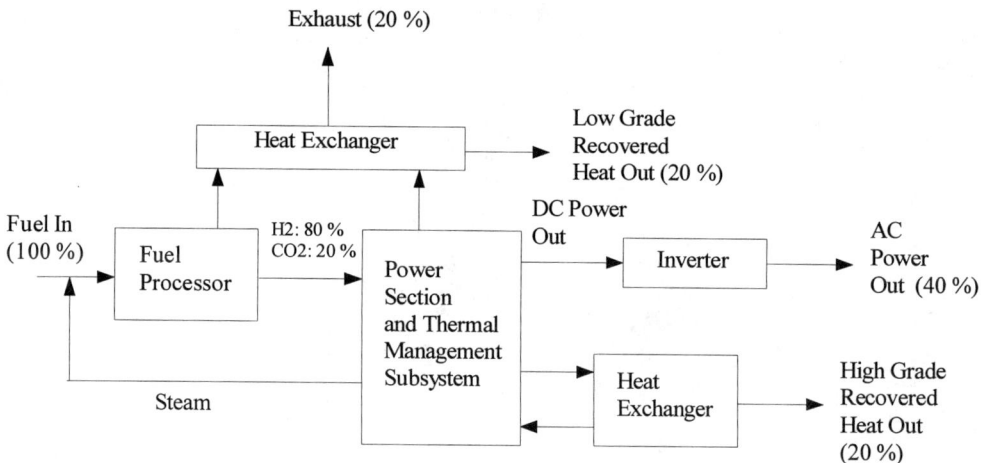

Figure 4-42. Power Plant Simplified Block Diagram [125]

4.4.3. Electrodes and Manufacturing

The Foil Technology

The production technologies for low-, middle, and high temperature fuel cells have one thing in common. This is the foil concept. As shown in Figure 4-43 schematically, Fuel cells consist of foils which can represent both the precursor or the finished single component for the cell (anode, cathode, matrix). With exception of the electrolyte all the foil components or the parts made by the foil forming procedures are porous. The thickness amounts only to a fraction of a millimeter.

As shown in Figure 4-44, the fuel cell is completed by the bipolar plate (BP), which connects the anode of one cell sequence with the cathode of the next cell sequence (A/M/K/BP//A/MK/BP//). The structure of the bipolar layer guarantees the uniform distribution of the reactant gases over the electrode surfaces and the collection and transmission of the current from the cathode from one cell unit to the anode of the next unit. The bipolar plate is also formed with foil technology processes.

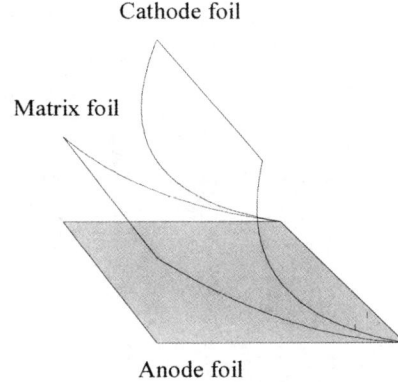

Cathode foil

Matrix foil

Anode foil

Figure 4-43. Foil Concept for the Structure of Fuel Cells

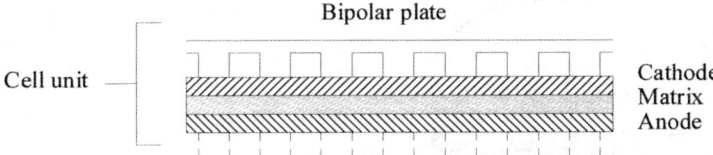

Bipolar plate

Cell unit

Cathode
Matrix
Anode

Figure 4-44. Cell Unit with Bipolar Plate

Using the foil technology there are two advantages. First, the possibility to produce the cell components and their precursors in large surfaces with relatively low-cost procedures and suitable large scale production technologies (tape casting, foil casting, calandering). Second, one can keep the material flow-through very low by precise adjustment of the thickness and the porosity of a foil. From all components the matrix has the lowest grain size of < 1 µm. The reason for this is that a difference pressure of 1 bar can be kept without blowing the capillaries of the matrix empty (according to the law of Laplace).

$$\Delta p = \frac{2\sigma}{r} \tag{4-6}$$

σ = Surface tension of the electrolyte
r = Pore radius

Both the anode and cathode are gas diffusion electrodes. Porous graphite paper serves as support for the catalyst layer, consisting of PTFE-bonded high surface area carbon black, activated with small platinum particles. The choice is determined by corrosion behavior, the specific surface area and the suitability for manufacturing a PTFE-bonded electrode layer. Acetylene blacks are commonly used, for instance, the Vulcan XC-72 and Carbon Black Pearls 2000-quality. The amount of platinum used in PAFCs is approx. 0.1 mg/cm^2 at the anode and 0.5 mg/cm^2 at the cathode. The diameters of the carbon particles are in general not exceeding a few tenths of nanometers in order to have an optimal utilization of the platinum by a surface to volume ration of the small Pt-crystallites which exceeds some 10^6/cm equivalent to a specific surface area of more than 5 m^2/g Pt.

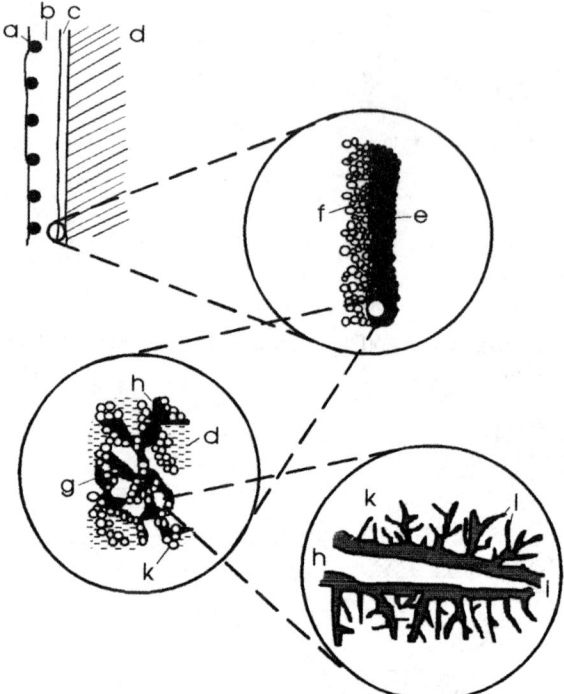

Figure 4-45. Gas diffusion electrode. a) gas side, b) hydrophobic backing layer, c) working layer, d) electrolyte, e) PTFE bonded catalyst support, f) electronically conductive layer of coarse porosity, g) PTFE particle, h) gas channels, i) electrolyte film, k) catalyst, l) micro- and mesopores [126]

The manufacturing process of the electrodes consists of a tape casting procedure: PTFE and activated platinum are worked into highly porous (P > 60%) felts of only some tenths of micrometer thickness, the "working layer", and applied by the doctor blade procedure on an approx. 0.3 mm thick graphite paper. The graphite paper is composed of long glassy carbon fibers (fiber-diameters of the order of 10 μm) which are glued to each other by a graphitised phenolic resin. The carbon substrate is glued by pyrolysis of a resinous material to the grooved bipolar plate. A SiC-containing matrix is applied on one or both electrodes with the doctor blade procedure. The entire electrode is compressed when mounted into the stack in order to ensure a low contact resistance (see Figure 4-16).

Although this technology seems to be mature for the introduction into the market, problems still exist and show weak features of the PAFC technology [127]. The used carbon or graphite is inherently and thermodynamically instable in the gas atmosphere and under the electrode conditions of the PAFC cathode. The consumption of the carbon cathode is an electrochemical reaction because the dissolution rate of the carbon support of the porous cathode depends on the cathode potential and increases exponentially with increasing cathode potential.

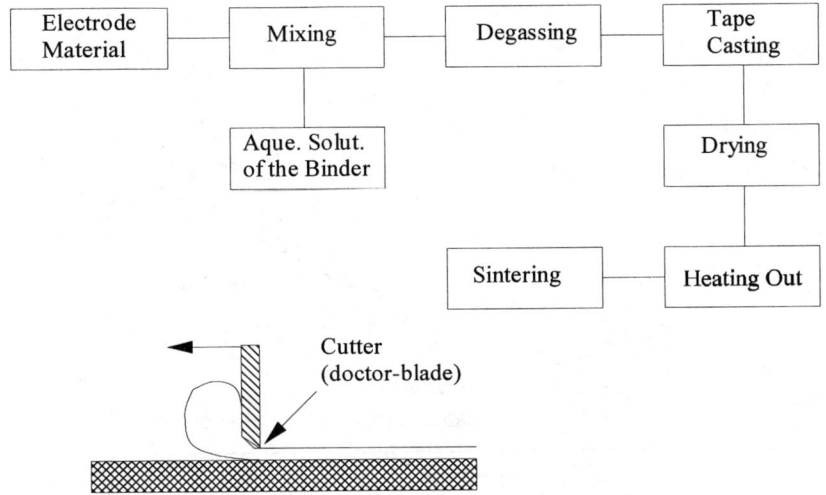

Figure 4-46. Production of "green" foils with the "doctor-blade" technique

The two principal reactions that describe the electrochemical oxidation of carbon in aqueous solutions are:

$$C + 2 H_2O \longrightarrow CO_2 + 4 H^+ + 4e^- \qquad (4\text{-}7)$$

and

$$C + H_2O \longrightarrow CO + 2 H^+ + 2e^- \qquad (4\text{-}8)$$

which have standard electrode potentials (vs. SHE) at 25 °C of 0.207 and 0.518 V, respectively. Carbon monoxide is thermodynamically unstable with respect to CO_2. The CO–CO_2 reaction

$$CO + H_2O \longrightarrow CO_2 + 2 H^+ + 2e^- \qquad (4\text{-}9)$$

has a standard electrochemical potential of –0.103 V. When the potential-pH diagrams for the C–H_2O system are examined, it is apparent that the stability domain of carbon is practically non-existent in aqueous solutions that are free of oxidizing agents, because of the overlap in the stability domain of carbon and the electrochemical reactions involving water to form H_2. In practice, however, carbon exhibits reasonable corrosion stability in aqueous systems because the above reactions are kinetical highly inhibited.

Figure 4-47 shows the dependence of corrosion current and electrode potential for various carbon materials. At 1.0 V vs. RHE, the corrosion current ranges from more than 0.1 mA/cm^2 for non-heat treated carbon blacks to about 1 µA/cm^2 for 100% dense graphites. The Tafel slope varies from 90 mV/Dec. for heat treated materials up to 330 mV/Dec. for non heat treated materials.

Cathode lifetime and carbon stability is one of the most important problems of PAFC technology, because it is reported by producers of PAFCs that after a lifetime of approx. 40 000 h, the carbon content of a PAFC cathode is less than 20% of the initially present

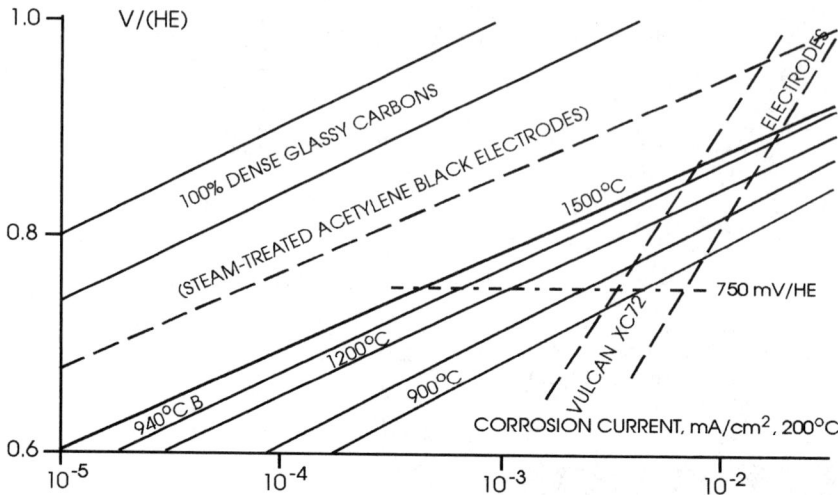

Figure 4-47. Dependence of Corrosion Current and Electrode Potential for Various Carbon Materials (after 1000 Minutes at 200 °C in PA Electrolyte). Dashed Lines: Electrodes; Solid Lines Bipolar Plate Components. Temperatures Refer to Carbon Heat-Treatment 940 °C (B = 940 °C with Boron Treatment). Pure Graphite Lies Close to Treated Acetylene Black [128].

carbon. Therefore, a search for an alternative, more stable catalyst support which would allow to extend simultaneously the lifetime and the current densities at equal or even enhanced cell voltage is of interest [129, 130].

4.4.4. The Electrode Materials

The evolution from 1965 to the present day in the development of cell components for PAFCs is shown in Table 4-7. In the mid-1960s, the conventional porous electrodes were polytetrafluoroethylene (PTFE)-bonded Pt black, and the loadings were about 9 mg Pt/cm^2. During the past two decades, Pt supported on carbon black has replaced Pt black in porous PTFE-bonded electrode structures as the electrocatalyst. A dramatic reduction in Pt loading has also occurred; the loadings' are currently about 0.10 mg Pt/cm^2 in the anode and about 0.50 mg Pt/cm^2 in the cathode. The operating temperature, and correspondingly the acid concentration, of PAFCs has increased to achieve higher cell performance; temperature of about 200 °C and acid concentrations of 100% H_3PO_4 are commonly used today.

Carbon is a critical material in fuel cells, particularly phosphoric acid fuel cells, since no other practical material has the necessary properties of electronic conductivity, reasonable corrosion resistance, surface properties and low cost-factors which make an inexpensive fuel cell a reality.

Carbon is a nonmetallic substance with a wide range of crystalline and amorphous structures different from diamond. Overlap of the valence and conduction bands

Table 4-7. Evolution of Cell Component Technology for Phosphoric Acid Fuel Cells

Component	ca. 1965	ca. 1975	Current Status[7]
Anode	PTFE-bonded Pt black	PTFE-bonded Pt/C Vulcan XC-72	PTFE-bonded Pt/C Vulcan XC-72
	9 mg/cm^2	0.25 mg Pt/cm^2	0.10 mg Pt/cm^2
Cathode	PTFE-bonded Pt black	PTFE-bonded Pt/C Vulcan XC-72	PTFE-bonded Pt/C Vulcan XC-72
	9 mg/cm^2	0.5 mg Pt/cm^2	0.5 mg Pt/cm^2
Electrode Support	Ta mesh screen	carbon paper	carbon paper
Electrolyte Support	glass fiber paper	PTFE-bonded SiC	PTFE-bonded SiC
Electrolyte	85% H$_3$PO$_4$	95% H$_3$PO$_4$	~100% H$_3$PO$_4$

Table 4-8. Typical Physical Properties of Carbon

Triple point	~4020 K and ~110 bar
Normal melting point	~3820 K
Normal boiling point	4470 ~ 5070 K
Specific gravity	
diamond	3.515
graphite	2.266
amorphous carbon	1.8–1.9
C-C bond length	0.154 nm
Electrical resistivity [μΩm]	
carbon, petroleum coke base	35 ~ 46
carbon, anthracite base	33 ~ 66
carbon, lampblack base	58 ~ 81
carbon lampblack graphitized	46 ~ 66
carbon fiber	> 600
glassy carbon	30 ~ 50
graphite, petroleum coke base	8 ~ 13
graphite fiber	~42

give it many of the properties of a metallic conductor. Table 4-8 lists some typical physical properties of carbon.

When assessing the chemical stability of carbon it is important to characterize the material clearly. Not only do different forms of carbon exhibit large variations in chemical stability, but the pretreatment history also has a strong effect. Reversible thermodynamics tells us that carbon should be oxidized to CO$_2$ or its derivatives (HCO$_3^-$, CO$_3^{2-}$, depending on pH) at only about 50 mV above the hydrogen potential in aqueous media. This is illustrated by the potential-stability diagram for carbon as a function of pH

---

7 1994/1995

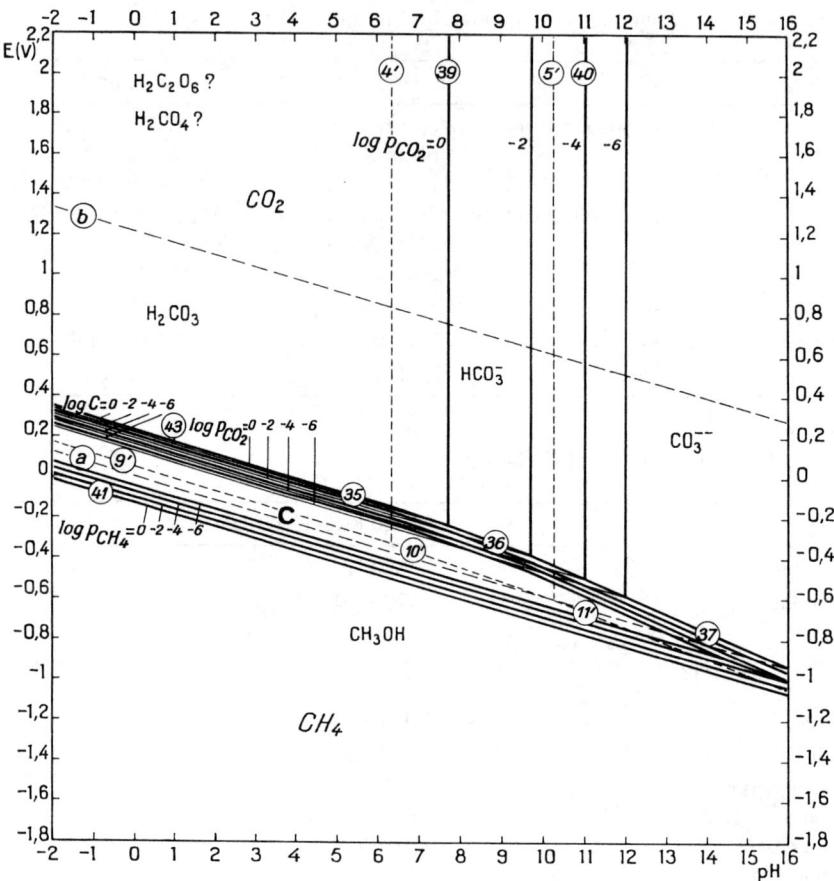

Figure 4-48. Pourbaix Potential pH Equilibrium Diagram for the System Carbon Water, at 25 °C (Considering carbon in the form of graphite). The portion of the diagram above the line log C = 0 refers to solutions containing 1mol of dissolved carbon per liter in the form of $H_2CO_3 + HCO_3^- + CO_3^{2-}$. The portion of the diagram below this line refers to solutions thermodynamically saturated with solid carbon.

(Pourbaix diagram [131]). In practice, however, the oxidation of carbon is kinetically so slow that carbon is considered to be very resistant to chemical attack except in oxidizing environments at elevated temperatures.

One of the major breakthroughs in PAFC technology that occurred in the late 1960s was the development of carbon blacks and graphites for cell construction materials; these developments are reviewed by Appleby [132] and Kordesch [133, 134]. It was shown at that time that carbon black and graphite were sufficiently stable to replace the more expensive gold-plated tantalum cell hardware. The use of high surface area carbon blacks to support Pt permitted a dramatic reduction in Pt loading, without sacrificing performance. However, carbon corrosion and Pt dissolution become problematic at cell voltages above ~0.8 V; consequently, low current densities with cell voltage about 0.8 V and hot idling at open circuit potential are to be avoided.

The phosphoric acid fuel cell contains noble metal electrocatalysts (mainly platinum or one of its alloys) for both H_2 oxidation and O_2 reduction, which are highly dispersed metal particles on a high-surface-area carbon powder. The mechanisms for introduction of the platinum into the carbon black support need consideration [135].

In the event that impregnation techniques are used, then the metal solution or solvent are evaporated within the carbon structure. Concentration processes occur as the solvent is lost until crystallization of the platinum salt is initiated, which then progresses until all of the solvent (water) is removed and crystals of the platinum salt are deposited within the carbon structure. It can be seen that such a process will lead to relatively large crystallites, since the precursor for the platinum crystallite is the salt deposit that has been grown out of solution. This process seems unsatisfactory for producing high surface area platinum crystallites.

Colloidal processes are now used to control the sizes of the latent platinum particles, but even these processes produce electrocatalysts that are dependent on the surface of the carbon supports, and the chemistries have been described in the patent literature [136, 137].

The essential chemistries for preparation of the colloidal platinum, outlined in these patents depends upon the formation of an insoluble white sodium platinum sulphite intermediate, which is hydrolyzed and thermally decomposed to form a stable platinum sole. The process steps described in a patent are given below [138]:

Starting with chlorplatinic acid

$$H_2PtCl_6 + Na_2CO_3 \longrightarrow Na_2PTCl$$
$\qquad$ sodium carbonate

Add $NaHSO_3 + Na_2S_2O_5 \xrightarrow{\text{pH4}} add$ $Na_2CO_3 \xrightarrow{\text{pH7}}$

Gives $Na_6Pt(SO_3)_4 \longleftarrow Na_2SO_3 + Pt(NH_3)_2Cl_2$
$\qquad$ white ppt. $\qquad\qquad$ sodium sulphite
$\qquad\qquad\qquad\qquad$ platinum aminochloride

or

$$H_2Pt(OH)_6 + Na_2SO_3 \xrightarrow{\text{boil}} Na_6Pt(SO_3)_4 + SO_2$$
$\qquad$ hydroxyplatinic acid $\quad$ white ppt.

Filter white ppt. slurry with water and io exchange on Dowex 50

Gives $Na_3H_3Pt(SO_3)_4 \xrightarrow{\text{Further ion exchange}} H_2Pt(SO_3)_4$

$$\text{Boil} \xrightarrow{\quad\quad} H_3Pt(SO_3)_2OH + SO_2 \xrightarrow{\ dry,\ heat\ to\ 135°C\ } \text{Platinum colloid}$$

Platinum colloid can be dispersed in water (black "Sole"). Pt is 15–25 A.

For applications as an electrocatalyst

The black "Sole" is added to carbon dispersed in water, it adsorbs preferentially on the carbon exterior. Colloid can be filtered out of solution by adsorption on the carbon.

For applications as a gas-phase catalyst

"Sole" is added to high surface area alumina, or other ceramic support (as pellets), dried at 200 °C, then calcinated in air at 600 °C for 15 mins.

This produces 20 Å platinum particles on the exterior of the porous pellet.

The choice of carbon for the support is critical to the successful performance of PTFE-bonded electrodes. The carbon support must have high chemical/electrochemical stability, good electrical conductance, reasonably high surface area, suitable pore-size distribution, and low impurity content. The carbons that appear to possess the best combination of these characteristics are carbon blacks. The number of commercially available carbon blacks is large and a considerable effort is needed to determine the optimum one for use in fuel cell electrodes. Much of this effort is proprietary information of the commercial fuel cell developers, and thus the "know-how" used in the production of practical electrodes is not readily available.

The carbons, which are used, are predominantly carbon blacks. Channel blacks are no longer used because their availability is limited. The major emphasis on carbon supports for electrocatalysts is focused on furnace blacks because they are readily available and are generally low-cost materials, although further treatment (i.e., with heat or steam) may add considerably to the final cost. The carbon blacks for electrocatalyst supports have been developed for other commercial applications, and very little effort has been devoted to the development of carbon blacks specifically for PTFE-bonded electrodes. Instead, existing commercial carbon blacks are heat-treated or steam-activated to produce the properties that are desired for fuel cell applications.

The high-surface-area carbon blacks such as Vulcan XC-72 and Black Pearls 2000 are heat-treated at high temperature (~2700 °C) to reduce their corrosion rates in phosphoric acid fuel cells. It has also been found that heat treatment of Pt-catalyzed Vulcan XC-72 at 900 °C in nitrogen enhances its performance in phosphoric acid fuel cells, compared to that observed with catalyzed Vulcan XC-72 that was not heat treated [139]. This heat treatment produced an improvement in cathode performance of 10 to 25 mV at 200 mA/cm^2 in H$_3$PO$_4$ at 190 °C.

The two major problems that have been identified with Pt-catalyzed carbon blacks for gas diffusion electrodes in phosphoric acid fuel cells are carbon corrosion and loss of Pt surface due to sintering. The corrosion of carbon in phosphoric acid results in the structural degradation of the electrode and loss of Pt surface area. The recent trend in fuel cell technology has been to operate at higher temperatures and pressures, and both of these conditions increase carbon corrosion. A corrosion rate of three to four times larger is projected for carbon black as the temperature of the phosphoric acid fuel cell is increased from 190 to 210 °C. The corresponding cell voltage rises from 0.68 to 0.71 [140]. The corrosion rate of a carbon black, such as Vulcan XC-72, at 0.75 V and 200 °C is unacceptable if a lifetime of 40000 h is envisioned for commercial phosphoric acid fuel cells. At electrode potentials over 0.8 V, the cathode polarization losses are attributed in a large part to carbon corrosion. Inroads to overcome this serious limitation have been made; the two approaches are to heat-treat the carbon black or develop alternative supports [141]. A practical solution to overcome carbon corrosion is to prevent the cathode from reaching potentials where significant corrosion occurs by applying small loads or adding N_2 during standing times [142, 143].

4.4.5. Carbon Corrosion and Cathode-Catalyst Stability in the PAFC

Operations of the PAFC under pressure conditions and at atmospheric pressure provide greatly different environments for the cell components. At 190 °C and 1 atm, the acid concentration in typical cells corresponds to about 98–100 wt.-% of acid expressed as H_3PO_4. At 205 °C and 8.2 atm, the acid concentration corresponds to about 93 wt.-% H_3PO_4. The volatilities of the acids are not only different, but they also contain proportions of differently condensed PA, which behave as different chemical species as far as carbon corrosion is concerned. This fact is well illustrated by the work by Stonehart

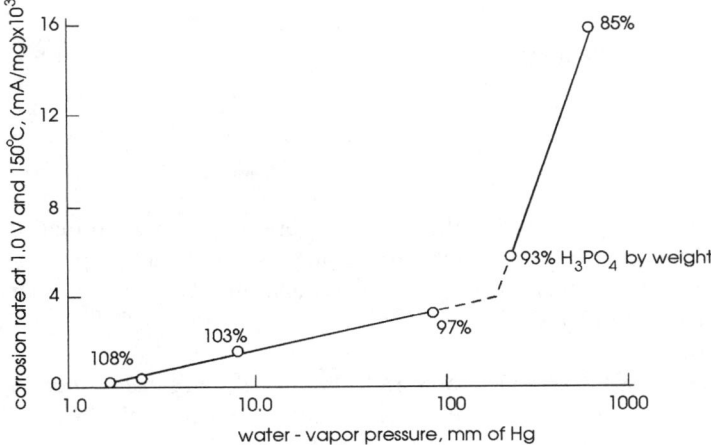

Figure 4-49. The effect of water-vapor pressure on the corrosion rate of Shawinigan acetylene black in concentrated phosphoric acid [145]

[144] (see also Figure 4-49). It shows the rate of corrosion of a carbon-catalyst support (Shawinigan acetylene black) as a function of PA concentration or water-vapor pressure at 150 °C and a potential of 1.0 V on the hydrogen scale. The curve shows a break at a concentration between 93 and 97 wt.-%, which represents a change in kinetics. In this concentration range, PA changes from the ortho- to the pyro-form. In the range 85 to 93 wt.-%, carbon corrosion is rapid, and the reaction order for water (the corrosion reactant) appears to be approximated unity.

In the more concentrated acid range, where condensed acids are present, the corrosion rates are lower, and the water-reaction order is close to zero. Arrhenius plots for the corrosion of carbon black as a function of acid concentration and temperature indicate that the corrosion rate at 205 °C in 93 wt.-% PA is perhaps a factor of 3 to 4 higher than at 190 °C in 98 wt.-% acid at the same electrode potential. However, cells operating under pressure do not only see this effect of acid concentration and/or water-vapor pressure but also the effect of increased cathode potential. Pressurized operation is used to improve system efficiency .

4.4.6. Performance

Cell performance for any fuel cell is a function of pressure, temperature, reactant gas composition and utilization. In addition, performance can be adversely affected by impurities in both the fuel and oxidant gases. Typical PAFC's will generally operate in the range of 100 to 400 mA/cm^2 at 600 to 800 mV/cell. Voltage and power constraints arise from the increased corrosion of platinum and carbon components at cell potentials above approximately 800 mV. The improvements of the Phosphoric Acid Fuel Cells are illustrated in Figure 4-50.

The goal is to maintain the performance of the cell stack during a standard utility application (~40,000 hours). Current state-of-the-art PAFCs show the following degradation with time:

$$\Delta V_{lifetime}(mV) = -3mV/1000 \text{ hours [151]} \tag{4-1}$$

Effect of Impurities

The concentration level of impurities entering the PAFC is very low relative to that of diluents or reactant gases, but their impact on the performance is significant. Some impurities (e.g., sulfur compounds) originate from the fuel gas entering the fuel processor and are carried into the fuel cell with the reformed fuel, where others (e.g., CO) are produced in the fuel processor.

Carbon Monoxide: The presence of CO in a H$_2$-rich fuel has a significant effect on the anode performance because CO poisons the electrocatalytic activity of Pt electrodes.

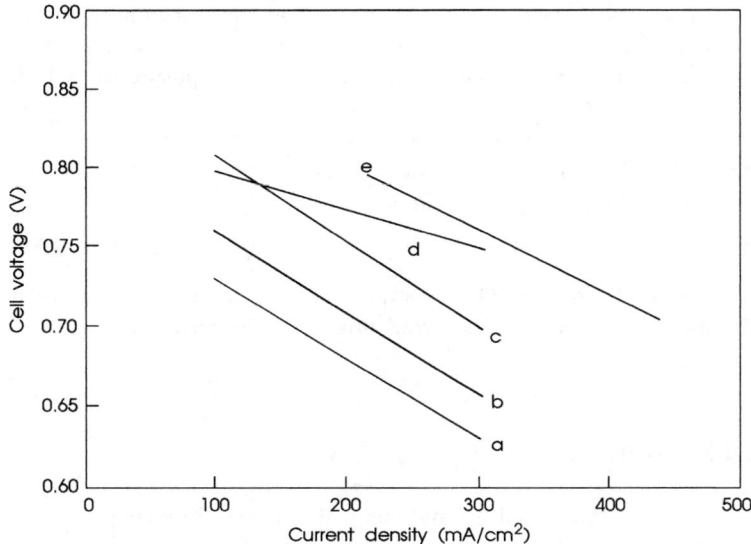

Figure 4-50. Improvements in the Performance of H_2-Rich Fuel/Air PAFCs.

a) 1977 – 190 C, 3 atm, Pt loading of 0.75 mg /cm^2 on each electrode [146]
b) 1981 – 190 C, 3.4 atm, cathode Pt loading of 0.5 mg/cm^2[147]
c) 1981 – 205 C, 6.3 atm, cathode Pt loading of 0.5 mg/cm^2 [148]
d) 1984 – 205 C, 8 atm, electrocatalyst loading was not specified [149]
e) 1992 – 205 C, 8 atm, 0.929 m^2, short stack, 200 hours, electrocatalyst loading not specified [150]

Both temperatures and CO concentration have major influence on the oxidation of H_2 on Pt in CO containing fuel gases. The following equation describes voltage loss resulting from CO poisoning as a function of temperature:

$$\Delta V_{CO} = k_{(T)} \, ([CO]_2 - [CO]_1) \, [152] \tag{4-11}$$

$k_{(T)}$ is a constant that is a function of temperature and $[CO]_1$, $[CO]_2$ are the percent CO in the fuel gas. The values of $k_{(T)}$ vary from -3.54 (mV/%, 190 °C) to -2.05 (mV/%, 204 °C).

Sulfur Containing Components. Hydrogen sulfide and carbonyl sulfide (COS) are impurities in fuel gases from fuel processors and coal gasifiers in PAFC power plants. Experimental studies by Chin and Howard [153] indicate that H_2S adsorbs on Pt and blocks the active sites for H_2 oxidation. The following electrochemical reactions involving H_2S are postulated to occur on Pt electrodes.

$$Pt + HS \rightarrow Pt\text{-}HS_{ads} + e \tag{4-12}$$

$$Pt\text{-}H_2S_{ads} \rightarrow Pt\text{-}HS_{ads} + H^+ + e \tag{4-13}$$

and

$$Pt\text{-}HS_{ads} \rightarrow Pt\text{-}S_{ads} + H^+ + e^- \tag{4-14}$$

Elemental sulfur is only expected on Pt electrodes at high anodic potentials, and at sufficiently high potentials, sulfur is oxidized to SO_2. The extent of poisoning by H_2S increases with increasing H_2S concentration, electrode potential and exposure time. H_2S like CO poisoning decreases with increasing cell temperature.

Other Compounds: Nitrogen compounds like molecular nitrogen act as a diluent but other nitrogen compounds (e.g., NH_3, HCN, NO_x) may not be as inocuous. NH_3 acts as a fuel. Oxidant gases react with H_3PO_4 to form a phosphate salt: $(NH_4)H_2PO_4$.

$$H_3PO_4 + NH_3 \rightarrow (NH_3)H_2PO_4 \qquad (4\text{-}15a)$$

which results in a decrease in the rate of O_2 reduction. A concentration of less than 0.2 mol-% $(NH_3)H_2PO_4$ must be maintained to avoid unacceptable performance losses [154].

4.4.7. Stacks and Systems

The fuel cell stack consists of a sequential arrangement of a ribbed bipolar plate, the anode, the electrolyte matrix and the cathode. The ribbed bipolar plate provides the gas supply to anode and cathode, respectively. The fuel is fed from one side of the stack, the residual gas leaves on the opposite side. A SiC based matrix filled with electrolyte is placed between each set of anode and cathode. The cathode is supplied with air or oxygen by the rectangularly ribbed rear side of the bipolar plate. A commercial stack consists of 30 up to approx. 500 single cells. The construction of an air-cooled stack is pictured in Figure 4-51. The reason for the usage of air instead of water for cooling is that it avoids the need to incorporate additional paths into the stack with resultant added series resistance and corrosion-complications.

The cell stack – built by ERC (Energy Research Corporation) – had a cross-flow configuration, and the process air stream was diverted into two types of channels in the stacks; channels in individual cells, with relatively small cross-sectional area; and channels in the cooling plates (approximately one for every five cells), with greater cross-section. The channel areas and distributions were calculated according to required process air flows, oxygen utilization, current density, and heat-transfer requirements. An advantage of this system is that the cooling portion of the air supply does not come into contact with the electrolyte and therefore does not contribute to its evaporation. The straight-through cross-flow manifolding is very simple.

B   – Bipolar plate with process air & fuel channels
CA – Anode DIGAS cooling plate
CC – Cathode DIGAS cooling plate

The Westinghouse Electric Corporation (WE) entered into a licensing agreement with ERC relative to their air-cooled PAFC technology. They produced a number of air-cooled systems, excluding however the DIGAS[R] concept, for which separately manifolded process-air and cooling-air streams were substituted. The WE approach involved a total redesign of the gas-distribution grooves on the bipolar plate to yield an arrangement

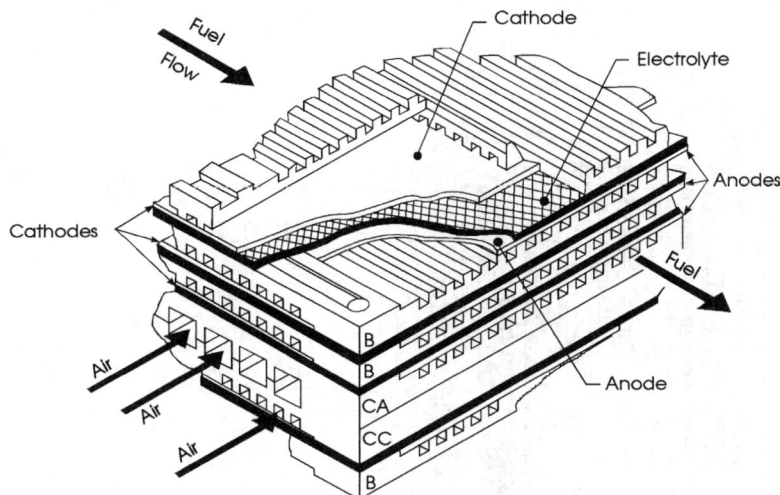

Figure 4-51. The ERC-WE DIGAS^R system [155]

which has become known as the Z plate. The new system manifolding required a change in the arrangement of the Fuel Cell stack within the pressure vessel. With pressure vessel dimensions dictated by practical considerations of size, wall thickness and availability, the most cost-effective diameter for 3,5 bar operation was found to be just over 1 m and will accommodate four 30 cm × 43 cm cell stacks arranged in a cruciform manner, with their longer sides forming a square in an internal duct. This duct serves to evacuate hot air downwards, after it has passed through the cells from segmental ducts between the outside of each stack and the pressure-vessel walls (Figure 4-52). The modules were planned to be operated at the following beginning-of-use conditions:

- Pressure   80 psia
- Temperature   205 °C
- Current density   267 mA/cm^2
- Fuel/air utilization   83/60%
- Cell voltage   720 mV

Despite large heat-exchanger area and piping-volume requirements, preliminary costs of components in the recirculatory coolant loop were estimated at only \$ (1979) 40/kW. This low additional investment cost, together with the intrinsic reliability of an air-cooled system (in the event of tube rupture in water-cooled tubes, cells will electrically short out with a catastrophic stack loss) made the concept look attractive.

Stacks developed by UTC (since 1985 International Fuel Cells) use water for the cooling. Prior to 1978 the stacks consisted of a sandwiched configuration as shown in Figure 4-51, in which the ribs in the cross-flow configuration are rotated through 90 ° for clarity. The phosphoric acid electrolyte is contained in a porous, hydrophilic matrix located between the catalyst layers of the anode and cathode. The electrolyte reservoir capacity of this so-called "conventional" design using multiple carbon paper layers was

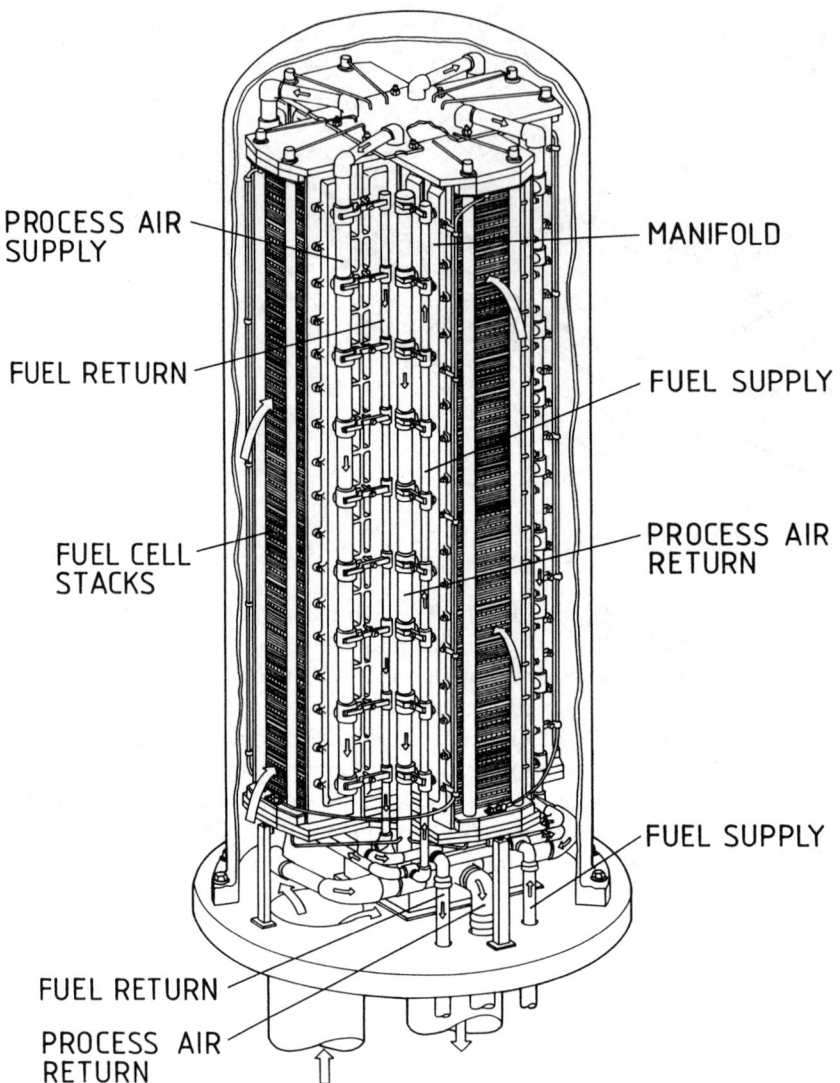

PROCESS AIR SUPPLY

MANIFOLD

FUEL RETURN

FUEL SUPPLY

FUEL CELL STACKS

PROCESS AIR RETURN

FUEL SUPPLY

FUEL RETURN

PROCESS AIR RETURN

Figure 4-52. Westinghouse 375 kW fuel cell module [156, 157]

very limited, and it involved a large number of carbon-to-carbon contacts that could result in a high ohmic resistance. Accordingly, it was replaced after 1979 by the so-called "ribbed substrate" design, shown in Figure 4-53, in which the ribs are formed on the reactant gas side of the electrode backing. A catalyst layer and a half-as-thick matrix is applied to the non-ribbed (flat) side of the porous substrate to form each electrode.

Unlike the earlier design, the ribbed substrate material is amenable to continuous processing since the ribs on each backing sheet run only in one direction. Separately manu-

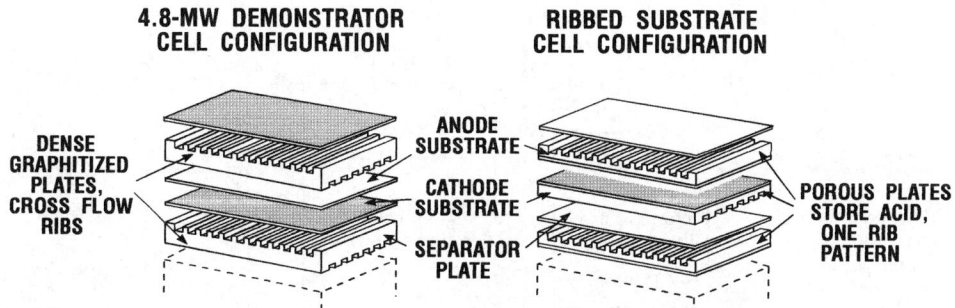

Figure 4-53. UTC's "conventional" and "ribbed substrate" stack design [158]

factured impermeable flat graphite sheets, slightly more than 1 mm thick, serve as bipolar gas separator plates contacting the substrate ribs for the anode and cathode of adjoining cells in a cross-flow arrangement. The ribbed substrates typically have a thickness of about 1.8 mm, giving a total cell weight (without cooling plates and electrolyte) of about 5.4 kg/m². As in the conventional stack arrangement, cooling plates are placed approximately one every five to eight cells, so that the pitch is five cells per inch (about 5 mm/cell).

The thick ribbed substrates allow about a five-fold increase in effective electrolyte volume compared with the conventional stack configuration, when both anode and cathode substrates are used as reservoirs. Approximately 1.5 kg of electrolyte per m² can be stored and it is estimated that this will be sufficient for more than 40 000 h of operation at 205–210 °C and 9.2 atm pressure. This increased storage has already been shown to greatly increase the long-term stability of cell performance in laboratory cells and short stacks. The ribbed substrate configuration was first used in the TEPCO 4.5 MW demonstrator plant, for which fabrication started in 1980, and in the UTC 40 kW on-site power plant demonstrator units.

4.5. Molten Carbonate Fuel Cells (MCFC)

4.5.1. Introduction

The Molten Carbonate Fuel Cell (MCFC), operating at a temperature between 600 and 650 °C, has been under intensive development for the last decade as a second generation Fuel Cell System. This label suggests both, the considerable advantages of the MCFC and the fact that it still needs substantial development before it can be successfully commercialised.

In Figure 4-54 the different Fuel Cell Systems are introduced together, however the melt chemistry and electrochemistry involved in the MCFC are unique. At the anode the net reaction is

$$H_2 + CO_3^{2-} \longrightarrow H_2O + CO_2 + 2e^-, \tag{4-15b}$$

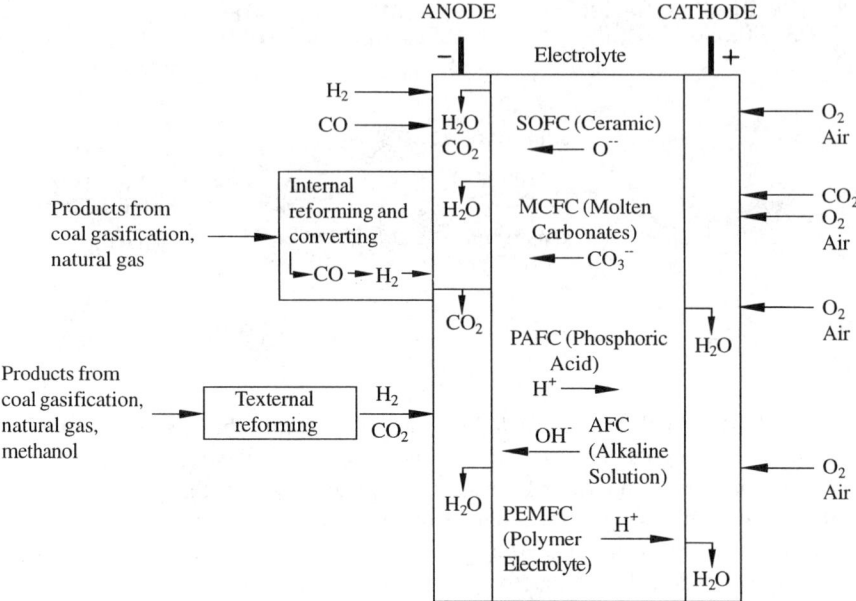

Figure 4-54. MCFC Technology in Comparison with other H_2-/O_2-Fuel Cell Systems

whereas at the cathode the net process is as follows:

$$1/2\,O_2 + CO_2 + 2e^- \longrightarrow CO_3^{2-} \tag{4-16}$$

The CO_2 has a function similar to Lewis acids (for example, the proton in middle-temperature cells such as the PAFC); however, protons are transferred in the PAFC via ion conduction in the electrolyte phase, whereas in the carbonate cell another shuttle mechanism is required. In the present concepts, this normally will be done by taking the CO_2-rich product gas from the anode, burning it completely to produce CO_2 and H_2O, condensing out the water, and adding the CO_2 to the cathode inlet air. However, in the future, the use of a CO_2 separation device or, in general, a "CO_2 transfer device" may be preferred, it will allow higher hydrogen utilisation in the cell and therefore greater system efficiency.

The reversible potential for a MCFC, taking into account the transfer of CO_2 is given by the equation:

$$E = E^0 + \frac{RT}{2F} \ln \frac{p_{H_2}\, p_{O_2}^{1/2}}{p_{H_2O}} + \frac{RT}{2F} \ln \frac{p_{CO_2,c}}{p_{CO_2,a}} \tag{4-17}$$

where the subscripts a and c refer to the anode cathode, respectively. When the partial pressures of CO_2 are identical at the anode and cathode, and the electrolyte is invariant, the cell potential depends only on the partial pressures of H_2, O_2 and H_2O. Typically, the CO_2 partial pressures are different in the two electrode compartments and the cell potential is affected accordingly.

The system efficiency and cost of the MCFC appear very attractive, even without a CO_2 separation device, in comparison with low-temperature cells. Although phosphoric acid fuel cell (PAFC) power plant systems are expected to gain commercial acceptance during the next few years, they will be characterised by relatively high material costs and heat rates. A major cost component will be associated with the noble metal catalysts that are needed for the anode and the cathode in acid cells. Another major cost component is the chemical plant necessary for conditioning the fuel by reforming carbonaceous fuels and largely removing CO (1,5% after the Shift Conversion) from the feed gas.

In high temperature systems, these shortcomings can be avoided because elevated temperatures accelerate the chemical and physical processes to such an extent that polarisation losses are less than with aqueous Fuel Cells, and less expensive materials can be used as electro-catalysts. Although the ideal efficiencies ($\Delta G / \Delta H$) of the overall cell reaction (oxidation of H_2 and CO) decrease at increasing operating temperatures, the higher operating temperature reduces polarisation losses to such an extent that the actual cell efficiency is increased.

The MCFC operating temperature (600–650 °C) is high enough to produce valuable waste heat, yet it is sufficiently low so as not to incur a too great free energy penalty (the theoretical open circuit is 100 mV higher than that of the 1000 °C SOFC). The waste heat can be used to provide compression work (to improve cell performance and particularly cell potential), to supply heat for bottoming cycles, and/or for cogeneration purposes (process or space heat). The most important use of process heat will be for the reforming of methane.

The MCFC appears to have a bright future for 21st century coal-fired baseload electric utility plants [159]. These, however, will be large (GW scale) and user confidence will be required before the MCFC is built in units of this size. Such user confidence will be best obtained if reasonably small MCFC units can be first built and field-tested under conditions where they can generate power economically. Dispersed units (a few MW in size) are unlikely to be coal-fuelled. However, the high temperature waste heat available in the MCFC can be used to advantage in two types of installation, notably in natural gas units and in applications where high quality waste heat is needed in industrial cogeneration.

The major advantage of the MCFC over the acid cell is that it can use internal reforming, i.e., its waste heat is directly available within the cell for the conversion of desulfurized methane directly to H_2 in a driven reforming reaction in the cell anode chamber. In the internal reforming molten carbonate cell (IRMFC), the space velocities are relatively low and reforming rates are quite adequate, provided that the reforming catalyst is protected from catalyst poisons (particularly) S and traces of carbonate [160, 161].

Summarising the status of its technology, the MCFC may be expected to enter commercial markets approximately 5 years behind the PAFC, provided that remaining technical problems can be overcome. The time lag is also important to produce general consumer acceptance for Fuel Cells. While the MCFC has a considerable number of technical and economic advantages over the PAFC and other low-temperature Fuel Cells for terrestrial stationary applications, its development has been restrained by problems of perception, as well as a lack of research effort. The high-temperature Fuel Cells, MCFCs and SOFCs, are presently perceived to have more disadvantages (corrosion and sealing problems) than advantages (rapid kinetics) resulting from the high operating temperature.

4.5.2. Operating Principle

In an operating MCFC, electrons are transferred from the anode through an external circuit to a cathode, where they participate in reduction reactions. Negative charges are conducted by carbonate anions (CO_3^{2-}) from the cathode through the molten electrolyte to an anode. At the anode, electrons are produced by oxidation. Figure 4-55 shows the essential features of an MCFC cell which uses reformate (e.g., a gaseous H_2 and CO mixture) as fuel. The electrode reactions are listed below.

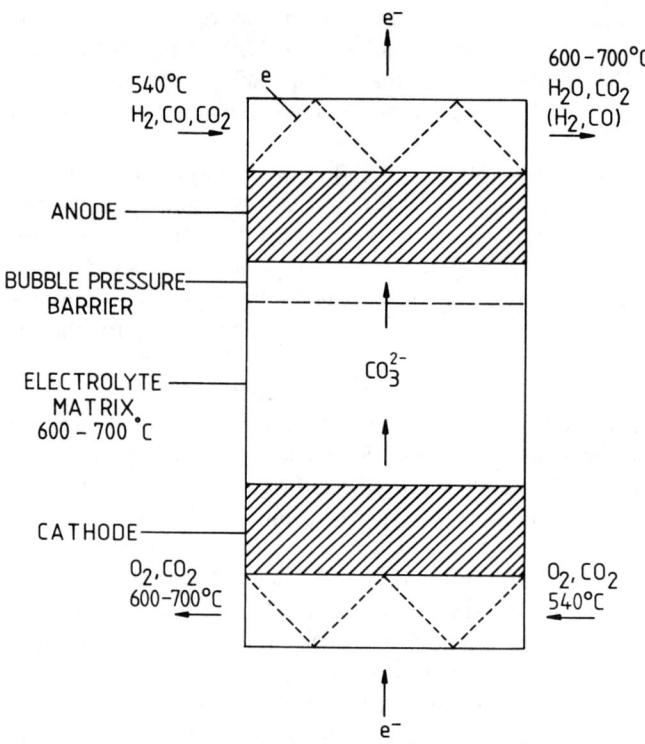

Figure 4-55. Schematic of a single cell in a MCFC cell stack, showing cell components, material flows, and current direction

Anode reactions:

$$H_2 + CO_3^{2-} \longrightarrow H_2O + CO_2 + 2e^- \qquad (4\text{-}18)$$

$$CO + CO_3^{2-} \longrightarrow 2\,CO_2 + 2e^- \text{ (minor)}; \qquad (4\text{-}19)$$

Shift reaction:

$$CO + H_2O \longrightarrow H_2 + CO_2 \qquad (4\text{-}20)$$

Cathode reaction:

$$1/2\,O_2 + CO_2 + 2e^- \longrightarrow CO_3^{2-}. \qquad (4\text{-}21)$$

Two Faradays of electricity generated involve the use of 1 mole of CO_2 at the cathode and the production of 1 mole of CO_2 at the anode. It is therefore desirable to recycle CO_2

in order to maintain the electrolyte composition invariant. This concept has led to a search for a practical "CO$_2$ transfer device". In practice, it has been found adequate to withdraw part of the anode exit gas, complete combustion and mix it with the cathode-gas stream. CO is not directly used by electrochemical oxidation, but produces additional H$_2$ when combined with water in the water gas shift reaction.

MCFCs differ in many respects from PAFCs because of their higher operating temperature and the nature of the electrolyte. The higher operating temperature of MCFCs provides the opportunity for achieving higher overall system efficiencies (potential for heat rates below 8000 kJ/kWh) and greater flexibility in the use of available fuels. On the other hand, the higher operating temperature places severe demands on the corrosion stability and life of cell components, particularly in the aggressive environment of the molten carbonate electrolyte. Another difference between PAFCs and MCFCs lies in the method used for electrolyte management in the respective cells. In a PAFC, PTFE serves as a binder and wet-proofing agent to maintain the integrity of the electrode structure and to establish a stable electrolyte/gas interface in the porous electrode. The phosphoric acid is retained in a matrix of PTFE and SiC between the anode and cathode. There are no materials available for use in MCFCs that are comparable to PTFE. Therefore, a different approach is required to establisch a stable electrolye/gas interface in MCFC porous electrodes. The MCFC relies on a balance in capillary pressures to establish the electrolyte interfacial boundaries in the porous electrodes. At thermodynamic equilibrium, the diameters of the largest flooded pores in the porous components are related by the equation:

$$\frac{\gamma_c \cos \theta_c}{D_c} = \frac{\gamma_e \cos \theta_e}{D_e} = \frac{\gamma_a \cos \theta_a}{D_a} \qquad (4\text{-}22)$$

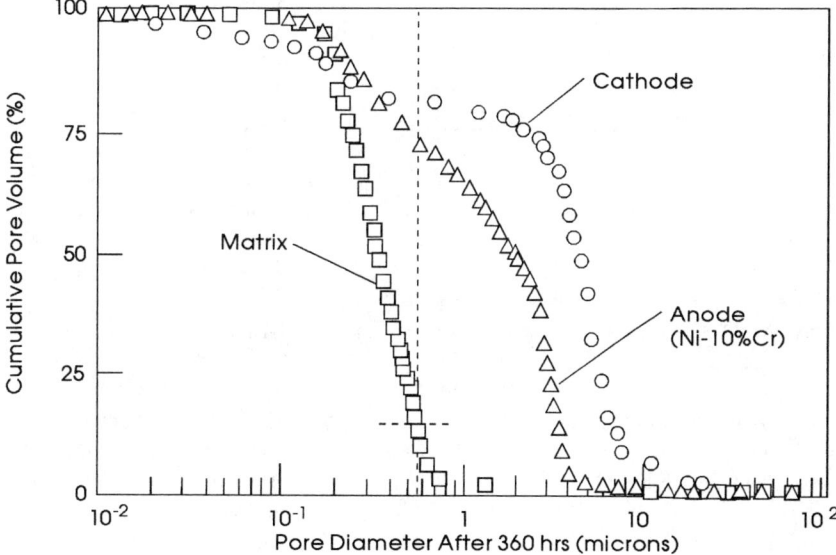

Figure 4-56. Pore size distribution requirements for well-matched anode, electrode matrix and cathode components [162]

where γ is the interfacial surface tension, θ is the contact angle of the electrolyte, D is the pore diameter, and the subscripts a, c and e refer to the anode, cathode and electrolyte matrix, respectively. By properly coordinating the pore diameters in the electrodes with those of the electrolyte matrix, which contains the smallest pores, the electrolyte distribution depicted in Figure 4-56 and described by Equation 4-22 is obtained.

This arrangement permits the electrolyte matrix to remain completly filled with molten carbonate, while the porous electrodes are partially filled, depending on their pore size distribution. The electrolyte content in each of the porous components will be determined by the equilibrium pore size ($<D>$) in that component; pores smaller than $<D>$ will be filled with electrolyte, and pores larger than $<D>$ will remain empty. A reasonable estimate of the volume distribution of electrolyte in the various cell components is obtained from the measured pore-volume-distribution curves and the above relationship for D.

Electrolyte management, that is, the control over the optimum distributuion of molten carbonate electrolyte in the different cell components, is critical for achieving high performance and endurance with MCFCs. Various processes (i.e. consumption by corrosion reactions, potential driven migration, creepage of salt and salt vaporization) occur, all of which contribute to the redistribution of molten carbonate in MCFCs [163].

Table 4-9. Evolution of Cell Component Technology for Molten Cabonate Fuel Cells

Component	ca. 1965	ca. 1975	Current Status
Anode	Pt, Pd, or Ni	Ni – 10 wt.-% Cr	Ni – 10 wt.-% Cr 3–6 μm pore size 50–70% porosity 0.5–1.5 mm thickness 0.1–1 m^2/g
Cathode	Ag$_2$O or lithiated NiO	lithiated NiO	lithiated NiO 7–15 μm pore size 70–80% initial porosity 60–65% after lithiation and oxidation 0.5–0.75 mm thickness 0.5 m^2/g
Electrolyte Support	MgO	mixt. of α-, β- and γ-LiAlO$_2$ 10–20 m^2/g	γ-LiAlO$_2$ 0.1–12 m^2/g 0.5 mm thickness
Electrolyte [mol-%]	52 Li$_2$CO$_3$–48 Na$_2$CO$_3$ 43.5 Li$_2$CO$_3$–31.5 Na$_2$CO$_3$, 25 K$_2$CO$_3$	62 Li$_2$CO$_3$–38 K$_2$CO$_3$ ~60–65 wt.-%	62 Li$_2$CO$_3$–38 K$_2$CO$_3$ 50 Li$_2$CO$_3$–50 Na$_2$CO$_3$ 50 Li$_2$CO$_3$–38 K$_2$CO$_3$ ~50 wt.-%
	paste	hot press "tile" 1.8 mm thickness	tape cast 0.5 mm thickness

4.5.3. Cell Components

The data in Table 4-9 provide a chronology of the evolution in cell component technology for MCFCs. In the mid-1960s, the electrode materials were, in many cases, precious metals, but the technology soon evolved to the use of Ni-based alloys at the anode and oxides at the cathode. Since the mid-1970s, the materials for the electrodes and electrolyte structure have remained essentialy unchanged.

A major development in the 1980's has been the evolution in the technology for fabrication of electrolyte structures. The conventional process used to fabricate electrolyte structures until about 1980 involved hot pressing (about 5000 psi) mixtures of $LiAlO_2$ and alkali carbonates (typically >50 vol.-% in liquid state) at temperatures slightly below the melting point of the carbonate salts (e.g., 490 °C for electrolyte containing 62 mol.-% Li_2CO_3-38 mol-% K_2CO_3). These electrolyte structures (also called "Electrolyte tiles") were relatively thick (1-2 mm) and difficult to produce in large sizes because large tooling and presses were required. The electrolyte structure produced by hot pressing are often characterized by: 1) void spaces (<5% porosity), 2) poor uniformity of microstructure, 3) generally poor mechanical strength, and 4) high iR drop. To overcome these shortcomings of hot pressd electrolyte structures, alternative processes such as tape casting and electrophoretic deposition for fabricating thin electrolyte structures were developed. The greates success to date has been reported with tape casting, which is a common processing technique used by the ceramic industry (see Section 4.6.6.).

4.5.3.1. Matrix Support Material

The matrix support material is the mixture of ceramic particles that forms the capillary network which contains the electrolyte. The support material provides structure to the matrix but does not participate in the electrical or electrochemical processes. The physical properties of the matrix, such as thermal cycle strength and differential pressure capability, are largely controlled by the support material. The size, shape and distribution of the support particles determines the interparticulate porosity and pore distribution. This in turn governs the electrolyte retention, volume and ohmic resistance of the matrix. The chemical and physical stability of the support particles is an important property for long term electrolyte retention. Particle instability will result in electrolyte loss and an overall degradation in matrix properties causing cell performance decay.

The matrix support is typically a mixture of fine and coarse particles as well as fibres. The fine particles which are currently used by all of the fuel cell developers are γ-$LiAlO_2$. Submicron sized γ-$LiAlO_2$ are used to provide a high porosity and small pore size for electrolyte inventory and retention. Coarse particles, in the size range of 100 μm, are added to the matrix for compressive strength and improved thermal cycle capability. The large particles serve as a second phase site for crack deflection and attenuation [164]. Some developers are also adding Al_2O_3-fibres to the matrix for additional tensile and bending strength to improve thermal cycle capability [165]. A typical composition for a matrix support employing fine and coarse particles and fibres is given in Table 4-10.

Table 4-10. Typical matrix support material composition [166]

Fine Particle	γ-LiAlO$_2$	0,1 µm	55%
Coarse Particle	α-Al$_2$O$_3$	100 µm	30%
Fiber	α-Al$_2$O$_3$	5 µm $\varnothing$	15%

The coarse particle is added for crack deflection and attenuation and the fiber is added for strength. The physical and chemical stability of the support materials in the electrolyte environment of the fuel cell is an important requirement. The particle and pore structure of the support material must remain stable to retain functionality over the life of the fuel cell. The fine particle, high surface area powders are particularly susceptible to changes with age. The γ-LiAlO$_2$ is the most stable crystal form of LiAlO$_2$ and has demonstrated superior phase stability compared to the alpha and beta forms of lithium aluminate. Recent 10,000 hours fuel cell tests at both ERC and Japanese developers have verified the phase stability of γ-LiAlO$_2$ in Li$_2$CO$_3$-K$_2$CO$_3$ electrolyte at 650 °C in both fuel and oxidant atmospheres [167, 168].

4.5.3.2. Cathode

The MCFC cathode, which in pre-1960 cells was made of silver and copper, has since the late 1960s been a porous nickel mass which, in the first hours of operation, is oxidised to nickel oxide and becomes lithiated. During this in situ oxidation and lithiation, the original structure changes drastically. The cathode mass acquires many very small pores in addition to original 5–10 µm size pores, which causes a characteristic wetting pattern, in which very small pores are preferentially filled but all to the internal surface area is easily wetted.

Cathodes show optimum performance when the electrolyte fills the small pores such that 20–25% of cathode-pore volume is occupied. Larger, gas filled, pores are then wetted by thin films. Even under optimum design conditions, cathode polarisation is much higher than that of the anode, indicating that oxygen reduction has slow and complicated kinetics even at 650 °C. Especially troublesome is the rapid increase in polarisation at lower temperatures. This is an obstacle to attempts to alleviate corrosion and electrolyte problems by a lower design temperature.

The solubility of the NiO cathode and the LiAlO$_2$ support material is another important consideration which is influenced by the electrolyte composition. The NiO solubility is a particular concern because the dissolved Ni ions are transported toward the anode and are reduced and deposited as metallic nickel particles within the matrix. This nickel deposition can lead to an electronic short circuit between the fuel cell electrodes and hence a loss in power output. In addition, the dissolution deposition process causes structural deterioration of the cathode due to material loss and electrolyte displacement in the matrix (see Figure 4-57).

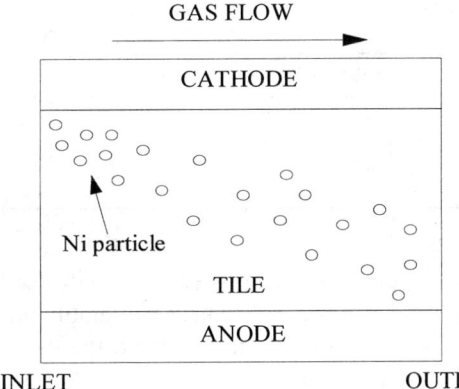

GAS FLOW

Figure 4-57. Outline of the Ni dissolution process in MCFC

NiO solubility is decreased in electrolyte compositions with a higher basicity or oxide ion concentration (O^{--}) [169–179]. This is because at the typical fuel cell operating condition NiO undergoes acidic dissolution according to the relation

$$NiO \longrightarrow Ni^{++} + O^{--} \tag{4-23}$$

Therefore higher oxide content lowers Ni^{++} concentration. The oxide ion concentration is governed by the CO_2 partial pressure and the dissociation constant of the alkali carbonate mixture according to

$$M_2CO_3 \longrightarrow M_2O + CO_2 \tag{4-24}$$

The relative basicity of the commonly used alkali carbonates is as follows

$$(basic)Li_2CO_3 > Na_2CO_3 > K_2CO_3(acidic) \tag{4-25}$$

Therefore the solubility of NiO is lower in the $Li_2CO_3 - Na_2CO_3$ mixture than in the $Li_2CO_3 - K_2CO_3$ mixture. For either binary system, the NiO solubility decreases as the lithium concentration increases beyond the eutectic composition. The lowest corrosion rates of the cathode side hardware have been noted for the 62% Li_2CO_3–38% K_2CO_3 and 52% Li_2CO_3–48% Na_2CO_3 eutectics. Addition of alkaline earth carbonates to either the $Li_2CO_3 - K_2CO_3$ or $Li_2CO_3 - Na_2CO_3$ systems has also been shown to reduce NiO solubility in out-of-cell tests. Additions of MgO, $CaCO_3$, $SrCO_3$ and $BaCO_3$ have all been shown to reduce dissolved Ni Concentrations in binary carbonate mixtures by increasing the electrolyte basicity [174]. Recently the effects of alkaline earth carbonates ($CaCO_3$, $SrCO_3$, $BaCO_3$) on fuel cell performance were systematically investigated. No influence was noted at small concentrations but a loss in performance was observed at concentrations in excess of 5 to 10 mol-%. Latest investigations show also a positive influence of addition some percent of potassium tungstate to the electrolyte [175].

To explain the relative slowness and other peculiarities of the cathode process, concepts of the complex oxide (peroxide, superoxide or hypothetical oxyanion) have already been proposed [177–186].

Table 4-11. Optimum amount of additives in mol-%[176]

	62 mol-% $Li_2CO_3 - K_2CO_3$	52 mol-% $Li_2CO_3 - Na_2CO_3$
$CaCO_3$	0–15	0–5
$SrCO_3$	0–5	0–5
$BaCO_3$	0–10	0–5

Another strategy is the development of cathodes made from alternative materials suitable to replace NiO. these materials should have good in-cell performance, preferably on the same level as that of NiO, and a very low dissolution-rate. In this respect, the thermodynamically stable ceramic oxides $LiFeO_2$ has a dissolution-rate equal to zero, but, according to tests in 3 cm² lab cells, a rather low performance. $LiCoO_2$ has a dissolution-rate which is almost an order-of-magnitude lower than that of NiO at atmospheric operation pressure [187–189]. As illustrated in Figure 4-58 the material B shows good in-cell performance, provided that the porous microstructure is optimized.

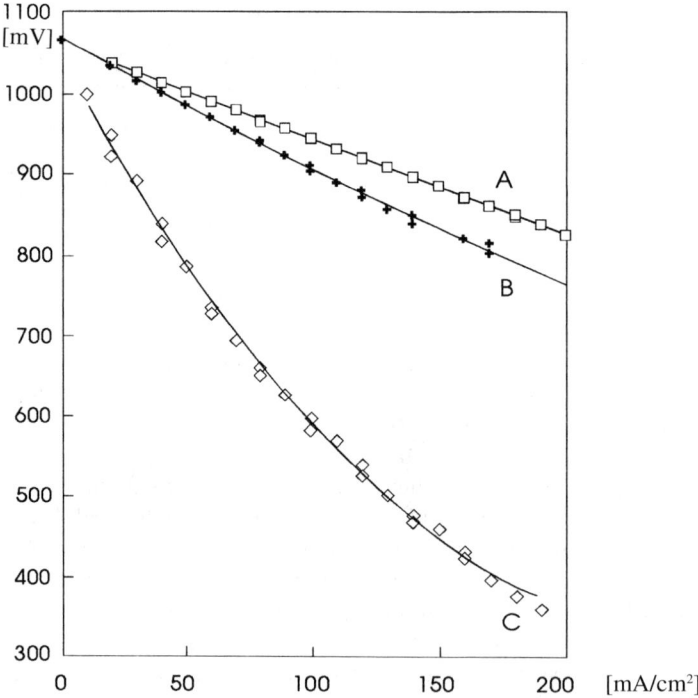

Figure 4-58. Voltage (mV) of 3 cm² Lab-Cells as a Function of Current Density [mA/cm²] [190]. A: with standard NiO cathode, B: with $LiCoO_2$ cathode, and C: with $LiFeO_2$ cathode. The lab-cells were operated at 650 °C and inlet gas conditions (70% air, 30% CO_2).

4.5.3.3. Anode

Ideally, MCFC anode pores are somewhat smaller than the gas-filled (large) pores of the cathode, but larger than the electrolyte-filled pores of the cathode. The matrix should have a pore size that is smaller than that of all other components, so that it always remains filled with electrolyte (see Figure 4-56).

Ceramic materials must be incorporated into the anode structure to stabilize it from further sintering, pore growth, shrinkage, and loss of surface area. Until recently, sintering has been controlled via the use of Ni-Cr alloy powder of Ni and Cr mixtures, usually in the 2–10 wt-% Cr range. The inital formation of Cr_2O_3, followed by surface formation of $LiCrO_2$, has been shown to be very effective in preventing drastic surface area loss by sintering. However, surface formation of $LiCr_2O_3$ has two disadvantages: First, its wettability by the electrolyte, changing anode surface area characteristics in respect to forming an ideal three-phase-boundary electrode. Second, a long-term growth of wettable fine-pore structure as chromium (in the form of chromite ion) leaches from the electrode. Recent work has concentrated on electrode structures containing oxidizable additives (e.g. aluminium [191]) that increase sinter resistance and are less leachable. These materials have also been shown to have high creep resistance [192, 193].

In addition to its electrocatalytic function, the anode may be required to perform other tasks. First, it may function as an electrolyte reservoir and gas cross-over barrier. The polarisation of the porous anode is relatively insensitive to its degree of filling by electrolyte, while that of the porous cathode must be maintained within a fairly narrow range. Therefore, the anode, but not the cathode, can function as a reservoir of electrolyte for the tile. The latter needs to be replenished, as a result of small but continuous electrolyte loss during long-term cell operation. However, care is needed to prevent too rapid electrolyte flow from the anode, which is ensured by a thin layer containing small, uniform pores at the anode-electrolyte interface; this layer is called the bubble-pressure barrier (BPB). Its primary purpose is that of preventing gas cross-over. The larger capillary retention ability associated with smaller pores increases the capability of the liquid electrolyte to resist pressure differentials and therefore prevents direct mixing of reactant gases (cross-over) through the electrolyte in the event of matrix failure. Furthermore, the electrolyte resides preferentially in small pores. Thus, proper electrolyte distribution between the anode and the tile may be achieved and the rate of electrolyte loss from the anode is reduced. In combination with the BPB, the anode is expected to provide major structural cell support in cell stacks, where the cathode and the very thin tile require this support.

MCFC anodes have been almost exclusively made from porous sintered nickel, with a thickness of 0,5–0,8 mm and a porosity of 55–70%, the mean pore size being approximately 5 μm. Because anode kinetics is faster than that of the cathode, less active surface area is required for the anodic process. Partial flooding of the comparatively thick anode is therefore acceptable, and it can provide a useful means of accommodating variations in the total carbonate content of the cell [196]. However, the optimal solution to the problem of electrolyte loss, i.e., an electrolyte reservoir in the MCFC, has yet to be found. This should be compared to the corresponding problem in the PAFC.

Hydrogen oxidation in molten carbonate has been studied at a variety of metal electrodes using cyclic voltammetry, AC impedance, potential step and chronocoulometric methods. There is considerable agreement about the kinetics, and a minor disagreement about the mechanism [195–197].

4.5.4. Internal Reforming Concept

The conventional Molten Carbonate Fuel Cell has the same reforming concept as the medium temperature Fuel Cells (PAFC), it uses an external reformer. A mixture of hydrocarbon fuel and steam is fed to a separate reformer and then hydrogen rich gas is fed to the anode of the Fuel Cell. The Internal Reforming Molten Carbonate Fuel Cell (IRMCFC) does not have an external reformer, but the reforming function is included in its stack.

The IRMCFC has the catalyst for the reforming reaction in the anode chamber or in a separate casing adjacent to the anode. In this case, at the anode, the following reactions occur:

$$CH_4 + H_2O \longrightarrow 3\,H_2 + CO \tag{4-26}$$

$$3H_2 + 2\,CO_3^{2-} \longrightarrow 3\,H_2O + 3\,CO_2 + 6e^- \tag{4-27}$$

$$CO + CO_3^{2-} \longrightarrow 2CO_2 + 2e^- \tag{4-28}$$

Net: $$CH_4 + 4\,CO_3^{2-} \longrightarrow 2\,H_2O + 5\,CO_2 + 8e^- \tag{4-29}$$

The major product from the reforming reaction, hydrogen, is electrochemically consumed in the anode reaction and the other product, water vapour, is consumed in the reforming reaction. Therefore, by coupling internally the reforming reaction with the cell reaction, the reforming reaction is shifted in the forward direction, which leads to a higher conversion rate. Figure 4-59 shows the different reforming concepts of the MCFC.

The most important question concerning this concept is whether the reforming reaction can take place with sufficiently high conversion at the relatively low operating temperature of the cell. Conventional external reformers are usually operated at around 800–900 °C and the thermodynamic equilibrium of methane conversion is then 95–99% (steam to carbon ration, S/C = 2,5–3,0). On the other hand, conversion at 650 °C – the typical MCFC operating temperature – is around 85% (S/C = 2,5). However, direct product exchange shifts the reaction forward to almost 100% under the high fuel utilisation conditions.

Figure 4-60 shows the relation between cell voltage and fuel utilisation. At higher utilisations, the cell voltage for simulated reformed gas and for methane conversion is driven to almost 100%. As shown in Figure 4-61 higher methane conversion is attained at higher current density, which may be understood in terms of higher consumption of hydrogen.

In an IRMCFC, the hydrocarbon fuel is typically methane or natural gas. However it is possible to use other fuels also. For example, Methanol may be used and other high car-

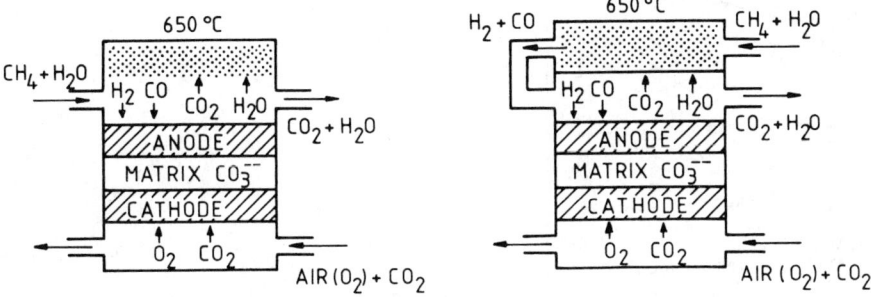

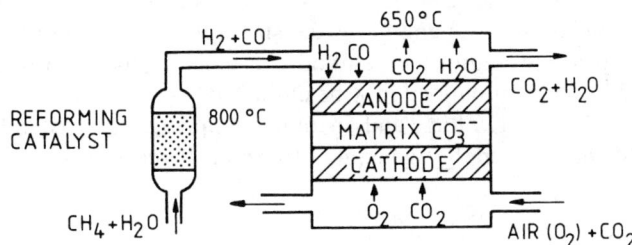

Figure 4-59. The different types of reforming for MCFC

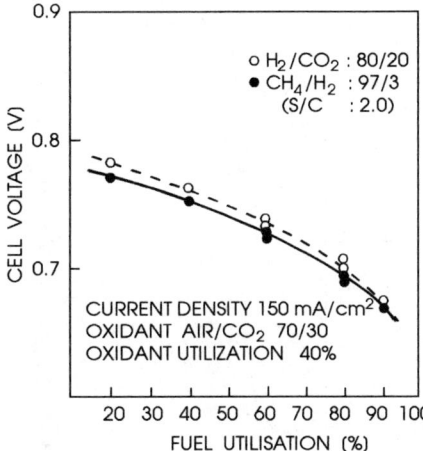

Figure 4-60. Dependence of Cell Voltage on Fuel Utilisation

bon fuels, for example propane and Naphta are possible fuels, although carbon depositi-on or cracking is a problem that must to be solved.

In the conventional Molten Carbonate Fuel Cell System (with external reforming), the heat required for the reforming reaction is usually supplied by burn-up (burn-out) of the anode exhaust gas followed by heat exchange. In the case of the IRMCFC, however, heat

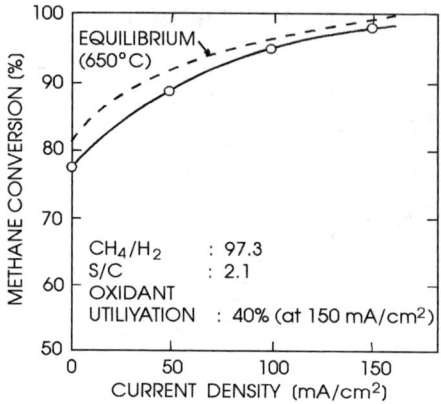

Figure 4-61. Methane conversion as a function of current density [198]

is provided by the fuel cell reaction. The fuel cell reaction itself is exothermic and removal of heat is necessary to control the temperature of the stack. If the heat from the cell reaction is automatically transferred within the cell and consumed in the reforming reaction, a significant fraction of the cooling load may be eliminated.

Knowing fuel utilisation, cell output voltage and operating temperature, the calculation of the electrical output and the heat energy generated by the cell reaction is possible.

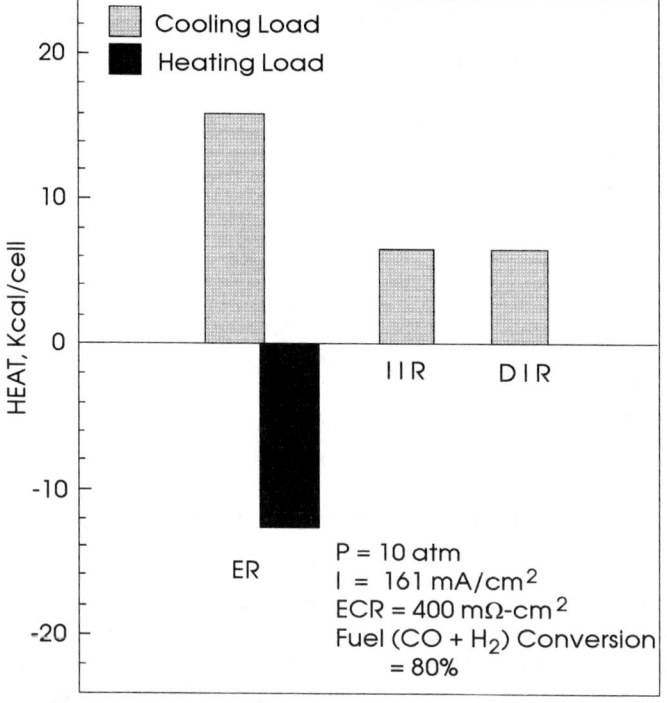

ER: External Reforming
IIR: Indirect Internal Reforming
DIR: Direct Internal Reforming

Figure 4-62. Heat rates in Internal and External Reforming

Although the result of this calculation depends on the heat release from the stack to the surroundings, the heat generated in the cell reaction is approximately equal to the cooling load, i.e., the heat to be removed from the stack. In the case of the IRMCFC, however, since part of the heat is utilised in the reforming reaction, the cooling load may be substantially reduced. Figure 4-62 shows a comparison of the heating/cooling loads in external and internal reforming system. As shown, the cooling load for the IRMCFC may be reduced to approximately one half of that for the external reforming system. This results in a reduction of the number and area of the heat exchangers, which in turn leads to plant simplicity and reduced cost for the IRMCFC [199, 200].

4.5.5. Performance of MCFCs

Over the past twenty years, the performance of single cells has improved from about 10 mW/cm^2 to > 150 mW/cm^2. During the 1980s both the performance and endurance of MCFC stacks showed dramatic improvements. The data in Figure 4-63 illustrate the progress that has been made in the performance of single cells, and in the cell voltage at 170 mA/cm^2 of small stacks at 650 °C, with low-Btu fuels (17% H$_2$ and CO) at 4.4 atm.

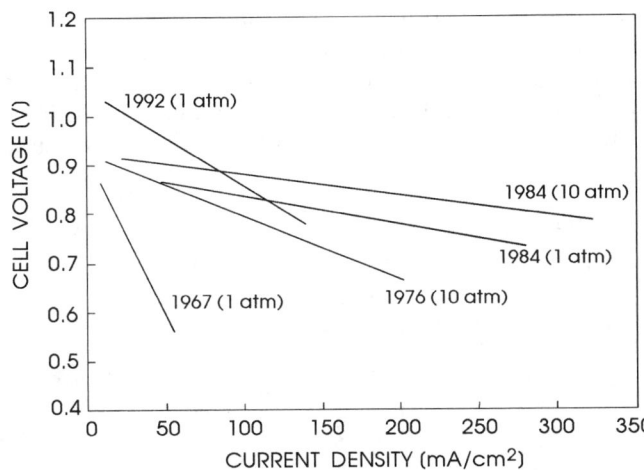

Figure 4-63. Progress in generic performance of MCFCs on reformate gas and air

The factors involved in choosing the operating condition for a MCFC are the same as those for the PAFC. These factors include stack size, heat transfer rate, voltage level, load requirement and cost. The performance curve is defined by cell pressure, temperature, gas composition and utilization. Typical MCFCs will generally operate in the range of 100 to 200 mA/cm^2 at 750 to 950 mV/cell.

Typical cathode performance curves obtained at 650 °C with an oxidant composition (12.6% O$_2$/18.4% CO$_2$/69% N$_2$), which is anticipated for use in MCFCs, and a common baseline composition (33% O$_2$/67% CO$_2$) are presented in Figure 4-64.

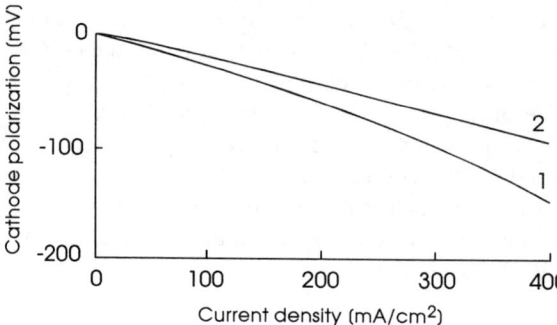

Figure 4-64. Effect of Oxidant Gas Composition of Cathode Performance in MCFCs at 650 °C
1) 12.6% O_2/18.4% CO_2/69.0% N_2),
2) 33% O_2/67% CO_2 [201]

The baseline composition contains the rectants, O_2 and CO_2, in the stoichiometric ratio that is needed for the electrochemical reaction at the cathode. With this gas composition, little or no diffusion limitations occur in the cathode because the reactants are provided primarily by bulk flow. The other gas composition, which contains a substantial fraction of N_2 yields a cathode performance that is limited by gas phase diffusion from dilution by inert gas.

Today cells as large as 1.0 m^2 are being tested in stacks. The focus has been on achieving performance in a stack equivalent to single cells. Several stacks have undergone endurance testing in the range of 7000 to 10000 hours. The voltage and power as a function of current density after 960 hours for a 1.0 m^2 stack consisting of 19 cells is shown in Figure 4-65. The data were obtained with the cell stack at 650 °C and 1 atm.

Endurance of the cell stack is a critical issue in the commercialisation of MCFCs. Adequate cell performance must be maintained over the desired length of service. Current MCFCs show an average degradation rate of:

$$\Delta V_{\text{lifetime}} \, [\text{mV}] = -5 \text{ mV}/1000 \text{ h} \qquad (4\text{-}30)$$

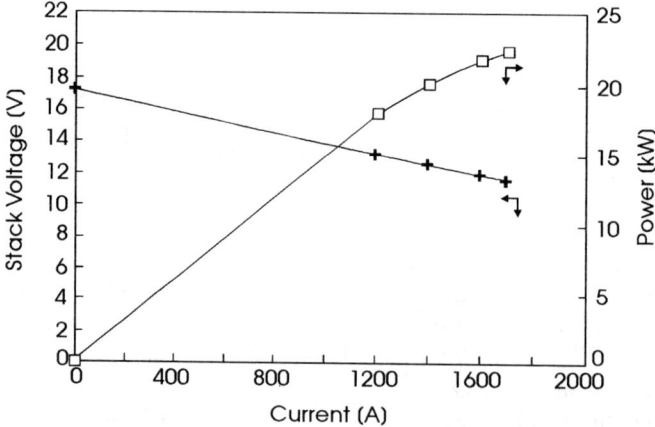

Figure 4-65. Voltage and power output of a $1.0/\text{m}^2$ 19 cell MCFC stack after 960 hours at 75% fuel utilisation [202]

The remainder of this section will review mainly the operating parameters like pressure and temperature which affect MCFC performance. Parameters such as Reactant Gas Composition, Reactant Gas utilization, Gas Impurities [203], and Current density are considered only generally.

Effect of Pressure

The dependence of the reversible cell potential of MCFCs on pressure is evident from the Nernst equation. For a change in pressure from P_1 to P_2, the change in reversible potential (ΔV_P) is given by:

$$\Delta V_P = \frac{RT}{2F} \ln \frac{P_{1,a}}{P_{2,a}} + \frac{RT}{2F} \ln \frac{P_{2,c}^{3/2}}{P_{1,c}^{3/2}} \tag{4-31}$$

where the subscripts a and c refer to anode and cathode. In a MCFC with the anode and cathode compartments at the same pressure (i.e., $P_1 = P_{1,a} = P_{1,c}$ and $P_2 = P_{2,a} = P_{2,c}$):

$$\Delta V_P = \frac{RT}{2F} \ln \frac{P_1}{P_2} + \frac{RT}{2F} \ln \frac{P_{2,c}^{3/2}}{P_{1,c}^{3/2}} = \frac{RT}{4F} \ln \frac{P_2}{P_1} \tag{4-32}$$

At 650 °C,

$$\Delta V_P(\text{mV}) = 20 \ln \frac{P_2}{P_1} = \left[46 \log \frac{P_2}{P_1} \right] \tag{4-1}$$

Therefore, a ten-fold increase in cell pressure corresponds to an increase of 46 mV in the reversible cell potential at 650 °C. Increasing the operating pressure of MCFCs results in larger cell voltages because of the increase in the partial pressure of the reactants, increase in gas solubilities, and the increase in mass transport rates. Opposing the benefits of increased pressure are the effects of pressure on undesirable side reactions such as carbon deposition (Boudouard reaction):

$$2\,CO \longrightarrow C + CO_2 \tag{4-34}$$

and methane formation (methanation),

$$CO + 3\,H_2 \longrightarrow CH_4 + H_2O \tag{4-35}$$

In addition, decomposition of CH_4 to carbon and H_2 is possible,

$$CH_4 \longrightarrow C + 2\,H_2 \tag{4-36}$$

but this reaction is suppressed at higher pressure. According to the Le Chatelier principle, an increase in pressure will favour carbon depsition and methane formation. The water gas shift reaction,

$$CO_2 + H_2 \longrightarrow CO + H_2O \tag{4-37}$$

is not expected to be affected significantly by an increase in pressure because the number of moles of gaseous reactants and products in the reaction is identical. Carbon deposition in a MCFC is to be avoided because it can lead to plugging of the gas passages in the anode. Methane formation is detrimental to cell performance because the formation of each mole consumes three moles of H_2, which represents a considerable loss of reactant and would reduce the power plant efficiency.

Effect of Temperature

The influence of temperature on the reversible potential of MCFCs depends on several factors, one of which involves the equilibrium composition of the fuel gas. The water gas shift reaction achieves rapid equilibrium at the anode in MCFCs, and consequently CO serves as an indirect source of H_2. The equilibrium constant (K),

$$K = \frac{P_{CO}P_{H_2O}}{P_{H_2}P_{CO_2}} \qquad (4\text{-}38)$$

increases with temperature, and the equilibrium composition changes with temperature and utilisation to affect the cell voltage, as Table 4-12 shows.

Table 4-12. Equilibrium Composition of Fuel Gas and Reversible Cell Potential.

Parameter	Temperature [K]		
	800	900	1000
P_{H2}	0.669	0.649	0.643
P_{CO2}	0.088	0.068	0.053
P_{CO}	0.106	0.126	0.141
P_{H2O}	0.137	0.157	0.172
E [V] *	1.155	1.143	1.133
K	0.247	0.48	0.711

* Cell potential calculated using Nernst equation and cathode gas composition of 30% O_2/60% CO_2/10% N_2.

Inspection of the result shows a change in the equilibrium gas composition with temperature. The partial pressures of CO and H_2O increase at higher temperature because of the dependence of K on T. The result of the change in gas composition, and the decrease in E_0 with increasing T, is that E decreases with an increase in T. In an operating cell, the polarization is lower at higher temperatures. The net result is that a higher cell voltage is obtained at elevated temperatures.

The two major contributors responsible for the change in cell voltage with temperature are the ohmic voltage drop and electrode polarisation. It appears that in the temperature range of 575 to 650 °C, about 1/3 of the total change in cell voltage with decreasing temperature is due to an increase in ohmic voltage drop, and the remainder from electrode polarization at the anode and cathode. Most MCFC stacks currently operate at an average temperature of 650 °C. Most carbonates do not remain molten below 520 °C and, as can be seen by the previous equations, cell performance is enhanced by increasing temperature. Beyond 650 °C, however, there are diminishing gains with increased temperature. In addition, there is increased electrolyte loss from evaporation and increased material corrosion. An operating temperature of 650 °C therefore represents an optimization of high performance and stack life.

4.5.6. MCFC R & D Programs

United States

At the present time, two major industrial vendors are involved in the Department of Energy's development program: Energy Research Co. (ERC) of Danbury and Torrington, CT and M-C Power Corp. of Burr Ridge, IL.

In April 1993, ERC passed a major milestone with a 250-hour test of a 120 kW MCFC stack, the largest tested to date. The test stack, achieved a 50% overall efficiency level, also a record for MCFC test cells. The ERC technology is distinguished from other molten carbonate concepts by its use of "internal reforming". The 120 kW prototype stack contains 246 individual flat cells assembled in six cell groups. Each cell package has an area of 2 feet by 3 feet and is less than 3/8ths of an inch thick. Assembled, the full-size unit is about 10-feet high and can be shipped by truck.

The 120 kW stack is the prototype for the 2 MW, natural gas-fueled power plant. The Department of Energy will fund two eight-stack modules and related hardware with the balance of plant financed by a consortium of five California utilities, the National Rural Electric Cooperative Association, the Electric Power Research Institute, and ERC. Previously, in 1992, ERC operated a 70 kW stack at the Pacific Gas and Electricity test facility in San Ramon, California, that logged more than 2000 hours of cumulative running time.

The San Ramon and Santa Clara units are the technological forerunners of ERC's expected commercial product line. ERC has announced that after these initial demonstrations, and a marketing run of about 100 megawatts of plants, it expects its molten carbonate fuel cell power plants to be commercialized by 1998.

Over the longer term, large molten carbonate central station power plants may be fueled by coal-derived gas. ERC has shipped a 20-kilowatt stack to Destec Energy Systems in Plaquemin, LA for integration with a coal gasifier. Also, in May, 1993, the Department of Energy selected a coal gasification-combined cycle project proposed by Duke Energy Co. for siting in Camden, NJ. The project will include a test of an ERC manufactured 2.5 MW MCFC to be fueled by cleaned, coal-derived gas (see Figure 4-66).

ERC has also completed a pilot manufacturing facility in Torrington which has a projected capacity of two to three megawatts per year [204, 205].

M-C Power has tested its configuration of a molten carbonate fuel cell in two full-area (11 square feet – the largest size known to have been fabricated to date), 20 cell, 20 kW stacks, with the second, slightly improved version accumulating 1500 hours of testing. A third full-area stack is now being tested.

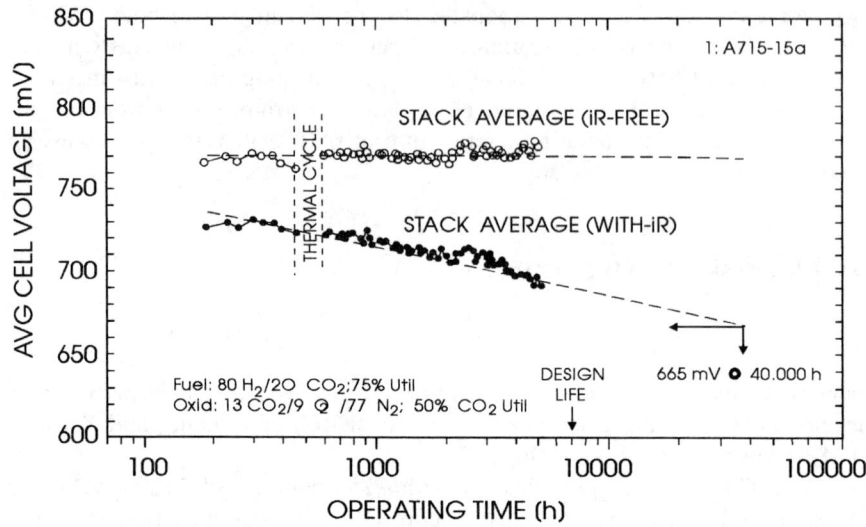

Figure 4-66. Lifegraph of Accelerated Stack Test and Performance Projection to 40 000 hours (300 cm^2, 5 Cell Stack; 160 mA/cm^2 Operation at the MW-Class Integrated System Condition): A Decay of –75 mV in 40 000 Hours is projected.

M-C Power is fabricating a 250 kW stack consisting of 250 full size cells for testing at the UNOCAL Science and Technology Centre in Berea, CA, beginning in early 1994. The company will also fabricate two stacks for a 250 kW test at the Kaiser Permanente Medical Centre in San Diego, CA in 1995. Co-sponsors of the project include the Gas Research Institute, the Electric Power Research Institute, Kaiser Permanente Medical Centre, San Diego Gas and Electric, and M-C Power.

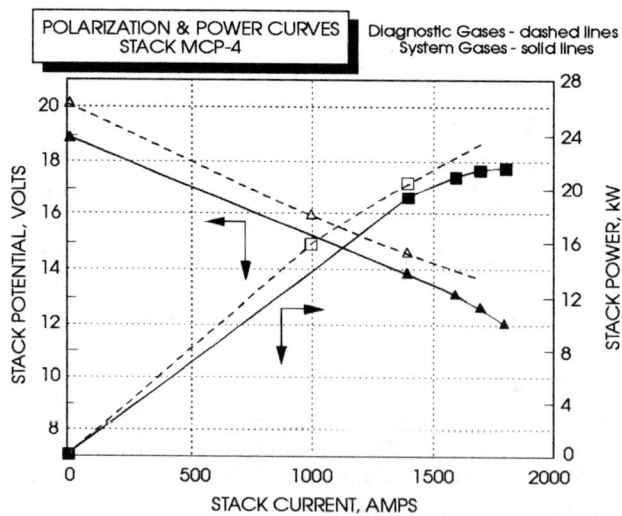

Figure 4-67. Polarization and power curves for stack MCP-4 diagnostic gases – dashed line system gases – solid lines (Courtesy of M-C Power)

M-C Power is currently producing full-size stack components in its pilot manufacturing facility in Burr Ridge with a manufacturing capacity of two to three megawatts per year. It expects to have market entry plants of 500 kilowatts to 3 megawatts ordered by 1995 and installed by 1997.

Additional to the development efforts of ERC, M-C Power, IFC has had also a long-term commitment to the development of MCFC technology for electric utility applications. As part of that development activity, IFC has until 1992 completed several power plant conceptual designs that provide utilities with flexibility in selection of fuels at minimum environmental impact and at competitive bus bar electricity costs for a range of ratings and utility applications.

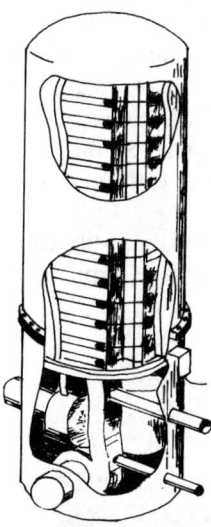

Figure 4-68. IFC's Product Integrated Stack Unit [206]

Japan

In Japan, Research and development work will be conducted on large-size stack technology, plant equipment, system technology, and fuel gas-related technology for coal gasification and total system. The goal is to develop a 1000 kW system by fiscal year 1997 after successful R & D for a 100 kW stack.

Stack Development

Studies are being conducted on a composite, large-capacity stack and an internally reforming stack: A 100 kW composite, large-capacity stack will be manufactured, and tested at the Akagi Laboratory. This stack will comprise two cascaded 50 kW modules, each module consisting of 44 laminated cells of 1.2 m^2 (22 cells each in the upper and lower potions, with an intermediate heater in between). A modified 100 kW large-capacity stack will also be manufactured and tested at Akagi. This stack will comprise 100 laminated cells of 1.0 m^2 area (50 each in the upper and lower portion of the stack, with an intermediate header), R&D for an internally reforming, 30 kW stack with 60 cells of 0.5 m^2 area will be conducted [207–210]. Research and development of a 1000 kW class power generation system of the externally reforming type was be initiated in the latter part of fiscal year 1993.

Development of Balance of Plant and System

In order to develop a suitable control system for the plant, simulation technology and a conceptual design of a 1000 kW class power generation system will be conducted. A high-temperatur multifunctional valve will be studied under operation. Regarding the stack-related peripheral system, studies will be made of the manufacture and operation of molten carbonate scrubbers and high-temperature blowers. Research will be conducted on the operation of a heat exchanger-type reformer and a two-stage catalytic reformer. Piping work will be done for the equipment that supplies fuel to the system. Studies will be made of the manufacture and operation of a waste heat recovery heat exchanger and on the operation of a turbine compressor.

In 1994 the following companies are engaged in all these projects: Hitachi, Ltd., Ishikawjima-Harima Heavy Industry Company (IHT), Mitsubishi Electric Corporation (MELCO), Sanyo and Toshiba [211–230].

Figure 4-69. 100 kW class Indirect Internal Reforming Molten Carbonate Fuel Cell Stack of Mitsubishi Electric (Courtesy of Mitsubishi Electric Corporation)

Europe

Ansaldo in Italy is pursuing MCFC development and has joined other European organizations in jointly sponsored projects. Ansaldo is building an automatic fabrication facility to initially produce two or three megawatts of molten carbonate stacks per year with the potential to ultimately triple the output. A 100 kW stack is planned for manufacturing in 1994 and is to be installed in Spain in 1995.

The Dutch government has enlisted two Dutch companies, Stork and Royal Schelde, in its MCFC development program. The arrangement has created Brandstofcel Nederland (BCN) which is planning a 50 kW demonstration unit in 1994 and two 250 kW demonstrations (one fueled by coal gas, the other by natural gas) in 1994 and 1995 [231].

4.6. Solid Oxide Fuel Cells (SOFC)

4.6.1. Introduction

Several features special to solid oxide electrolyte fuel cells may make them very attractive for utility and industrial applications. The very high operating temperature (~1000 °C) ensures that all fuel compositions, when combined with the necessary amount of water vapor, will oxidize rapidly and reach thermodynamic equilibrium if sufficient air is provided on the cathode side. The high temperature of reaction makes expensive catalysts unnecessary and permits direct processing of the fuel in the fuel cell itself. Solid oxide systems are able to attain at least 96% of the theoretical voltage at open circuit, (since gas crossover and electronic conductivity in the electrolyte are small). In the absence of major internal resistance drop they can operate at much higher current densities than molten carbonate fuel cells with high fuel conversion efficiency. The SOFC requires no CO_2 recycling, leading to a further simplification of the system compared with that of the MCFC.

High temperature operation and tolerance to impurities in the fuel streams make solid oxide systems especially attractive when combined with coal gasification plants. The heat released in the fuel cell reactions can be efficiently transferred and used for coal gasification or hydrocarbon reforming.

Because the solid oxide electrolyte is normally very stable, no electrolyte migration problems under cell operating conditions exist as in the MCFC. In addition, the composition of the electrolyte is not a function of fuel or oxidant composition. In the absence of liquid phases, there are no three-phase interfaces to maintain, no problems with pore flooding, and no catalyst wetting problems. Solid oxide cells also have a good tolerance to overload, underload, and even to short-circuiting. The progress until 1993 is shown in Figure 4.70.

One disadvantage of the SOFC is the less negative free energy of formation of water at the 1000 °C operating temperature, compared with the MCFC temperature of 650 °C. This lowers the SOFC open circuit potential by about 100 mV compared to the MCFC. Therefore, unless polarization and IR drop are correspondingly lower, the SOFC will

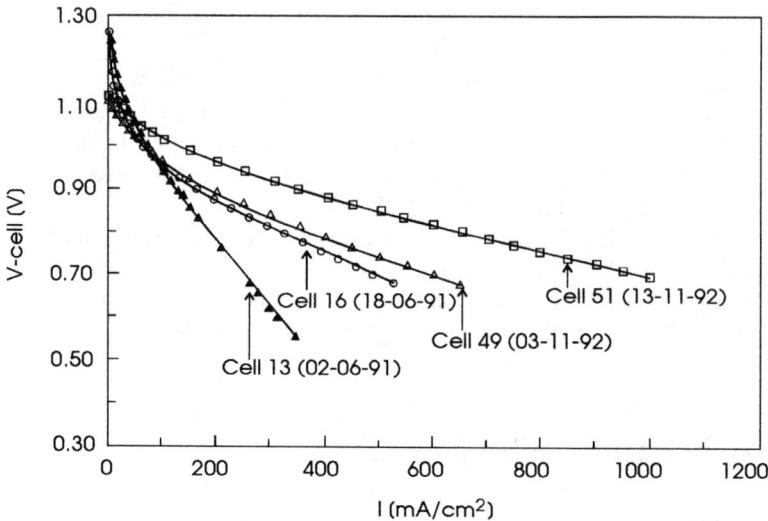

Figure 4-70. Examples for performance and progress of SOFC electrodes (1991–1993) at ECN [232]

always be less efficient than the MCFC, typically by about six percent. Some of this may be made up by the use of its higher quality waste heat in a topping cycle. 1994 concepts propose high performance catalyzed-reaction layers for SOFC applications already at 800 °C [233].

Because of the high operating temperature of the SOFC, sulfur tolerance levels are claimed to be one to two orders of magnitude higher than for other types of fuel cells, and this permits the use of high temperature sulfur removal methods, which are more energy efficient than the low temperature methods required for the MCFC to lower the sulfur content of the gas to less than 10 ppm. This tolerance to fuel with a high impurity content may make the SOFC very attractive for operation on heavier fuels such as diesel oils and, most importantly, coal gas.

4.6.2. Reactions, Electrolyte and Cell Materials of SOFCs

Cell Reactions: In an operating SOFC, negative charges are conducted by electrons from the anode through an external circuit to the cathode; it is conducted by negative ions (O^{2-}) from the cathode through the electrolyte to the anode.

Figure 4-71 shows the operating principle, the anode and cathode reactions, and the overall reactions for an SOFC operating on a reformed fuel (mixture of CO and H_2). The SOFCs typically operate at 1000 °C; at this temperature, the electrolyte is an oxygen-ion conductor, and the free energies (and therefore the associated Nernst potential) of the overall reactions are substantially lower than at lower temperature (e.g. 43.3 kcal/mole at 1200 K versus 54.6 kcal/mole at 300 K for H_2). The heat of reaction is nearly independent of temperature, therefore the potential (ideal efficiency) is reduced by the high-temperature operation.

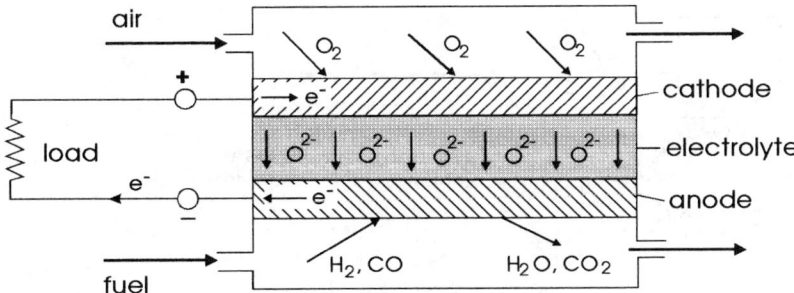

Figure 4-71. The schematic shows the essential features and the overall electrode reactions of an SOFC unit

anode reaction: $\qquad$ $H_2(g) + O^{2-} \longrightarrow H_2O(g) + 2e^-$ (4-39)

anode reaction: $\qquad$ $CO(g) + O^{2-} \longrightarrow CO_2(g) + 2e^-$ (4-40)

combined anode reaction:

$$aH_2(g) + bCO(g) + (a+b)O^{2-} \longrightarrow aH_2O(g) + bCO_2(g) + 2(a+b)e^- \quad (4\text{-}41)$$

overall cathode reaction:

$$1/2(a+b)O_2(g) + 2(a+b)e^- \longrightarrow (a+b)O^{2-} \quad (4\text{-}42)$$

overall cell reaction:

$$1/2(a+b)O_2(g) + a\,H_2(g) + b\,CO(g) \longrightarrow a\,H_2O(g) + b\,CO_2(g) \quad (4\text{-}43)$$

The actual efficiency and electrical energy produced will be lower if the voltage is reduced by IR losses or polarization. Maintaining the temperature level requires cooling of the cell to remove a quantity of heat equal to the difference between the heat of reaction and the electricity produce. The heat removed is the difference in enthalpy between the products of the reaction leaving the generator at high temperature and the cooler reactants supplied to the generator. Because the heat is available at high temperature (1000 °C), it can be used in an efficient bottoming cycle to produce more electricity or in applications requiring high grade heat. Thermodynamic analysis show that the electric efficiency is over 80% [234].

Electrolyte: The electrolyte is ZrO_2 (zirconia) doped with 8–10 mole-% of Y_2O_3 (yttria). Pure zirconia is an insulator. When yttria is mixed with zirconia, some Y^{3+} replaces Zr^{4+} in the fluorite-type crystal structure. A number of oxide-ion sites become vacant because three O^{2-} replace four O^{2-} when two Zr^{4+} are replaced by two Y^{3+} in the lattice. The mobile ionic species within the electrolyte in the SOFC system are the O^{2-}-ions as compared to protons (H^+) in the PAFC and PEFC and CO_3^{2-}-ions in the MCFC. At high temperatures, the O^{2-}-ions move across the electrolyte via vacant lattice sites, conducting charges ionically, as does the carbonate ion in the MCFC. In contrast, the H^+ ion in the PAFC travels in the reverse direction from the anode to the cathode. All three current carrying species arrive at the appropriate electrolyte-electrode interface, or three phase boundary, and react with the gas phase within the porous electrode. The

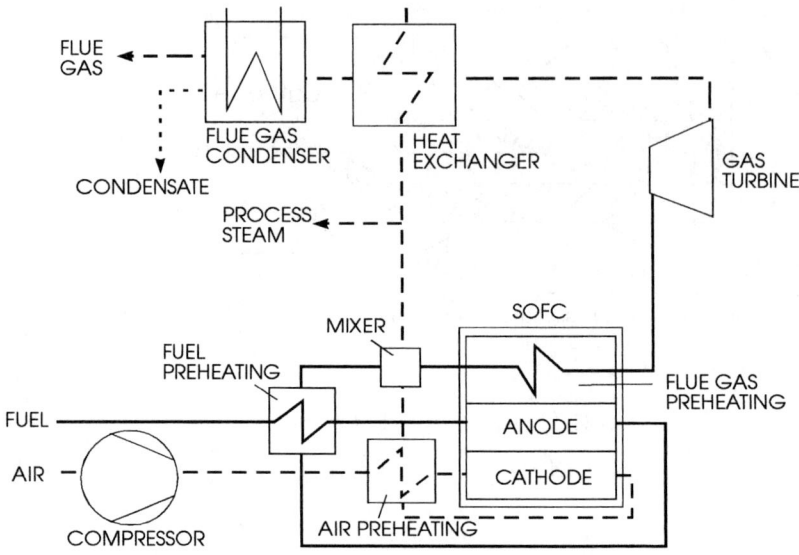

Figure 4-72. Proposal for SOFC Cogeneration power plants

micro-structural characteristic of this interfacial three phase boundary is a critical parameter controlling the gas dynamics and the electrochemical efficiency of the fuel cell.

Doped CeO_2 is one of the candidates to replace zirconia as electrolyte and would allow lower operating temperatures [235, 236]. Consequently, some SOFC components can be made out of low cost metals and metal shaping technologies can be applied. Gas sealing problems might be overcome much easier than at high operating temperatures.

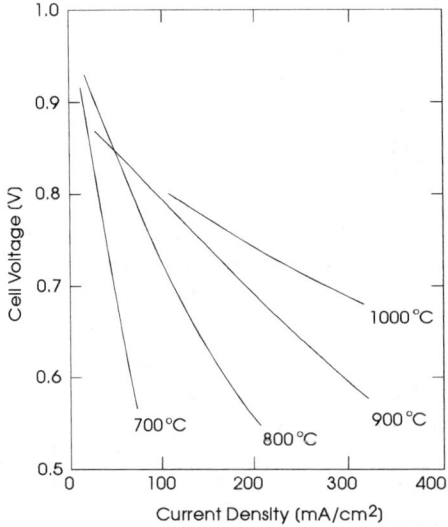

Figure 4-73. Dependence of cell performance on temperature. Fuel (67% H_2/22% CO/11% H_2O) utilization is 85% and oxidant (pure O_2) utilization is 25% [237] (see also Figure 4-76).

Materials: Table 4-13 summarizes the properties of conventional SOFC materials as well as CeO_2 as intermediate temperature electrolyte and Fe-Cr-alloy as intermediate temperature interconnecting material.

Table 4-13. Properties of SOFC materials

	Electrolyte		Anode		Cathode	Interconnect	
	YSZ, Y_2O_3 [8 mol%]	CeO_2 Gd_2O_3 [11 mol%]	Ni/ZrO_2	LSM	LSC	$LaCrO_3$	FeCr alloy
Spec. Gravity [g/cm³]	5.97	7.24		6.8–6.9		6.74	7.22
Porosity	< 5	<5	30			< 5	0
El. conductivity $\Omega^{-1}cm^{-1}$ at 950 °C	0.1	0.12	3000	100	950	30	7690
at 700 °C	0.03	0.04	3000	98	1080	30	
Ionic transference number at 950 °C	1	0.7	–	–	–	–	–
at 700 °C	1	0.85	–	–	–	–	–
Thermal expansion [cm/cm K x 10⁻⁶]	10.5	12.2	12–14	12	18–22	9.5–10.7	12..5
Thermal conductivity	3.8	–	–	–	–	–	23
Bend strength [MPa] at RT	368	250	42	26	–	–	tensile strength 520–720
Bend strength [MPa] at high temperature	280	150–200	–	–	–	–	–
Fracture toughness [MPs $\sqrt{m}$] at RT	3	3	–	–	–	–	–

Porous gas-diffusion electrodes are used because diffusion of reactant species through solid plate electrodes becomes the limiting rate process at low currents and sufficient power outputs are attainable only with porous electrodes. The processes occurring at the electrodes are not well understood. There are no quantitative models for the effects of electrode structure and composition on electrode performance for solid electrolyte systems [238]. For liquid electrolyte systems, the straight-pore, thin-layer, and hybrid models are useful in describing the influence of electrode morphology on electrode performance, while variants of the double-layer theory describe the effects of the electrolyte-electrode interface on electrode performance. Studies of the mechanism and rate-determining step for electrode reactions should be useful in future electrode development.

The successful commercialization of SOFC technology during the next years will depend on the ability to produce cost competitive systems that can operate for five to ten years before replacement of the SOFC modules. These requirements are very common for all Fuel Cell systems for stationary applications.

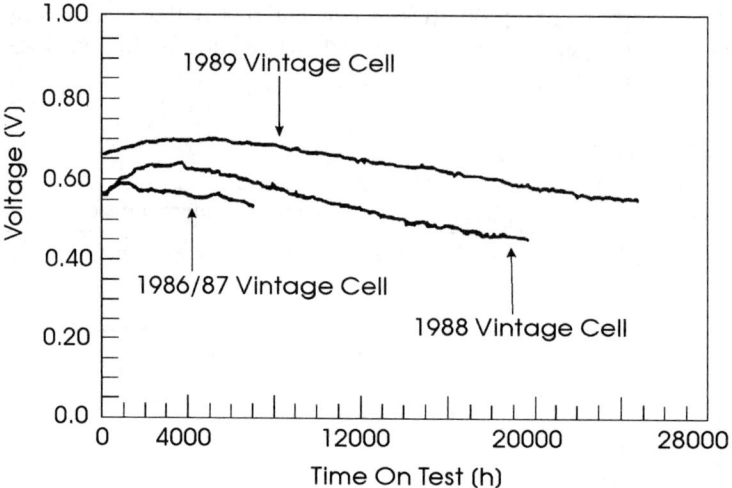

Figure 4-74. Life expectancy of different Westinghouse Electric SOFC cells during a long-term test run [239] (Courtesy of Westing house Electric)

In mid 1992 several test cells from SOFC modules of WE surpassed 30 000 hours of operation during a long term test run. WE's goal is to develop cells which can operate for 50 000 to 100 000 hours with a voltage degradation rate of less than 0.1 percent per 1000 hours (see Figure 4-74).

4.6.3. Variations of SOFC Developments

Basically four different designs shown in Figure 4-75 are under development. The designs and materials for SOFC stacks were extensively reviewed by Minh [240–242].

- Tubular design
  Seal Less Tubular Design
  Segmented Cell In-Series Design
- Planar design
- Monolithic design

The above four SOFC designs differ only in cell geometry. The 'cell' is the repetitive electrochemical building block, connected in series and parallel, forming the 'stack' or unit of fabrication. The basic SOFC 'cell' consists of the following common parts:

(a) The anode. (b) The electrolyte. (c) The cathode. (d) The interconnect or bipolar plate. (e) The support tube (in the tubular design only).

A brief outline of the materials of construction of the above five components will highlight their common features, for they are all based upon the same materials selection

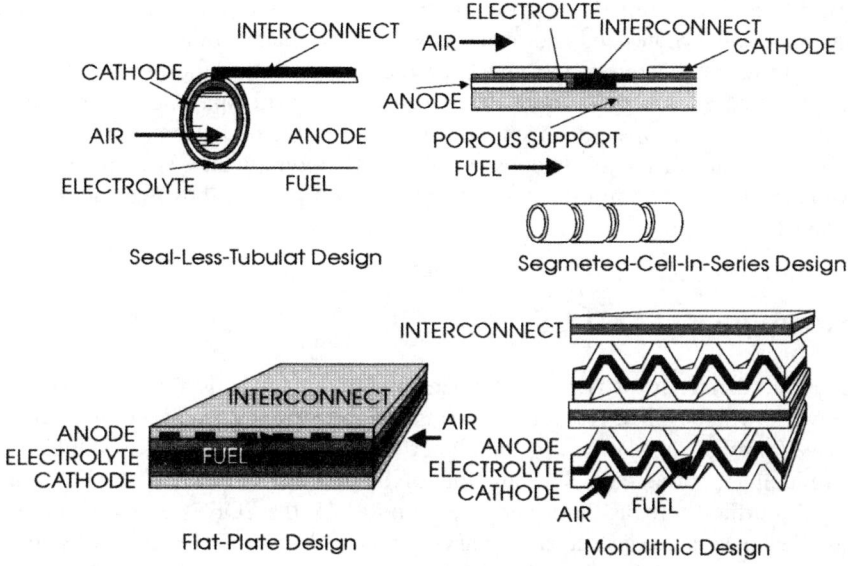

Figure 4-75. SOFC designs

Table 4-14. List of solid oxide fuel cell components and their compositions using the Westinghouse concept as an example.

Component	ca. 1965	ca. 1975	ca. 1985 *
Anode	Porous Pt	Porous-Pt	Ni-ZrO$_2$ cermet** Deposit slurry, EVD ~150 μm thick 20–40% porosity
Cathode	Porous-Pt	Stabilized ZrO$_2$ impregnated with prawsodymium oxide	Sr-doped lanthanum manganite Deposit slurry, sinter ~ 1 mm thick 20–40% porosity
Electrolyte	Yttria-stabilized ZrO$_2$ 0.5 mm thick	Yttria-stabilized ZrO$_2$ EVD	Yttria-stabilized ZrO$_2$ EVD ~ 40 μm thick
Cell Interconnect	Pt	Mn-doped cobalt chromite	Mg-doped lanthanum chromit
Support tube	Yttria-stabilized ZrO$_2$	Yttria-stabilized ZrO$_2$	Calcia-stabilized ZrO$_2$, 34-35% porosity, 12.8 mm diameter 2 mm wall thickness 457 mm length

* Specifications for Westinghouse 5 kW fuel cell
** Y$_2$O$_3$ stabilized ZrO$_2$

with minor dopant variations depending upon the fuel cell type and mode of fabrication. All are ceramic materials synthesized and formed by conventional ceramic processes.

In each of the three SOFC types: the *anode* or fuel electrode is a porous cermet of nickel and an inert phase such as zirconia or yttria-stabilized zirconia; the anode is an electronic conductor with a projected current density in the range of 1 A cm^{-2}; it is fabricated usually as a mixture of nickel oxide and stabilized zirconia which is converted to the conductive cermet *in situ* within the cell. The fabrication process, however, will differ with the respective designs

4.6.4. Performance of SOFCs

The thermodynamic efficency of SOFCs at open circuit voltage is lower than that of MCFCs and PAFCs which utilize H$_2$ and O$_2$ because of the lower ΔG at higher temperatures (as discussed previously). However, the higher operating temperatures of SOFCs is beneficial in reducing polarization. The amount of data sets compared to PAFC and MCFC is limited. Furthermore, as described in Section 4.6.5. the SOFCs being developed have unique designes, are constructed of varying materials and are fabricated by differing techniques (chapter 4.6.6.). Flat plate SOFCs will undergo considerable development in materials, design, and fabrication techniques. As SOFC technology progresses, it will mature towards more standardized cells as has happened with PAFCs and MCFCs which are closer to conformity with technical properties.

Effect of Pressure:

The reversible potential of cathodes and their operating voltage in SOFCs are expected to increase with an increase in oxidant pressure.

Effect of Temperature

An operating temperature that is lower than 1000 °C would be desirable for SOFCs owing to the lower interdiffusion and thermal expansion mismatch, for example. The dependence of SOFC performance on temperature is illustrated in Figure 4-76 for oxygen (25% utilization) and a fuel of 67% H$_2$/22% CO/11% H$_2$O) (85% fuel utilization). The sharp decrease in cell voltage as a fuction of current density at 800 °C is a manifestation of the high ohmic polarization (low ionic conductivity) of the solid electrolyte at this temperature. The ohmic polarization decreases as the operating temperature increases to 1000 °C, and, correspondingly, the current density at a given cell voltage increases dramatically. The data in this Figure show a larger decrease in cell voltage with decreasing temperature between 800 and 900 °C than that between 900 and 1000 °C, at constant fuel utilization and current density. The Nernst potential decrcases with an increase in temperature and it counteracts the performance improvement that is observed as the temperature increases.

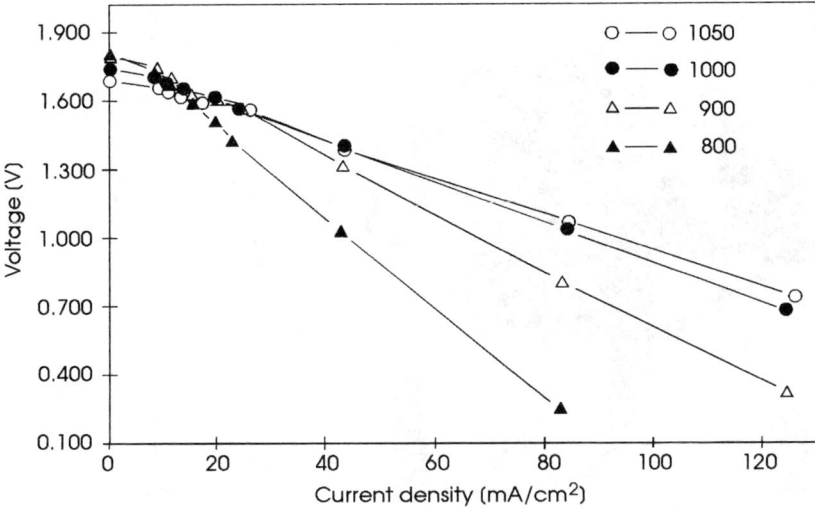

Figure 4-76. Dependence of cell performance on temperature in °C.
Fuel (67% H_2/22% CO/11% H_2O) utilization is 85% and oxidant (pure O_2) utilization is 25%[8]

4.6.5. SOFC Programs

USA: Commercial Solid Oxide development today is proceeding along tree paths, differentiated by the geometric configuration of the cells: Westinghouse Electric Corp., is developing a tubular arrangement in which each cylindrical, one- to two-meter long cell contains a cathode on the inside, the anode on the outside, an electrolyte layer of yttria stabilized zirconia sandwiched in between. Allied Signal Corp. is developing a monolithic design made up of a corrugated arrangement of electrodes and electrolyte. Ztek, Ceramatec and others are developing a planar configuration. In the Ztek concept, the electrodes and electrolyte are configured in a disc so thin that 16 of them stacked on top of each other stand only one inch high. Hundreds of discs would be packed together to form the company's basic 25-kilowatt unit.

To date, the Westinghouse configuration has received the most attention. In 1991, the Department of Energy and Westinghouse signed a US-$ 140 million, five-year cooperative agreement to bring the tubular solid oxide configuration to commercial readiness. The Department of Energy is funding US-$ 64 million of the cost, while Westinghouse and other industrial sponsors are providing US-$ 76 million. The agreement calls for Westinghouse to test progressively larger single cells and cell bundles, culminating in test of two 25-kilowatt generators, three to five 100-kilowatt modules, and a 2-megawatt module. The first test units have been installed at Kansai Gas and Osaka Gas in Japan. Smaller Westinghouse test units have passed 35000 hours of running time, with about

---

8 Allied-Signal Aerospace Company, 1992

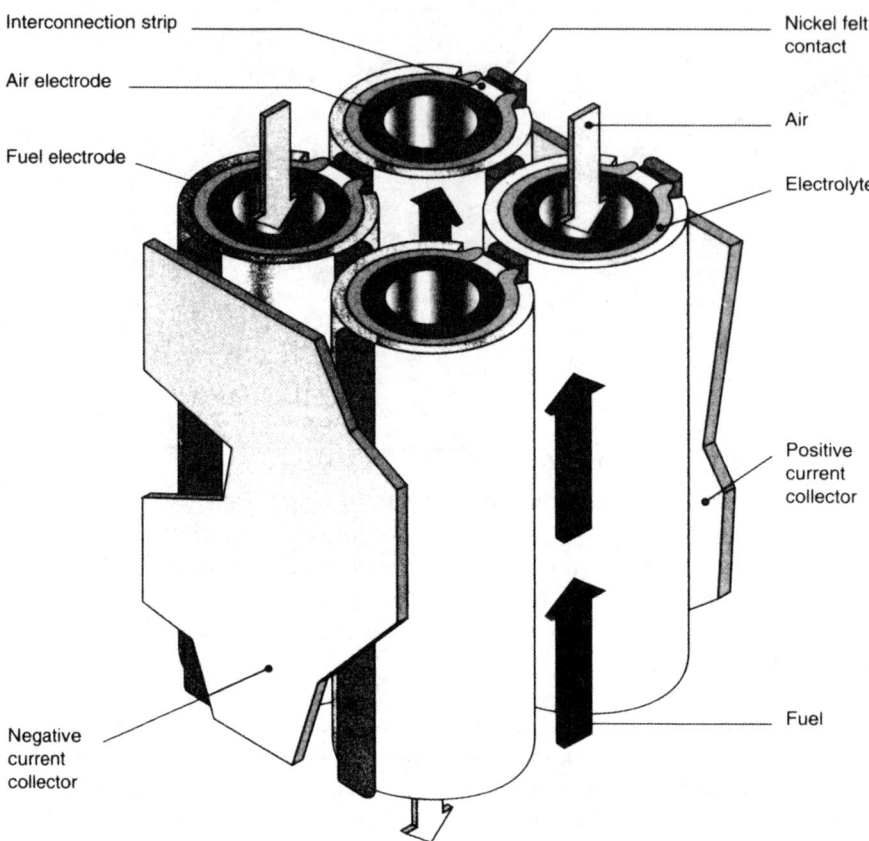

Figure 4-77. Westinghouse Tubular Design (Courtesy of Westinghouse)

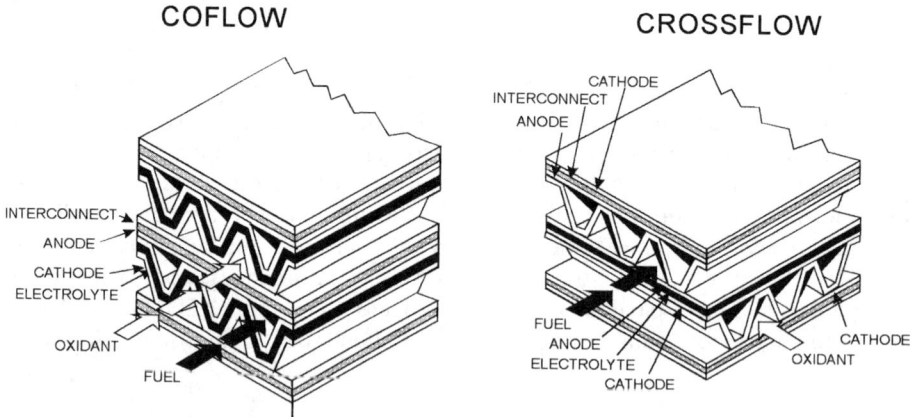

Figure 4-78. Monolithic Design designed by Allied-Signal (Courtesy of Allied-Signal)

50 000 hours being required to support commercial operation. Westinghouse plans to test a megawatt-scale prototype in 1996.

In the supporting research component of this project, Westinghouse is optimistic to develop solid oxide tubes nearly 80 inches (2.032 m) long, or roughly double the current length and power output. The company also plans to boost the capacity of its Monroeville pilot manufacturing plant from 500 kilowatts of annual capacity to two megawatts per year. The Department of Energy has also provided support to other solid oxide concepts, most notably the Allied Signal monolithic configuration [243, 244].

International Efforts

Japan's R & D will be conducted on elements of a power generation system, including materials, key technologies, and studies of basic designs and system of several kilowatts capacity. The project will then proceed to develop a module of several tens of kilowatts capacity in fiscal years 1996–97. At later date, R & D will focus on a power generation system at an appropriate scale [245].

Germany has increased its solid oxide development efforts with government cost-sharing being provided to Siemens AG, Power Generation; Dornier GmbH; Fraunhofer Gesellschaft e.V.; and Messerschmidt-Bolkow-Blohm [246, 247]. Australia is also a new entry into the international solid oxide development effort. A joint venture partnership, Ceramic Fuel Cells Ltd., was formed in 1991 and includes both government and private research organizations, a state electric commission, and an Australian company [248].

Summarizing: Of all SOFC configurations proposed so far only the Westinghouse tubular design has reached the maturity needed to extrapolate performance potentials from experimental evidence. Other concepts are still at an early stage of development. The technologies have not yet solidified enough to provide a base for assessing potentials or weaknesses. Nevertheless, later entries may establish themselves as winning SOFC technologies. Comparative Evaluations of the performance potentiale of SOFCs have been already made and should help identify the most promising SOFC concepts [249, 250].

4.6.6. Processing Techniques of SOFC Components

A special interest exists in the fabrication methods of ceramic SOFC components. This is the reason why detailed descriptions of these technologies are given in this chapter. The ceramic fabrication processes involved in the different designs of SOFC can be grouped into two categories: Those leading to mechanically stiff parts with predominantly structural functions and those leading to porous or dense but relatively thin coatings and parts which have mainly electrical functions with minor mechanical requirements.

Table 4-15. SOFC component manufacturing processes [251]

stiff, structural ceramic parts	functional ceramic parts
extrusion	screen printing
dry pressing	slurry coating
tape casting	EVD, PVD
calandering	plasma- and flame spraying

Dry Pressing

Some components such as the interconnect plates or electrolyte tubes for the segmented-cell-in series design can be shaped by the dry pressing process of powder granules.

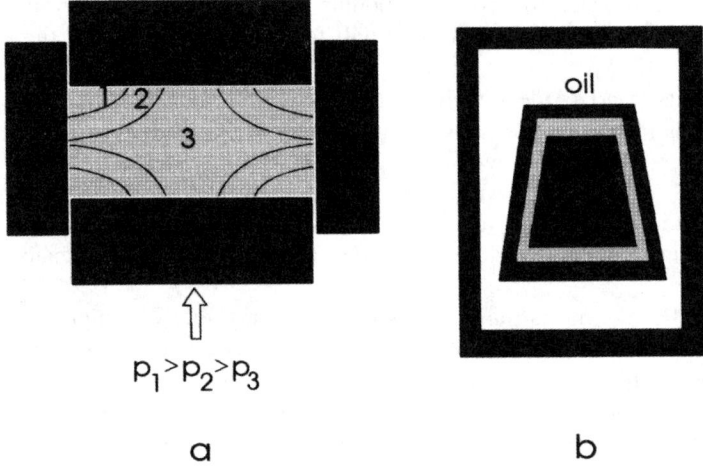

$$p_1 > p_2 > p_3$$

a   b

Figure 4-79. Dry Pressing a) axial b) isostatic

Dry pressing can be performed in the axial mode using a rigid die with moving punches or in the isostatic mode employing the pressure via an oil bath to a rubber mold filled with a powder. Usually dry pressing is used for parts thicker than 0.5 mm and parts with surface relief in the pressing direction.

Feedstocks for dry pressing are spray-dried powder granules containing ceramic powder, deflocculant, binder plasticizers and lubricants.

The maximum pressure is commonly in the range of 20–100 MPa. The quality (flatness, dimensional accuracy, microstructure, etc.) of the pressed parts is good in case of simple geometries, however limited when the part has several surface contours different in height.

Powder filling degree, pressure and granule density determine the green density in the pressed part. The green density is inhomogeneous over the cross section due to pressure gradients in the powder compact caused by frictional effects (see Figure 4-79). This might lead to warpage and possible cracking of the part during sintering.

Typical green densities (after pressing) range from 50 to 60 Vol.-% which is only slightly higher than the green density of the spray dried granules.

Dry pressing is a low cost, high-capacity forming technique ranging around 2 to 5 US-$/kg (powder excluded) in mass production of simple shaped parts.

Tape Casting

Tape casting is a suitable processing technique to produce thin ceramic sheets with smooth surfaces and precise dimensional tolerances. This process is widely used for multi-layer ceramic substrates and capacitors for the electronic industry. Casting is performed by applying a ceramic slip with a movable doctor blade on a temporary support (see Figure 4-80). After casting, the ceramic slip consisting of powder dispersed in an organic binder/solvent pasta is dried and the plastic tape can be handled for shaping, such as corrugating, coating or laminating. Multi-layer tapes are fabricated by subsequently casting several layers on top of each other. Tape casting is a low cost, high capacity forming technique. Forming costs range between 2 to 5 US-$/kg. Costs for organic processing materials range around 1 US-$/kg tape. If water-based chemicals are used, costs may range between 5 to 10 US-$/kg.

After evaporating the solvent, the tapes can be fired. During sintering to full density, shrinkage of about 50 Vol.-% occurs. During the sintering process the materials' inherent strength is low and the component is crack sensitive. Co-firing of different materials require the total matching of shrinkage as well as shrinkage rate throughout the whole binder burn-out and sintering temperature regime. This is the major draw-back of the monolithic stack design. The most important factors determing the results are the properties of the ceramic powder and the composition as well as the preparation of the tape casting slurry. Self supporting electrolytes commonly are 150 to 200 μm in thickness. However, also non self supporting layers as thin as 2–10 μm can be fabricated.

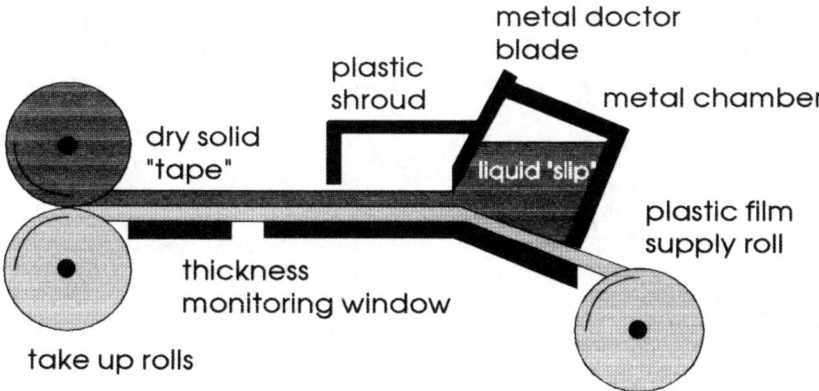

Figure 4-80. Tape casting

Tape Calandering

The tape calandering is a technique to form a continuous thin sheet or tape of controlled size as shown in Figure 4-81. Ceramic powder, binder and plasticizer mixed in a high-intensity mixer. The friction resulting form the mixing heats the batch and softens the binder to form a plastic mass. The mass is rolled into a thin, flat tape using a two-roll mill, and the tape thickness is controlled by the spacing of the rolls. This technique has been applied to fabricate monolithic fuel cells via a co-sintering process, using multi-layer tapes. In this case individual tapes are laminated in a second rolling operation. For co-firing processes, the starting powder characteristics, especially surface area and particle size should be optimized to match the shrinkage.

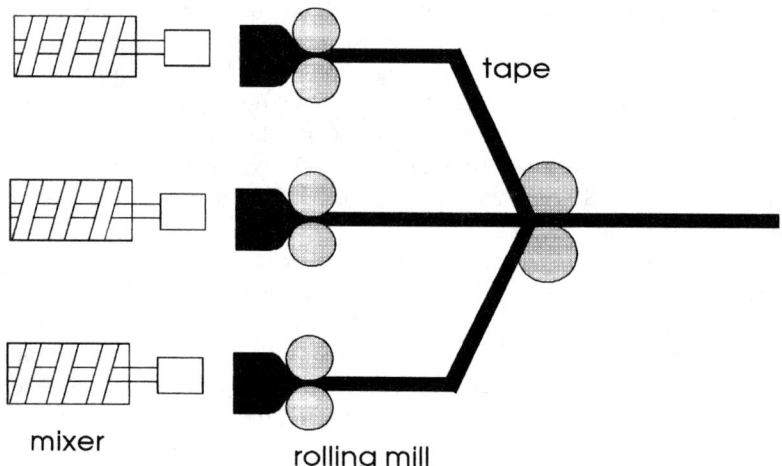

Figure 4-81. Tape calandering

The thermal expansion coefficients of each component should also be the same. This tape calandering process has been applied to fabricate fuel cells with very thin electrolyte layer, which enables to operate fuel cells based on YSZ at temperatures as low as 600 °C.

Screen Printing

The screen printing technique is used to produce porous or dense layers with a thickness from a few microns to ten microns. This technique has widely been used to fabricate hybrid circuits up to several ten microns.

A paste from a mixture of starting powder, an organic binder and plasticizer is screen-printed in a desired pattern onto a substrate and is then dried and fired at an elevated temperature to produce films. This process is schematically described in Figure 4-82. The patterned screen is affixed to the printer, the paste with a high viscosity is placed on

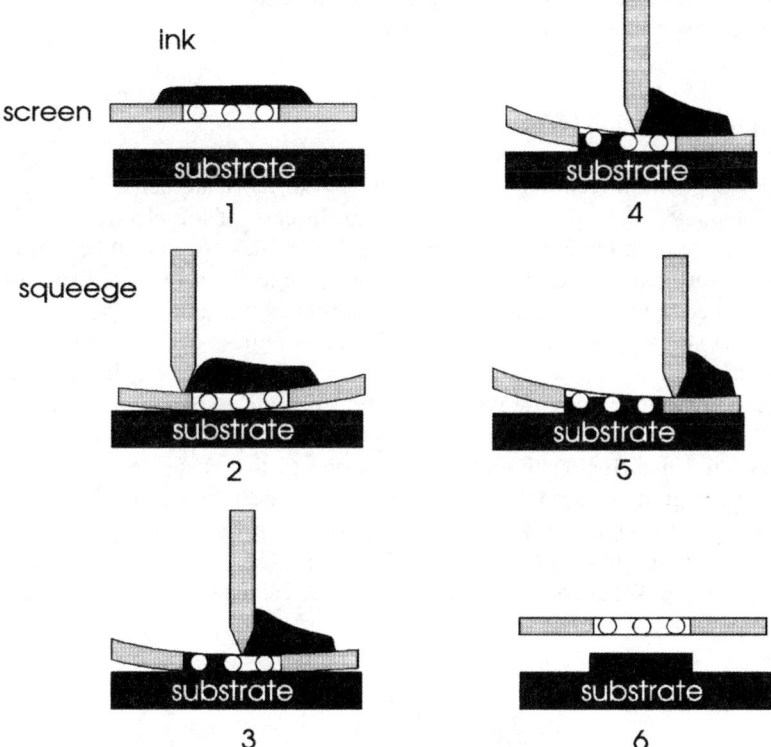

Figure 4-82. The screen printing process

top of the screen. The paste is then forced through the openings in the screen by a squeegee (rubber roller) which depresses the screen to contact the substrate, as it traverses the pattern. The kind and amount of organic binder as well as the mixing rate of binder and powder should be optimized to obtain a continuous and homogeneous microstructure. The thickness of the green film is determined by the thickness of the screen wires. During sintering, shrinkage will occur. Controlled porosity is desirable as well as a good attachment of the layer to the substrate which is achieved by careful control of the powder characteristics and sintering conditions. During sintering some shrinkage parallel to the substrate will occur. This might lead to cracks in the sintered film. Thick, crack-free layers may be build up by repeated coating and sintering operations of thin layers.

Slurry Coating

Thin porous layers, but also dense films may be fabricated by the slurry coating method. In case of a dense electrolyte (YSZ) a stabilized water suspension with a small amount of powder (ca. 5 wt.-% YSZ) is coated on a substrate, dried a room-temperature, pre-heated at elevated temperatures, and finally sintered. This cycle is repeated 5 to 10 times. Thin and dense films with no observable cracks are produced.

Electrochemical Vapor Deposition (EVD)

Electrochemical vapor deposition techniques have been used to fabricate dense thin layers on porous substrates, especially for the tubular design. In case of the electrolyte for tubular cells, steam and/or oxygen is fed to the interior of the seal-less support tube while metal chloride vapor, hydrogen and argon are fed to the outside. Hydrogen is used to remove the chlorine formed, and argon is used as a flow-regime conditioning gas.

The EVD consists of two steps, schematically described in Figure 4-83. The first step is the same as in the normal chemical vapor deposition. The reactant metal chloride and steam (or oxygen) react to form oxides, depositing on the substrates. Once the pores are completely closed, the reactants are no longer in direct contact. Further growth of the film proceeds by oxygen ion diffusion through the film due to the presence of a large oxygen chemical potential gradient across the deposited film. The metal chloride reacts with oxygen ions diffusing through the deposited dense oxides in the presence of electronic conduction. The growth rate is controlled by solid state diffusion in the oxide film and can be described like the Wagner oxidation of metals [252]. The layer thickness is proportional to the square root of the reaction time.

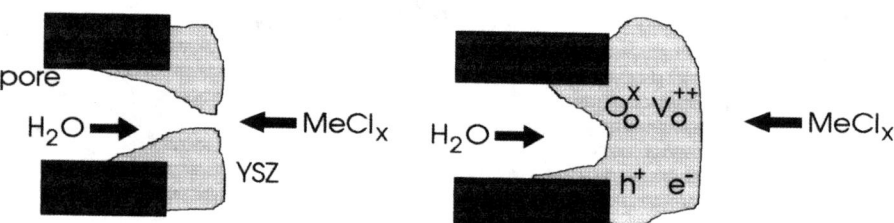

Figure 4-83. Model for the layer growth in the EVD Process

Physical Vapor Deposition (PVD)

Physical vapor deposition (PVD) is a generic term for a variety of sputtering techniques. Radio frequency (rf) sputtering has been most widely used for YSZ thin deposition in production of SOFC cells. YSZ deposition rates during rf sputtering are relatively low due to low sputtering yield of YSZ (ca. 0, 25 μm/h). In addition, magnetron sputtering provides ion irradiation of the film during deposition, which has been shown to be crucial for obtaining high density films at low temperature.

The ion-beam-assisted deposition (IBAD) technique resembles the reactive magnetron sputtering technique, where an ion beam and a reactive background gas are the primary components of the coating. Thin films may also be deposited by vacuum evaporation of a target on an electron beam heating device. The achievable deposition rates are in general two orders of magnitude higher than those of the above mentioned methods (typically 1 μm/min.).

Thermal Spraying

In this deposition process the coating material is fed in the form of a rod or powder into a zone of very high temperature and is projected at high velocity onto a substrate surface. The flame spray technique (FS) uses a combustion flame (H_2-O_2 or C_2H_2-O_2) to melt the spraying material and usually air as a carrier gas. The plasma jet from the combustion gas which passes through a dc arc provides the heat source in plasma arc spraying. The temperature of the plasma flame can reach up to 16000 °C and is capable of melting all refractory oxides. The substrate might be heated in order to clean the surface from adsorbed volatile species increasing the adherence of the oxide coating on the substrate.

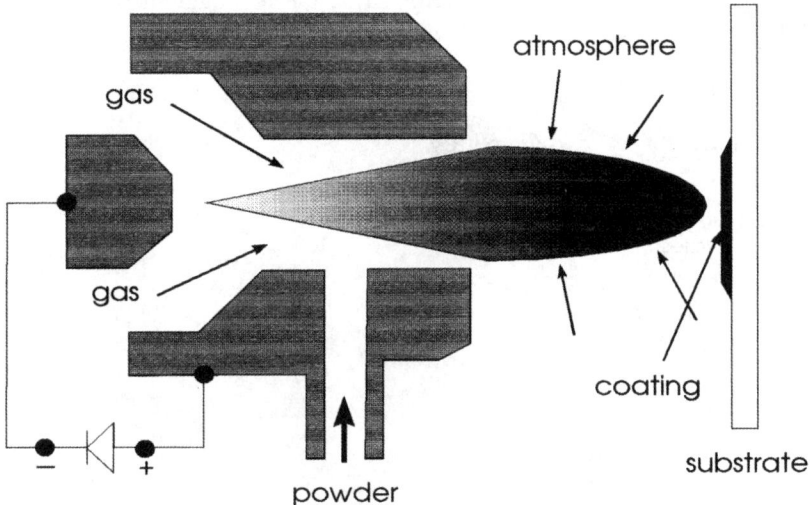

Figure 4-84. The plasma spray process

Common to all thermal spraying processes are the high temperatures involved, the rapid quenching of the melted ceramic particles on the substrate during deposition and the thermal treatments required to improve the coating quality. To achieve dense layers, plasma spraying in vacuum is highly promising. A new technique for depositing films of almost any material has become the pulsed laser deposition (PLD). Laser pulses 20–30 ms short vaporize the target material. Pulsed laser evaporation produces controllable amounts of high-kinetic-energy particles, preserve the stoichiometry of multi-component compounds and allow high growth rates.

Sintering

In all processes where powder particles are involved in the green body forming technique, the final processing step requires sintering. The sintering process provides bonds between particles and between particles and a substrate.

During sintering at increasing temperature, accelerated diffusion of the chemical species takes place. Thereby material is transported within the microstructure of the powder compact from energetically unfavorable to more favorable places. Due to the difference in chemical potential, the concentration of vacancies beneath a concave surface is higher than beneath a convex surface. The transport of vacancies from a concave surface can occur by lattice and boundary diffusion resulting in mass transport. The driving force for sintering therefore is reduction in surface energy of the powders. A finer grain size powder sinters at lower temperatures and to a higher final density. However, finer powders tend to agglomerate and therefore are more difficult to handle. Controlled packing of the fine powders during the forming stage is more difficult. The sintering of grains in agglomerates occurs at a lower temperature than that of agglomerates among each other, as described in Figure 4-85. As a result, pores between agglomerates remain, even after sintering at a very high temperature [253]. Therefore, fine but non-agglomerated powders are needed to fabricate ceramic components with homogenous microstructure.

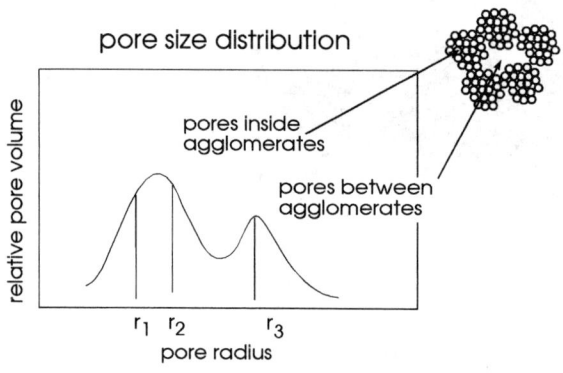

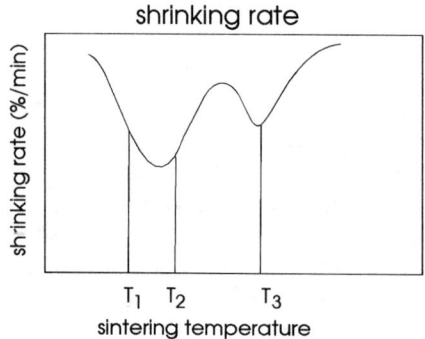

Figure 4-85. Typical pore size distribution of a ZrO$_2$ powder compact and the corresponding shrinkage rate

High quality starting powders may be prepared via wet-chemical processes such as co-precipitation and spray pyrolysis.

In case electrode layers are fabricated via powder sintering, high final densities are not required. The green electrode layer may be sintered on pre-sintered substrates or electrolytes. Therefore, the shrinkage of the electrode layers should be 1-dimensional rather than 3-dimensional at the interface. Sintering perpendicular to the dense substrates is preferred with a minimum of shrinkage parallel to the substrates. Furthermore, the perpendicular shrinkage itself should be well controlled. To achieve this goal, the thickness of the green layer should not exceed a critical thickness, their green density of the layer should be very high and sintering activity should be controlled by particle size, particle distribution and sintering conditions.

4.7. Direct Methanol Fuel Cells (DMFC)

Electrochemists are challenged to use a fuel which is easily transported and converted into energy from the liquid state. Methanol is the favored fuel [254]. The direct methanol/air cell (DMFC) uses a sulfuric-acid electrolyte because of the need to reject CO_2 produced during the electro-oxidation of methanol. In an acid medium, the overall reaction of methanol oxidation can be written as follows:

$$CH_3OH + H_2O \longrightarrow CO_2 + 6\,H^+ + 6e^-\qquad(4\text{-}44)$$

with $E^0 = 0.029$ V/SHE [255].

The advantages of the methanol liquid-feed fuel cell over the cells designed for gas-feed may be summarized as follows:

– elimination of fuel vaporizer and its associated heat source and controls,
– elimination of complex humidification and thermal management systems,
– dual-purpose use of the liquid methanol/water as fuel and as an efficient stack coolant, and
– significantly lower system size, weight and temperature than existing fuel cell systems.

In spite of these advantages, however, the DMFC remains poorly developed compared to other fuel cells, and it has been suggested that the obvious advantages of methanol as a fuel should be harnessed by prior reforming of the methanol to yield hydrogen and CO_2 [256]. Great difficulties have also been encountered in improving the efficiency of the DMFC itself. These problems may be summarized as:

– acid electrolytes must be used because carbonate formation is a serious problem in alkaline solution, particularly at the current densities regarded as commercially desirable. The acid electrolyte causes problems of corrosion and, more fundamentally, is responsible for the slow electrode kinetics of the reduction of oxygen at the air cathode.

–   similar electrocatalysts have been proposed for both anode and cathode, leading to the
    problem of mixed potentials at both electrodes and a marked reduction in efficiency. This
    is a particular difficulty at the cathode and partitioning the cell with a membrane to avoid
    "chemical short circuits" will introduce a further source of inefficiency and resistance.
    The anode reaction is very slow near the thermodynamic potential, at least on the cata-
    lysts currently used. A large overpotential loss is therefore encountered [257].
–   the catalysts currently used for the anode are all based on a high platinum content.
    These catalysts are easily poisoned both by impurities and (more seriously) by pro-
    ducts of the anodic reaction itself [258].

These difficulties have rendered the DMFC an apparently far less attractive option
than other cell designs, particularly in view of the fact that both anode and cathode suffer
poor kinetics, which is significantly worse than polymer electrolyte fuel cells (PEFCs).
The DMFC will only play an important role in future developments if substantially better
performance at both anode and cathode can be realized.

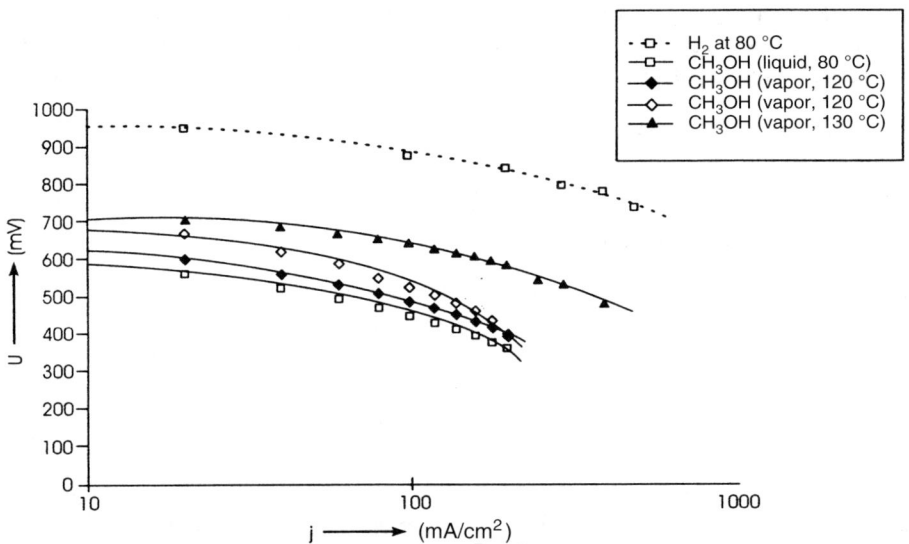

Figure 4-86. V-I characteristics of a DMFC under different operating conditions [259]

The development of "in-situ" infrared spectroscopic techniques allowed significant
progress in identification of the intermediates adsorbed at the electrode catalysts, caused
by residues of aldehyde, carboxylic acid and other intermediates produced during the
galvanic-oxidation of methanol. Electro-Modulated Infrared Reflectance Spectroscopy
(EMIRS) led to the first "in-situ" unambiguous proof of the presence of adsorbed CO on
platinum electrodes during the oxidation of methanol [260, 261].

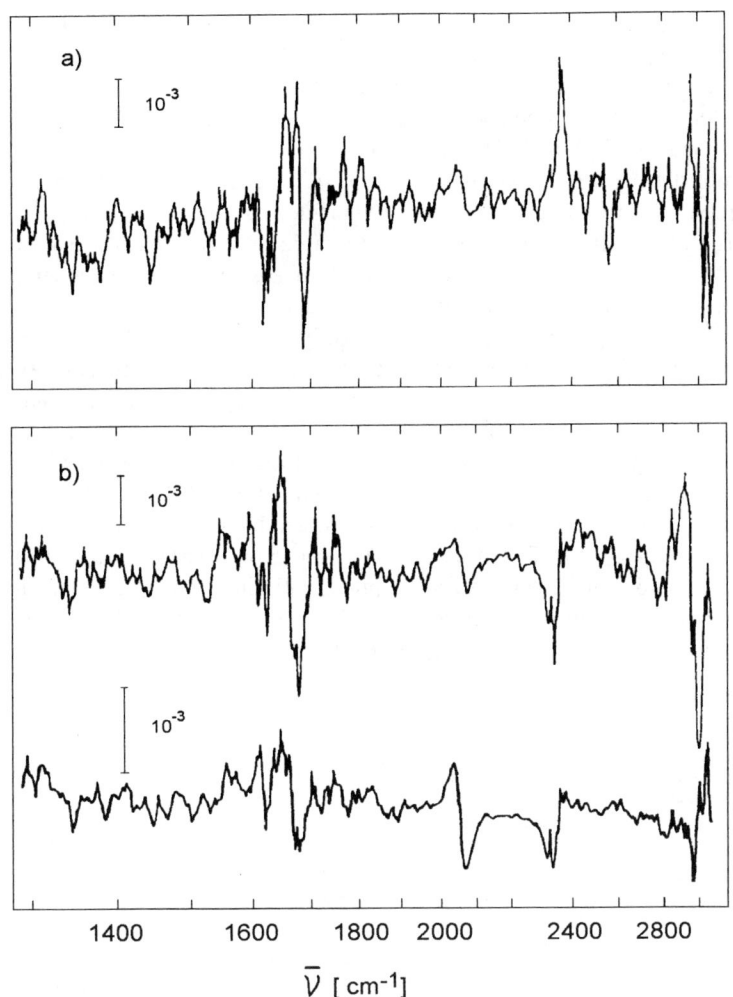

Figure 4-87. EMIRS Spectra of Species Resulting from Methanol Absorption on Polycrystalline Platinum as a Function of the Number of Scans: 0.5 M $HClO_4$ + 5×10^{-3} M CH_3OH; E = 0.4 V, E = 0.2V/RHE, f = 13.5 Hz, room temperature, a) first scan, b) 10th and 25th scans [262]

CO was mainly adsorbed linearly bonded to the electrode surface (band at around 2050 cm^{-1}), but bridge-bonded species were also present (very small band at around 1850–1900 cm^{-1}). These two species are responsible for the poisoning phenomena observed during the working of a direct methanol fuel cell [263]. Such results were widely confirmed by different groups with EMIRS and similar methods, such as SNIFTIRS or IRRASS, using Fourier transform spectrometers [264–266].

However, these conclusion were partially refuted by other observations made with different methods using electrochemical measurements coupled with mass spectroscopy, such as Differential Electrochemical Mass Spectroscopy (DEMS) or Electrochemical Thermal Desorption Mass Spectroscopy (ECTDMS) [267–269]. These techniques, which need large-area electrodes made with the catalytic material mixed with Teflon powder, allowed the detection of adsorbed CO, but the main species identified were claimed to be either adsorbed COH [270–272] or adsorbed CHO [273, 274].

These apparently contradictory observations made by IR spectroscopies and mass spectroscopies could probably be related to different experimental conditions – in particular the electrode structures, which are different in the two sets of experiments. It appears to be clear that the chemical nature and distribution of adsorbed species depend on the electrode structure and on the degree of coverage by the species.

In an acid medium, platinum appears to be the most efficient catalyst for the electro-oxidation of methanol. Various attempts have been made to increase the catalytic activity of the electrode and to decrease the poisoning phenomena by modifying the surface nature.

Two kinds of modification are generally proposed – bulk alloys and foreign adatoms [275]. In the case of alloys, various systems were investigated; Pt-Sn [276] Pt-Pd [277] and Pt-Ru [278–281]. Only Pt-Ru and possibly, Pt-Sn electrodes display a greater activity than pure platinum. Apparently, the Pt-Ru alloy is the most promising one, with a negative shift of the polarization curves compared to pure platinum, and a decrease in poisoning [281–284]. However, superficial enrichment of one metal of the alloy could arise under working conditions, and modify the activity of the electrodes.

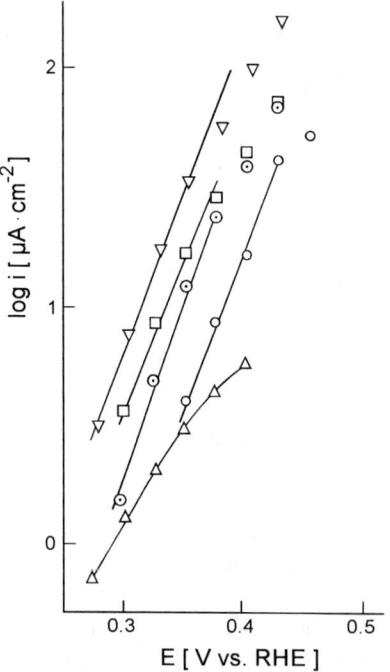

Figure 4-88. Tafel-plots i vs. V for CH_3OH electro-oxidation on various Pt alloy-SPEs in 1 M CH_3OH + 0,5 M H_2SO_4 at 50 °C; (∇) cPtRuSn; ($\square$) aPtSn; ($\triangle$) cPtSn, ($\odot$) aPtAuRu, (o) cPtRu, where 'a' and 'c' denote amorphous and crystalline, respectively [285].

The second possibility for modification of electrocatalytic properties is to adsorb a foreign metal on its surface by electrodeposition at a more positive potential than the Nernst potential, i.e. by "underpotential deposition" or "upd". The atoms adsorbed on the electrode surface, the so-called adatoms, greatly modify the adsorption of organics. Adatoms generally lead to positive effects with significant enhancement of the electrocatalytic activity of platinum. However, methanol seems to be a special case, the influence of adatoms being very limited [286]. Only Pb, Ru, Bi, Sn, or Mo adatoms increase slightly the platinum activity, and mainly at lower potentials [287].

Work at the Jet Propulsion Laboratory (JPL) [288]

The Jet Propulsion Laboratory (JPL) published a breakthrough in DMFC technology in 1994, using a 3% aqueous solution of methanol (20 psig, anode-side) and oxygen (20 psig, cathode-side) for their solid polymer (PEM) membrane. The water produced at the cathode is circulated back to the reservoir, where it is injected with methanol. The unused methanol/water solution from the anode chamber is also circulated back to the reservoir.

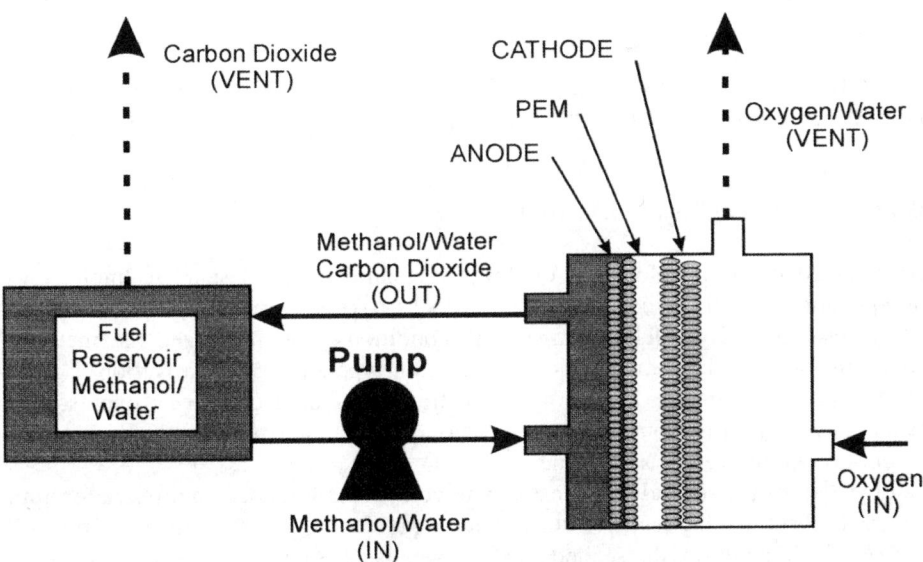

Figure 4-89. Diagram of the JPL Direct Methanol Liquid-Feed Fuel Cell [289]

Laboratory versions of the JPL-PEM cell, operating directly on a 3% methanol in water solution at a temperature of 60–90 °C, delivering an output of 0.50 V at 300 mA/cm² using oxygen. The performance has improved since March 1992.

This output at the cell level is quite high in relation to that of prior direct-oxidation methanol fuel cells (power density up by factor of 20). JPL plans a scaling-up of the cells

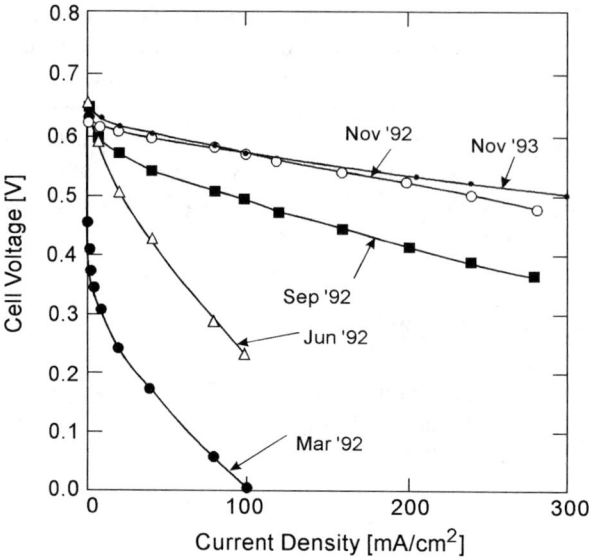

Figure 4-90. Improvements in Cell Performance [290]

to 4"×6" (50 amps continuously). An overall efficiency > 40% is being predicted by JPL for the next generation fuel cell stacks (8"×8"×17").

Work at International Fuel Cells (IFC)

IFC is engaged in the development of direct methanol fuel cell stack technology for DOD applications [291]. Two developmental stacks have been produced. The first two-cell stack test focused on defining operating conditions that minimized the methanol crossover, using partially hydrophobic gas diffusion type anodes and a Nafion®117 polymer electrolyte membrane. Figure 4-91 illustrates methanol crossover (expressed as equivalent ASF) as a function of current density. The data are superimposed on performance curves obtained at the same time.

The objective of the second stack test was to correct the low methanol oxidation limiting current observed in the first stack which precluded operation above 200 ASF (215 mA/cm²). In addition, the second stack of 4 cells (258 cm²) was actively cooled by circulating the water-methanol fuel through the anode and an external cooling loop. Figure 4-92 illustrates the best cell performance. It achieved 0.5 V at the 250 ASF (270 mA/cm²) design point.

IFC plans to continue its DMFC program concentrating on alternative technology approaches to their hydrophilic anode concept. New membrane technologies and the development of new cathode catalysts that do not oxidize the crossover methanol are considered possible solutions.

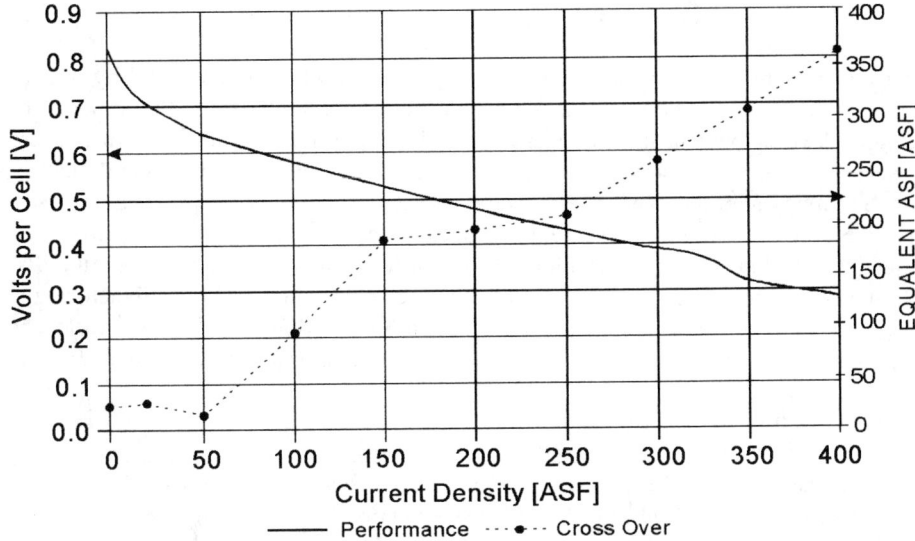

Figure 4-91. Methanol Crossover versus Current Density [292]

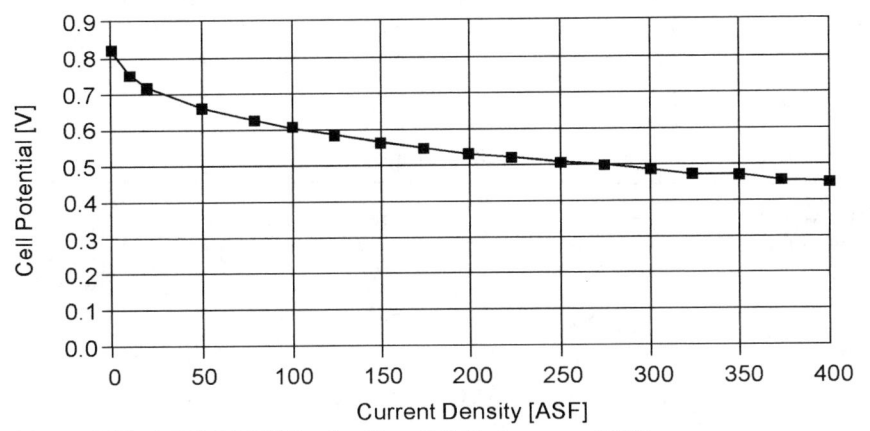

Figure 4-92. 4 Cell DMFC Stack – Best Cell Performance [292]
Operating Conditions: 104 °C, 50 psia, anode catalyst: 50/50% atomic percent Pt/Ru, cathode catalyst: 4 mg/cm^2 Pt black; electrolyte: Nafion$^®$117.

4.8. Hybrid Cells

Here a hybrid cell is defined as a galvanic electrochemical generator, in which only one of the active reagents is in the gas phase. Hybrid cells occupy an intermediate position between the closed-galvanic cells and fuel cells, in which both cathodic and anodic reactants are supplied continuously (usually in gaseous form) from sources external to the cell. Hybrid cells take advantage of both battery and fuel cell technologies.

4.8.1. Metal – Air Batteries

Metal-air batteries have a very favourable energy density, which is achieved by not incorporating positive, active components within the cell. To a large extent, the impetus for research and development in this field has arisen from possible EV applications, where energy density is a critical parameter. These considerations have long intrigued electrochemists, and many attempts have been made over the years to develop practical systems. Zinc has received by far the most attention, because it can efficiently be electroplated from aqueous solution. However problems of dendrite formation, non-uniform zinc dissolution and deposition, limited solubility of the reaction product and unsatisfactory air-electrode performance made it difficult to demonstrate this battery type as an electric vehicle power source. A survey of Metal/Air batteries can be found in Reference 293.

Table 4-16. Comparison of Metal/Air Batteries

Metal Anode	Electrochemical Equivalent of Metal [Ah/g]	Theoretical Cell Voltage* [V]	Valence Change	Theoretical Specific Energy (of Metals) [Wh/g]	Practical Operating Voltage [V]
Li	3.86	3.4	1	13	2.4
Ca	1.34	3.4	2	4.6	2.0
Mg	2.20	3.1	2	6.8	1.4
Al	2.98	2.7	3	8.1	1.6
Zn	0.82	1.6	2	1.3	1.2
Fe**	0.96	1.3	2	1.2	1.0

  * Cell voltage with oxygen cathode
** Fe^{2+} (1.8 Wh/g for Fe^{3+})

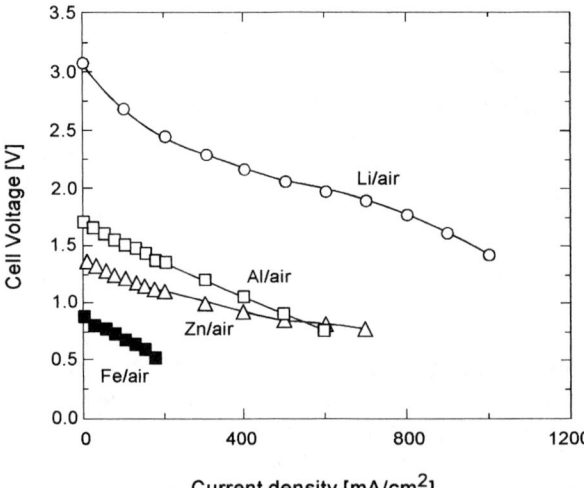

Figure 4-93. Typical Cell Voltage-Current Density Performance Curves for Alkaline Metal/Air Cells

Table 4-16 provides a comparison of metal/air cells with various metal anodes, listed in order of their decreasing electrochemical equivalent weight. The theoretical values for the cell voltage and specific energy are based on the overall cell reaction for metal/air cells containing alkaline or neutral electrolytes. The large difference between the theoretical cell voltage and the typical operating voltages for these metal/air cells is an indication of the electrode polarization in these systems, with the air electrode contributing a large fraction of the polarization. Promising results have been obtained with metal/air cells containing anodes of Li, Al, Mg, Zn, or Fe (see Figure 4-93). Of these four metals, only zinc and iron are being considered for secondary (electrically recharged) metal/air batteries. Zinc, Magnesium, Aluminum [294, 295] and Lithium are useful electrodes in batteries, which can also be made "mechanically rechargeable". "Fresh" metal is added to the cell to replace the electrode material consumed by electrochemical reaction.

4.8.2. Zinc – Air Batteries

This system is used for powering warning lights (navigation aids), remote signals, railroad track circuits and hearing aids. The batteries used in these applications are primary batteries, which are usually discharged at low current densities over long periods. Extensive research efforts have been carried out to develop secondary zinc/air battery systems [296]. Various designs have been proposed for secondary zinc/air batteries to improve their performance and life. The overall cell reaction may be represented as:

$$Zn + \tfrac{1}{2}O_2 + H_2O + 2\,OH^- \longrightarrow Zn(OH)_4^{2-} \qquad E_0 = 1.62\ V \qquad (4\text{-}45)$$

The major problems associated with the zinc electrodes in secondary zinc/air cells are: (1) shape change arising from charge-discharge cycling, (2) formation of dendrites that penetrate the separator, (3) anode passivation during discharge, and (4) detrimental effect of zinc discharge products on the performance of the air electrode.

The development of an efficient high-rate bifunctional air electrode remains a formidable challenge. The problem is that the oxygen which starts to evolve from the cathode during charging oxidizes the carbon material and the strongly alkaline electrolyte in the porous structure is slowly converted into carbonate, causing physical damage. Cycle-life time is therefore short. Inspite of discouraging results in the very long history of rechargeable air-zinc batteries attempts have been made again to produce longer lasting electrically rechargeable air-zinc batteries [297]. A special version for computers was developed by AER Energy Resources, Smyrna, GA, USA [298].

At Kyushu University in Japan, new Perovskite catalysts for high current performance, have been developed. 240 cycles at 100 mA/cm^2 with 38% Zn utilisation were obtained [299].

The carbon oxidation can be avoided if the oxygen gas is evolved on a third electrode placed between tubular carbon-air electrodes and zinc plate electrodes [300].

In a more recent construction [301] the zinc electrode is positioned between the air electrode for O$_2$-reduction and the electrode for oxygen evolution (0.1 mm nickel plate) (see Figure 4-94). Air electrodes and circulating powder zinc in alkaline solution we used

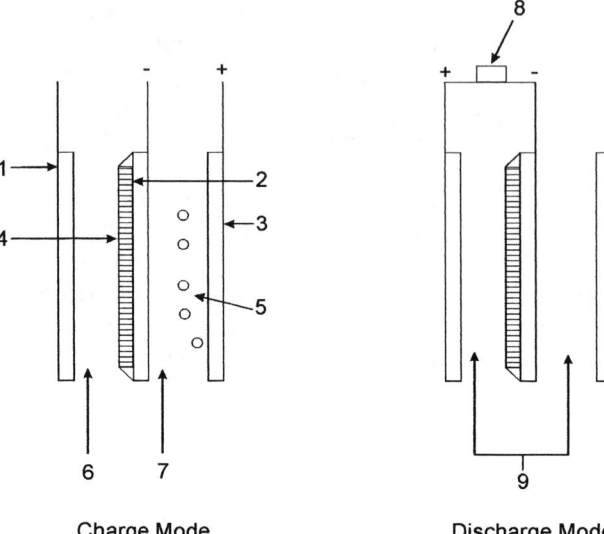

Charge Mode   **Discharge Mode**

Figure 4-94. Schematic Diagram of the Charge and Discharge Mode of Operation of a Zinc/Air Cell with a Third Electrode. 1) air electrode; 2) anode base plate; 3) third electrode; 4) Zinc deposit; 5) O_2; gas evolution; 6) electrolyte containing zincate species; 7) zincate-free electrolyte; 8) discharge load, 9) electrolyte

Figure 4-95. A Sony Zinc-Air Battery System powered an experimental car (Hideo Baba in the driver, SAE paper No. 710237 (1971) (Courtesy of Sony)

in an experimental car already in 1971 (Figure 4-95). A Ni–Cd battery was used for peak power. The used up electrolyte was externally negenerated.

The zinc electrode, especially in plate designs shows often shape change and short circuiting dendrite formation. Also passivation occurs when the cells are deeply discharged. To circumvent these difficulties, and also avoid the electrical charging, mechanically exchangeable anodes were introduced. This technology is rather old and has been tried for military applications, e.g. for radio batteries in the mid 1970's with alkaline electrolytes [302–304].

"Reserve Batterie Designs", with a possibility to make up the battery electrolyte later by water addition have a long shelf-life. All mechanically rechargeable systems (the discharged metal electrode is replaced) permit short recharge times and allows mass-production of battery fuel at facilities dedicated to that function. This approach makes use of simple "unifunctional" air electrodes, which need operate only in a discharge mode. Sanyo manufactures a series of 14-pack batteries, ranging from 20 to 200 Ah at voltages from 1 to 12 V. They were designed as water-activated reserve batteries, but are also mechanically refuelable and have provisions for replacement of the zinc electrode cassettes.

In January 1994, the German Post Office presented a new Zn/air project. The battery is produced by the Electric Fuel Corporation (EFL) in Israel. As it is also mechanically rechargeable, ELF's new battery is the zinc equivalent of aluminum-air. ELF's concept is to package each zinc anode in a 'cassette' holding the zinc/zinc oxide slurry in tight contact with a zinc anode plate. It thus gets over the problem of change in shape of the zinc anode, and simplifies the air electrode. The cassettes are then withdrawn in a machine, which if located at the roadside, can act like a refuelling (gasoline) station [305–307].

Figure 4-96. Mercedes-Benz 180E Van. Prominently carrying the Postdienst Logo as it appeared on the cover of the US share holder's prospectus, issued by EFL, 1994

The net energy-efficiency data are given in Table 4-17. A transporter like the 180E will cover 250–300 km on one charge, with a maximum speed of 100 km/h and an ability to restart on a gradient of 20%. In general terms, EFL concludes that zinc-air makes it possible to carry 10 times the amount of fuel on board a vehicle as is possible with conventional lead-acid batteries.

Table 4-17. Performance of the 157 kWh Battery for the 4.6 Ton GVW MB410 being Used in Postdienst Tests. (2 Strings of 4 Modules with a Total of 526 Cells)

Specific energy	208 Wh/kg
	243 Wh/dm^3
Peak power density at end of discharge	101 W/kg
	118 W/dm^3
Total capacity, at C/4 rate, to 80% zinc consumption[9]	492 Ah
Nominal battery voltage	320 V
Battery energy	157 kWh
Battery weight	758 kg
Battery volume	649 dm^3

4.8.3. Iron – Air Batteries

Iron-electrode batteries, that could utilize the virtually inexhaustible resources of iron, have been examined as futuristic, electrochemical power sources [308, 309]. Alkaline, rechargeable iron/air and nickel/iron rechargeable batteries are the two, most-discussed systems [310–312].

The net cell reaction in a secondary nickel/iron cell is:

$$2\,NiOOH + 2\,H_2O + Fe \longleftrightarrow 2\,Ni(OH)_2 + Fe(OH)_2 \tag{4-46}.$$

The system has an open-circuit voltage (OCV) of 1.33 V. Oxidation of active iron to $Fe(OH)_2$ comprises the first discharge step of the iron electrode. During subsequent discharge, conversion to Fe_3O_4 takes place. The cell reaction for a secondary iron/air cell, which has an OCV of 1.28 V, is:

$$Fe + \tfrac{1}{2}O_2 + H_2O \longleftrightarrow Fe(OH)_2 \tag{4-47}.$$

The iron/air battery looked especially attractive. It has a theoretical energy density value of 764 Wh/kg. In practice, however, this performance has not been achievable. This is due to the low charge/discharge efficiencies of iron electrodes. Another important

---

9 A common method for indicating the discharge as well as the charge current of a battery is the C rate, expressed as: I = M×C$_n$; where I = discharge current [A], C = numerical value of rated capacity of a battery [Ah]; n = C rate at which cell or battery was rated; M = multiple or fraction of C. For example, the C/4 or 0.4 C discharge rate for a battery rated at 492 Ah is 123 A

scientific problem is the non-availability of an efficient, rechargeable air electrode. During recharge of an iron/air cell, the electrode reaction occurring at the air electrode is the anodic evolution of oxygen through the reaction:

$$4\,OH^- \longrightarrow O_2 + 2\,H_2O + 4e^- \tag{4-48}.$$

At the high positive potential required for this reaction to proceed at a significant rate, undesirable side effects occur, i. e. corrosion and/or erosion of the electrode support and poisoning of the electrocatalyst. The oxygen has to be disposed safely.

The charge/discharge cycle life of nickel/iron batteries under normal conditions of use is about 2000 cycles at 80% depth-of-discharge (DOD). This provides a long calendar life, extending up to 20 years. By comparison, typical lead/acid accumulators have much lower cycle-lives, with a calendar life ranging from 3 to 6 years. A nickel/cadmium battery has a cycle-life of about 1500 cycles at 80% DOD. Iron/air batteries under development are expected to exhibit a performance of close to 1000 cycles.

Iron electrodes also have environmental advantages compared with other battery electrode materials, such as nickel, cadmium and lead, which are substantially toxic. Furthermore, iron electrodes can withstand mechanical shocks and vibrations, as well as over-charge and deep-discharge. Large surface iron is pyrophoric when dried out.

Between 1910 and 1950, nickeloxide/iron batteries were produced in large numbers in the USA and other countries for industrial traction applications. Further development and perfection of these batteries was, however, retarded by the emergence of the nickel/cadmium system. The latter has a lower degree of gas evolution, and therefore less water-loss problems. Nickel/cadmium batteries have proved attractive for several applications, leading eventually to the present-day sealed, maintenance-free versions. In a parallel development, the advent of large-scale automobile manufacture, especially in the USA and Japan, stimulated the evolution of cost-effective lead/acid batteries. The situation has, however, changed since then. The limited resources, high cost and toxicity of cadmium are such that the nickel/cadmium system will be almost certainly relegated to special applications that justify the extra investment, e.g. for service at sub-zero temperatures, in emergency power supplies, satellite power systems and defense equipment. As for the lead/acid battery, its relatively poor cycle-life, together with the high toxicity and corrosive effects of sulfuric acid vapours, point to the eventual preference for the nickel/iron counterpart. The latter is safe, economical and reliable in applications such as traction, lighting and fans in trains, shunting yard operations and mining locomotives. Nickel/iron and iron/air batteries have also been projected as potential candidates for electric vehicle (EV) applications ranging from family cars to commercial vans. It is noteworthy, that power densities generally required for EV applications (~ 100 W/kg) can be met by either of these battery systems (see Table 4-18), however, energy densities are poor compared to Zn-Air.

Figure 4-97 shows a cross-section of a cell developed by the Swedish National Development Co. (SNDC) in pile form with air and electrolyte flow. With a good distribution of air within the air element, 50% of the oxygen introduced to the system is consumed at the air electrode at the 5 h rate. The 1978 prototype battery contained piles, each consisting of 190 air and 95 iron electrodes (30 kWh). The electrodes (117×205 mm) of each pile have been divided into 19 cells connected in series [314].

Table 4-18. Nickeloxide/Iron and Iron/Air Batteries [313]

	Energy Density [Wh/kg]	Power Density [W/kg]	Efficiency [%]	Cycle-life
1991 status				
Ni/Fe	50–60	75–110	60	1000
Fe/air	52–109	102–146	50	500
Projected status				
Ni/Fe	75–80	100–130	72	2000
Fe/air	98–195	181–309	68	1000

The Ni-Cd-System uses Nioxide-Cd, the Fe-Air System uses Fe metal-Air, (Language as used commonly).

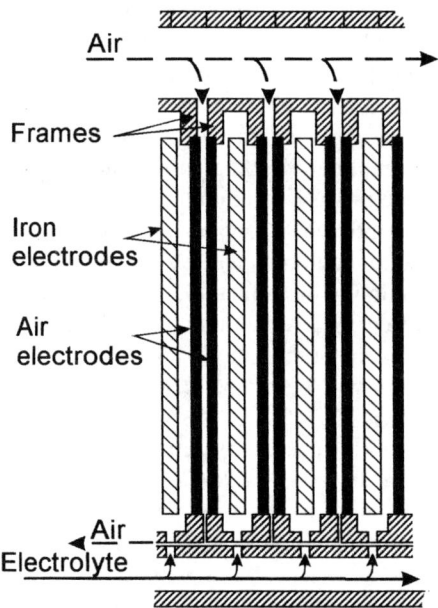

Air

Frames

Iron electrodes

Air electrodes

Air

Electrolyte

Figure 4-97. Cross-Section of SNDC Iron-Air Battery Pile [314]

Problems that adversely affect the performance of iron electrodes are spontaneous corrosion in the charged state (leading to a high rate of self-discharge), a low faradaic-efficiency for the anodic dissolution reaction, poor low-temperature performance, and low charge acceptance at high ambient temperature due to a parasitic, hydrogen-evolution reaction (HER). In addition, the improved 'Edison' process of electrode manufacture is expensive [315]. Efforts have been made to circumvent these problems. However the cycle of the air electrolytes as reported in Table 4-18 [313] has not been verified.

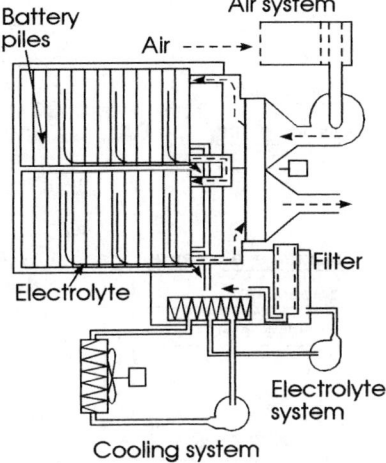

Figure 4-98. Schematic Cross Section of Swedish National Development Corp. Iron/Air Battery Including Auxiliary System.

4.8.4. Metaloxide – Hydrogen Batteries

As with the metal-oxygen system, metaloxide-hydrogen cells can be considered as closed galvanic/fuel cell hybrids. They make use of the hydrogen-electrode technology of fuel cells, and that of nickel- or silveroxide electrodes from alkali secondary cells [316]. Some advantages of the nickeloxide-hydrogen compared to the nickeloxide-cadmium battery are (1) higher specific energy, (2) inherent protection against overcharge and overdischarge (reversal) and (3) cell pressure can be used as an indication of state of charge. Some of the disadvantages are (1) relatively high initial cost due to limited production, which could be offset by cycle-life costs and (2) self discharge, which is proportional to hydrogen pressure. Nickeloxide-hydrogen batteries have captured a large share of the space-battery market in recent years. They are rapidly replacing nickeloxide-cadmium as the energy-storage system of choice (see Figure 4-98).

The charged hydrogen-nickeloxide cell is represented as:

$$Pt(s), H_2(g) / KOH(aq) / NiO(OH)(s),$$

and has an OCV of 1.5–1.6 V in its fully charged state. The associated cell reactions are:

$$(\tfrac{1}{2})H_2(g) + NiO(OH)(s) + H_2O(1) \longleftrightarrow Ni(OH)_2 \cdot H_2O \qquad (4\text{-}49).$$

Hydrogen pressure rises from about 0.5 MPa in the fully-discharged state to 3–10 MPa when charged, and the pressure in this vessel can be used to monitor the state of charge. Direct reaction between hydrogen and nickel oxide is relatively slow, but 6–12% of capacity is lost per day. This system is extensively reviewed by Smithrick [317].

Energy densities of metaloxide-hydrogen systems can be further improved by the association of a more positive electrode with the hydrogen electrode. The silveroxide electrode has a much greater energy density than the nickeloxide electrode. The practical energy density of Agoxide-H$_2$ is about 80 to 100 Wh/kg, i.e. 2–5 times that of Nioxide–Cd [318, 319].

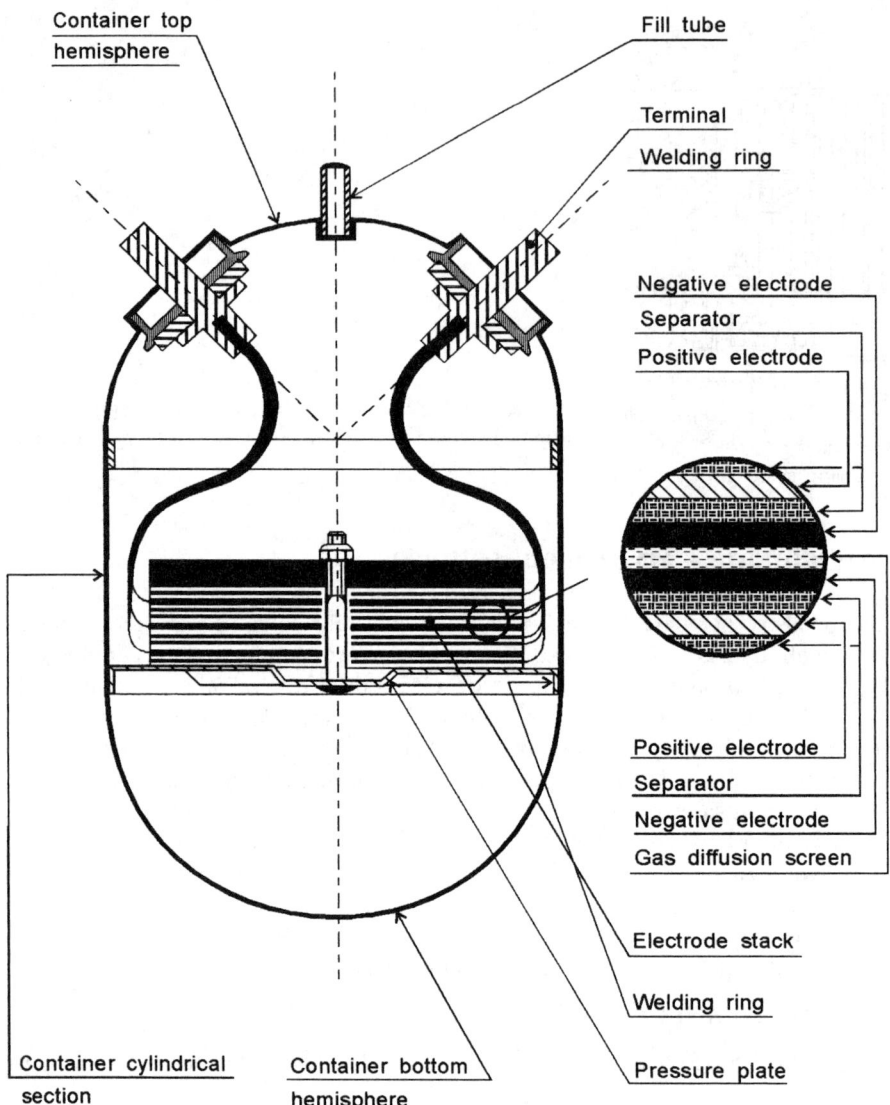

Figure 4-99. Schematic Configuration of a Metal Oxide-H_2 Cell.

4.8.5. Metaloxide – Metalhydride Batteries

This type of batteries contains no gas diffusion electrodes like fuel cells, but they are mentioned because they involve hydrogen as gas, filling the cell space, which is free of electrolyte. The gas pressure is determined by the state of charge of the cell, the temperature and the type of the metalhydride used. Metalhydrides may also be the means of storing hydrogen for fuel cells. The system has become the replacement for Nickel(oxide) – Cadmium cells and may also replace the Nickel(oxide) – Hydrogen cells using catalysed fuel cell anodes. The system and its applications in the consumer and perhaps in the electric vehicle area is discussed extensively in a survey article [320] and in the Handbook of Batteries by D. Linden [321].

References for Chapter 4

[1] Kinoshita, "Electrochemical Oxygen Technology", John-Wiley & Sons, Inc., New York, 1992, Chapter 2, pp. 19–112

[2] Kordesch, J. Gsellmann, B. Kraetschner, Power Sources **9**, (J. Thompson, ed.), p. 379, Academic Press, New York, Copyright 1983, Academic Press, Inc.

[3] O. Lindström, W. Lavers, "Cost Engineering of Ammonia and AFC power plants", 1994 – Fuel Cell Seminar, San-Diego (CA), pp. 196–198

[4] E. Gülzow, M. Schulze, G. Steinhilber, K. Bolwin, 1994 – Fuel Cell Seminar, San-Diego (CA), pp. 319–322

[5] A. Al-Saleh, et al.: "Performance of Ni/PTFE, and Ag/PTFE Electrodes in Alkaline Fuel Cell in the Presence of Carbon Dioxide Impurity in the Feed Gases", Paper from the Chemical Engineering Department, King Fahd University of Petroleum and Minerals, Dhahran 31261, Saudi Arabia 1992

[6] K. Kordesch, "Gibt es Moeglichkeiten zur Herstellung und Verwendung" von "Low Cost – Low Tech" Zellen?, in Brennstoffzellen: Stand der Technik-Enwicklungslinien-Marktchancen, (ed. H. Wendt, V. Plzak), Duesseldorf 1990, VDI-Verlag, pp. 57–63

[7] K. Kordesch, "The Choice of Low-Temperature Hydrogen Fuel Cells: Acidic – or Alkaline?", Hydrogen Energy Progress – IV, Proc. of the 4th World Hydrogen Energy Conference, California, USA, 13–17 June 1982, ed. T. N. Veziroglu, W. D. v. Vorst, J. H. Kelley. Volume 3, pp. 1139–1148

[8] Pratt & Whitney Aircraft, Design and Devolopment of H_2–O_2 Fuel Cell Power Plants for Apollo Command Module, NASA/Houston, Final Report 6/16/1962

[9] Richard B. Ferguson "Apollo Fuel Cell Power System", Proceedings 23rd Annual Power Sources Conference, 1969, pp. 11–13

[10] J. L. Platner, D. Ghere, P. Heiss, US-Army, 2 kW Fuel Cell System. 19th Annual Power Sources Conf. (Symp.), 1965, pp. 32–35

[11] M. B. Clark, W. G. Darland, K. V. Kordesch, Electrochem. Technol. **3** (1965), p. 166

[12] A. Winsel, "The Eloflux Fuel Cell System", DECHEMA-Monogr. **92** (1983), pp. 1885–1913

[13] Varta, US 3 597 275, 1971 (A. Winsel, R. Wendtland)

[14] H. van den Broeck, Commercial Development of Alkaline Fuel Cells, CEC-Italian Fuel Cell Workshop in Taormina, Italy, 1987, pp. 73–77. The Commission of the European Communities, Brussels, Belgium 1988

[15] Max Schautz, et al., 1992 – Fuel Cell Seminar, Tucson (AZ), oral presentation, no extended abstract in the proceedings

[16] M. Schautz, G. Dudley, "Testing of Space Fuel Cells at ESTEC", Reprinted article form ESA Bulletin No. 60

[17] F. Baron, "Fuel Cell for European Space Vehicle, Status of Activities", Proc. of the European Space Power Conference held in Graz, Austria, 23.–27. August 1993, pp. 635–646

[18] K. H. Tetzlaff, R. Walz, C. A. Gossen, Journal of Power Sources **50** (1994) 311

[19] K. Kordesch et al., "Fuel Cell Research and Development Projects in Austria", Proceedings of the 7th World Hydrogen Energy Conference, Moscow 1988. T. N. Vézirglu (ed.): Hydrogen Energy Progress – VII, Pergamon Press, Oxford 1988

[20] C. Myrén, S. T. Naumann, "Fuel Processor Systems for Small Scale AFC Power Plants", 1994 – Fuel Cell Seminar, San Diego (CA), pp. 192–195

[21] O. Lindström, S. Schwartz, Y. Kiros, M. Ramanathan, A. Sampathrajan, "Small Scale AFC Stack Technology", 1994 – Fuel Cell Seminar, San-Diego (CA), pp. 323–324

[22] Y. Kiros, "Tolerance of Gases by the Anode for the Alkaline Fuel Cell", 1994 – Fuel Cell Seminar, San-Diego (CA), pp. 364–367

[23] J. Kivisaari, "Influence of Impurities in the Anode and Cathode Gases, on Fuel Cell Performance and Life – A Literature Overview", 1994 – Fuel Cell Seminar, San-Diego (CA), pp. 436–437

[24] O. Lindström, "A Critical Assessment of Fuel Cell Technology", Department of Chemical Engineering and Technology, Royal Institute of Technology Stockholm 1993, ISRN KTH/KT/FR – 93/94 – SE, ISSN 1101-9271

[25] O. Lindström, "A Critical Assessment of Fuel Cell Technology", 1994 – Fuel Cell Seminar, San-Diego (CA), pp. 297–298

[26] M. S. Wilson, J. A. Valerio, S. Gottesfeld, "Endurance Testing of Low Pt Loading Polymer Electrolyte Fuel Cells", ECS, Spring Meeting, San Francisco (CA), May 22–27, 1994, Volume 94-1, Abstract Number 607, p. 966

[27] R. Cohen, "Gemini Fuel Cell System", Proceedings 20th Power Sources Conference, 24.–26th May 1966, pp. 21–24

[28] K. Prater, J. Power Sources **29** (1990)

[29] D. Watkins, "Solid Polymer Fuel Cell Technology", 4th Annual Battery Conference, Long Beach, California, January 1989

[30] M. S. Wilson, T. E. Springer, T. A. Zawodzinski, J. R. Davey, C. R. Derouin, J. A. Valerio, S. Gottesfeld, "Development of Components for a Polymer Electrolyte Fuel Cell of Low Cost and High Performance", 1994 – Fuel Cell Seminar, San Diego (CA); Nov./Dec. 1994, p. 281

[31] S. Kamasaki, Sh. Sekido, Y. Miura, "Characteristics of Fuel Cell Hydrgoen Electrodes Combined with SPE Nafion", Memoirs of Faculty of Technology, Tokyo Metropolitan University, No. 38, 1988, p. 141

[32] K. V. Kordesch, Electrochimica Acta **16** (1971) 597

[33] S. Srinivasan, E. A. Ticianelli, C. R. Derouin, A. Redondo, "Advances in Solid Polymer Electrolyte Fuel Cell Technology with Low Platinum Loading Electrodes", J. Power Sources **22** (1988) 359–375

[34] S. Srinivasan, E. A. Ticianelli, C. R. Derouin, A. Redondo, J. Power Sources **22** (1988) 359

[35] E. A. Ticianelli, C. R. Derouin, A. Redondo, S. Srinivasan, J. Electrochem. Soc. **135** (1988) 2209

[36] E. A. Ticianelli, J. G. Beery, S. Srinivasan, J. Appl. Electrochem. **21** (1991) 597

[37] E. A. Ticianelli, C. R. Derouin, A. Redondo, S. Srinivasan, J. Electroanal. Chem. **251** (1988) 275

[38] Z. Poltarzewski, P. Staiti, V. Alderucci, W. Wiezorek, N. Giardano, J. Electrochem. Soc. **139** (3) (1992) 761

[39] E. J. Taylor, N. R. K. Vilambi, E. B. Anderson, K. Donohue, "Electrochemical Catalyzation of High Platinum Utilization Gas Diffusion Electrodes for Proton Exchange Membrane Fuel Cells", received from PSI Technology Company, 20 New England Business Center, Andover, MA, 01810, 1993

[40] A. C. Ferreira, S. Srinivasan, "High Power Densities and Energy Efficiencies in PEMFCs using 0.05 mg Pt/cm^2 Gas Diffusion Electrode", ECS, Spring Meeting, San Francisco (CA), May 22–27, 1994, 969

[41] E. J. Taylor, E. B. Anderson, N. R. K. Vilambi, J. Electrochem. Soc. **139** (5) (1992) L45

[42] S. Srinivasan, D. J. Manko, H. Koch, M. A. Enayetullah, A. J. Appleby, J. Power Sources **29** (1990) 367

[43] S. Srinivasan, E. A. Ticianelli, C. R. Derouin, A. Redondo, J. Power Sources **22** (1988) 359

[44] M. S. Wilson, S. Gottesfeld, J. Electrochem. Soc. 139 (2) (1992) L28

[45] W-K. Paik, T. E. Springer, S. Srinivasan, J. Electrochem. Soc. **136** (3) (1989) 644

[46] P. G. Dirven, W. J. Engeln, C. J-M. van der Poorten, J. Applied Electrochem. **25** (1995) 122

[47] Ballard Advanced Materials, "Development and Evaluation of a Low-Cost Solid Polymer Electrolyte", IEA Workshop, Ottawa, Canada, Sept. 14–15, 1992

[48] T. D. Gierke, G. E. Munn, F. C. Wilson, J. Polym. Sci. Polym. Phys. **19** (1981) 1687

[49] Brochure DuPont Nafion® perfluorinated membranes; DuPont, Fayetteville, NC 28302, USA, 1994

[50] PEM Brennstoffzellen, Brochure, Siemens KWU FRN4, Erlangen, Germany, D-91050 (Dec. '92)

[51] K. Strasser, "PEM Fuel Cells for Energy Storage Systems", IECEC '91, 26th Intersociety Energy Conversion Engineering Conference, Boston, Massachusetts, August 4–9, 1991, p. 636

[52] A. J. Appleby "Fuel Cells for Traction Applications – An Overview", Proceedings from a symposium on Feb. 8, 1994 at the Royal Swedish Academy of Engineering Sciences, IVA, Stockholm, Sweden, pp. 13–60

[53] M. Wakizoe, J. Kim. S. Srinivasan, and J. Mc. Breen, "A Novel Methode for Determination of Transport Parameters of H_2 and O_2 in Proton Exchange Membrane Fuel Cells", in Proceedings of the Symposium on "Batteries and Fuel Cells for Stationary and Electric Vehicle Applications", ed. by A. R. Landgrebe, Z. Takehara, Proceedings Volume 93-8, pp. 310–323

[54] Yu-Min Tsou, M. C. Kimble, R. E. White, J. Electrochem. Soc. **139** (1992) 1913

[55] M. Wakizoe, J. Kim. S. Srinivasan, and J. Mc. Breen, "A Novel Methode for Determination of Transport Parameters of H_2 and O_2 in Proton Exchange Membrane Fuel Cells", in-Proceedings of the Symposium on "Batteries and Fuel Cells for Stationary and Electric Vehicle Applications", ed. by A. R. Landgrebe, Z. Takehara, Proceedings Volume 93-8, pp. 310–323

[56] A. Steck (Ballard Advanced Materials), "Development and Evaluation of a Low-Cost Solid Polymer Electrolyte as an Alternative to Perfluorinated Type Membranes in Fuel Cells", Proceedings IEA Workshop for the PEFC ANNEX, September 14 and 15, 1992, Ottawa (Canada)

[57] A. E. Steck, "Membrane Materials in Fuel Cells", Proceedings of the First International Symposium on New Materials for Fuel Cell Systems", Montreal, Quebec, Canada, July 9–13, 1995, p. 74

[58] G. G. Scherer, IEA Annex IV: "Collaborative Research & Development on the Polymer Electrolyte Fuel Cell", Subtask D: Membranes & Electrolytes, Annual Report for the Period 10/92–9/93, Paul Scherrer Institut, Switzerland, 22. 9. 1993

[59] F. N. Büchi, B. Gupta, O. Haas, G. G. Scherer, Electrochimica Acta **40** (5) (1995) 345

[60] F. N. Buechi, B. Gupta, J. Halim, O Haas, G. G. Scherer, "A New Class of Partial Fluorinated Fuel Cell Membranes", Extended Abstracts, ECS-Meeting San Francisco (CA), Volume 94-1, Abstract-No. 615, p. 978

[61] B. Gupta, F. N. Büchi, G. G. Scherer, A. Chapiro, Solid State Ionics **61** (1993) 213

[62] Guenther G. Scherer, "PEFC Research at Paul Scherrer Institute", Proceedings IEA Workshop for the PEFC ANNEX, September 14 and 15, 1992, Ottawa (Canada)

[63] M. S. Wilson, T. E. Springer, T. A. Zawodzinski, J. R. Davey, C. R. Derouin, J. A. Valerio, S. Gottesfeld, "Development of Components for a Polymer Electrolyte Fuel Cell of Low

Cost and High Performance", 1994 – Fuel Cell Seminar, San Diego (CA), Nov./Dec. 1994, p. 281

[64] H. P. Dhar, "Internally Humidified High Performance Proton Exchange membrane Fuel Cell", 1994 – Fuel Cell Seminar, San Diego (CA), Nov./Dec. 1994, p. 85

[65] A. B. Laconti, A. R. Fragala, J. R. Byack, Proceedings of the Symposium on Electrode Materials and Processes for Energy Conversion and Storage, edited by: J. D. E. McIntyre, S. Srinivasan, F. G. Will, The Electrochemical Society, Inc., Pennington, NJ, 1977, p. 354

[66] H. Koch, A. Nandi, N. K. Anand, O. Velev, D. H Swan, S. Srinivasan, A. J. Appleby, The Electrochemical Society, Inc., Volume 90-2, Seattle (Washington), October 14–19, 1990, Abstract No. 115, p. 176

[67] T. Springer, S. Gottesfeld, S. Randzinski, T. Zawodinski, The Electrochemical Society, Inc., Volume 90-2, Seattle (Washington), October 14–19, 1990, Abstract No. 118, p. 179

[68] T. E. Springer, T. A. Zawadzinski, S. Gottesfeld, J. Electrochem. Soc. **138** (8) (1991) 2334

[69] T. E. Springer, T. A. Zawodzinski, S. Gottesfeld, "Modeling Water Content Effects in Polymer Electrolyte Fuel Cells", received from Los Alamos National Laboratory, Los Alamos, NM 87545 (1991)

[70] D. M. Bernardi, J. Electrochem. Soc. **137** (11) (1990) 3344

[71] T. E. Springer, S. Gottesfeld, "Pseudohomogenous Catalyst Layer Model for Polymer Electrolyte Fuel Cell", Electronics Research Goup, Los Alamos National Laboratory, Los Alamos (NM) 87544, 1993 (private communication, extended paper without indication of the conference), 1994

[72] M. S. Wilson, T. E. Springer, T. A. Zawodzinski, S. Gottesfeld, "Recent Achievements in Polymer Electrolyte Fuel Cells (PEFC) Research at Los Alamos National Laboratory", 26th Intersociety Energy Conversion Engineering Conference, Boston (MA), August 4–9, 1991, Volume 3 "Conversion Technologies, Electrochemical Conversion", p. 636

[73] D. M. Bernardi, M. W. Verbruge, "Mathematical Model of A Gas Diffusion Electrode Bonded to a Polymer Electrolyte", AIChE **37** (8) (1991) 1151

[74] P. Staiti, et al. J. Applied Electrochem **22** (1992) 663

[75] D. M Bernardi, M. W. Verbrugge, J. Electrochem. Soc. **139** (9) (1992) 2477

[76] T. V. Nguyen, R. E. White, J. Electrochem. Soc **140** (8) (1993) 2178

[77] T. E. Springer, M. S. Wilson, S. Gottesfeld, J. Electrochem. Soc. **140** (12) (1993) 3513

[78] T. F. Fuller, J. Newman, J. Electrochem. Soc. **140** (5) (1993) 1218

[79] J.-T. Wang, R. R. Savinell, Electrochimica Acta **37** (15) (1992) 2737

[80] S. Gottesfeld, T. E. Springer, M. S. Wilson "Modeling and Experimental Diagnostics in PEFC", ECS, Spring-Meeting 94-1, Extended Abstracts, San Francisco (CA), May 22–27, 1994, p. 986

[81] D. K. Wilkinson, H. H. Voss, K. Prater, J. Power Sources **49** (1994) 117

[82] M. Watanabe, Y. Satch, Ch. Shimura "Easy Management of Water Content in Polymer Electrolyte Membrane with Mounted Porous Fiber Wicks", Batteries and Fuel Cells for Stationary and Electric Vehicle Applications, edited by: A. R. Landgrebe, Z.-I. Takehara, The Electrochemical Society, Inc., Proceedings Volume 93-8, p. 302

[83] A. Cisar, A. Gonzalez-Martin, O. J. Murphy, "Internally Humidified Membranes for Use in Fuel Cells", Proceedings of the First International Symposium on New Materials for Fuel Cell Systems", Montreal, Quebec, Canada, July 9–13, 1995, p. 104

[84] K. V. Kordesch, J. C. T. Oliveira, "Fuel Cells", Ullmann's Encyclopedia of Industrial Chemistry, Vol A 12, Weinheim, 1989

[85] M S. Wilson, S. Gottesfeld, J. Electrochem. Soc. **139** (2) (1992) L28

[86] D. M. Bernardi, J. Electrochem. Soc. **137** (1990) 3344

[87] J. Srmivason, et al., "High Energy Efficiency and High Power Density Proton Exchange Membrane Fuell Cells – Electrode Kinetics and Mass Transport", J. Power Sources **36** (1991)

[88] Kim Kinoshita, "Electrochemical Oxygen Technology", Berkeley (CA), John Wiley & Sons, Inc., New York, 1992, pp. 168–176

[89] J. H. Hirschenhofer, et al., Fuel Cells A Handbook (Revision 3), US DOE, January 1994, p. 6–9

[90] S. Srinivasan, E. A. Ticianelli, C. R. Derouin, A. Redondo, J. Power Sources **22** (1988) 359

[91] J. C. Amphlett, R. M Baumert, R. F. Mann, B. A. Peppley, P. R. Roberge, T. J. Harris, "A Performance Model for PEM Fuel Cells", 28th IECEC, Atalanta (Georgia), Aug. 1993, p. 1.1215

[92] D. R. Brighton, G. A. Clark, M.J.M. Rowan, P.L. Mart, "Analysis of the Performance of a commercial 5 kW PEM Fuel Cell System", Department of Defense – Defense Science and Technology Organisation, presented at the 1994 – Fuel Cell Seminar, San Diego (CA), Nov./Dec. 1994

[93] N. Keon, G. Simader, T. L. Ören, J. C.T. Oliveira, "Dynamic Modeling of PEFC" presented at the SCSC'95, Ottawa, Ontario, Canada, July 24–26, 1995

[94] "Fuel Cell Test at Siemens", Polymer Electrolyte Membrane-Brennstoffzellen (PEMBZ), Siemens KWU FRN 4, Erlangen, D-91050, December 1992

[95] R. A. Lemons, J. Power Sources **29** (1990) 251

[96] M. S. Wilson, T. E. Springer, T. A. Zawodzinski, S. Gottesfeld, "Recent Achievements in Polymer Electrolyte Fuel Cell (PEFC) Research at Los Alamos National Laboratory", IECEC 1991 – 26th Intersociety Energy Conversion Engineering Conference, Bosteon (MA), Aug. 4–9, 1991, pp. 636–641

[97] M. Waidhas et al., "Research and Development of Low Temperature Fuel Cell at Siemens", 1994 – Fuel Cell Seminar, San Diego (CA), Nov/Dec. 1994, p. 477

[98] S. Gottesfeld, J. Pafford, J. Electrochem. Soc. **135** (1988) 2651

[99] R. A. Lemons, J. Power Sources **29** (1990) 251

[100] J. H. Hirschenhofer, D. B. Stauffer, R. R. Engleman, "Fuel Cells: A Handbook (Revision 3)", US DOE Morgantown, West Virginia, Contract No. DE-AC01-88FE61684, January 1994

[101] K. Strasser, J. Power Sources **37** (1992) 209

[102] K. B. Prater, J. Power Sources **51** (1994) 129

[103] K. Prater, J. Power Sources **29** (1990) 239

[104] K. Prater, J. Power Sources **37** (1992) 181

[105] Ballard Power Systems Inc., Information Summary, 1993

[106] D. Watkins et al., "Solid Polymer Fuel Cell Technology", 4th Annual Battery Conference, Long Beach (CA), January 1989

[107] D. Watkins et al., "Canadian Solid Polymer Fuel Cell Development", 5th Annual Battery Conference, Long Beach (CA), January 1990

[108] A. E. Steck, "Membrane Materials in Fuel Cells", First International Symposium on New Materials for Fuel Cell Systems, July 9–13, 1995, Montreal (Canada)

[109] A. E. Steck, Ballard Advanced Materials, Progress Report, IEA, Amsterdam, April 1994

[110] Ballard Power Systems, News Release, "Ballard Receives $ 3.3 Million in Orders from Dow for Fuel Cells", February 1, 1995

[111] D. S. Watkins, B. C. Peters, "Development of a PEM Power Plant for Stationary Power Applications", 1994 – Fuel Cell Seminar, San Diego (CA), Nov./Dec. 1994, p. 250

[112] F. Mitlitsky, N. J. Colella, B. Myers, C. J: Anderson, "Regenerative Fuel Cells for High Altitude Long Endurance Solar Powered Aircraft", Intersociety Energy Conversion Engineering Conference, Atlanta (Georgia), August 9–13, 1993

[113] F. Mitlitsky, N. J. Colella, B. Myers, "Unitized Regenerative Fuel Cells for Solar Rechargeable Aircraft and zero Emission Vehicles", 1994 – Fuel Cell Seminar, San Diego (CA), Nov./Dec. 1994, p. 624

[114] H. P. Dhar, J. Appl. Electrochem **23** (1993) 32 - 37

[115] K. Ledjeff, A. Heinzel, "Regenerative Fuel Cell for Energy Storage in PV Systems", 26th Intersociety Energy Conversion Engineering Conference IECEC91, Boston (MA), August 4-9, 1991, p. 538

[116] A. J. Appleby, J. Power Sources **22** (1988) 377

[117] J.-A. Charleston, J. Reed, "Hydrogen-Bromine Fuel Cell Advance Component Development", National Technical Association Conference, Chicago (Illinois) July 13–16, 1988, NASA Technical Memorandum 101345

[118] R. F. Savinell, S. D. Fritts, J. Power Sources **22** (1988) 423

[119] S. D. Fritts, R. F. Savinell, J. Power Sources **28** (1989) 301

[120] D. R. de Sena, E. R. Gonzalez, E. A. Ticianelli, Electrochimica Acta **37** (10) (1992) 1855

[121] W. Vogel, J. T. Lundquist, J. Electrochem. Soc. **117** (1970) 1512

[122] H. C. Maru et al., "Superacid Electrolyte Fuel Cells for Transportation Applications", Proceedings Renewable Fuels and Advanced Power Sources for Transportation Workshop, June 17–18, 1982, Boulder, Colorado, p. 109

[123] A. J. Appleby, "Acid Fuel Cells; Overview of the State-of-the Art", Proceedings Renewable Fuels and Advanced Power Sources for Transportation Workshop, June 17–18, 1982, Boulder, Colorado, p. 55

[124] A. J. Appleby, "New Materials for Fuel Cell Systems", New Materials for Fuel Cell Systems 1, Montreal, Quebec, Canada, July 9–13, 1995, p. 2

[125] International Fuel Cells, Power Plant Description of the PC18 (UTC)

[126] H. Wendt, W. Jenseit, Chem.-Ing.-Techn. **60** (3) (1988) 180

[127] L. Christner, J. Ahmed, J. Faroque, "Corrosion of Phosphoric Acid Fuel Cell Components", Proceedings of the Symposium on Corrosion in Batteries and Fuel Cells and Corrosion in Solar Energy Systems, edited by C. J. Jonson, S. L. Pohlmann, Proc.-Vol. 83-1, pp. 140–158

[128] A. J. Appleby, "Acid Fuel Cells: Overview of the State-Of-The-Art", Proceedings Renewable Fuels and Advanced Power Sources for Transportation Workshop, 17–18 June 1982, Boulder (Colorado), pp. 55–73

[129] J. Scholta, H. Wendt, Internat. Symposium "Electrochemical Engineering and the Environment 1992", London, p. 237

[130] J. Scholta, H. Wendt, 1992–Fuel Cell Seminar, Tucson (AZ), p. 281

[131] M. Pourbaix, "Atlas of Electrochemical Equilibria in aqueous solutions", Pergamon press, Brussels, 1966, p. 449

[132] J. Appleby, in Poceedings of the Workshop on the Electrochemistry of Carbon, Edited by S. Sarangapani, J. R. Adridge, B. Schumm, The Electrochemical Society, Inc. Pennington, NJ, 1984, p. 251

[133] K. V. Kordesch, "Survey of Carbon and its Role in Phosphoric Acid Fuel Cells", BNL 51418, prepared for Brookhaven National Laboratory, December 1979

[134] K. Kordesch, J. Mc. Breen, H. Olender, S. Srinivasan, K. Tomatschger, "Selections of Carbon as Electrode Materials", Proceedings of the Symposium on Corrosion in Batteries and Fuel Cells and Corrosion in Solar Energy Systems, edited by C. J. Jonson, S. L. Pohlmann, Proc.-Vol.-83-1, pp. 124–139

[135] P. Stonehart, Ber. Bunsenges. Phys. Chem. **94** (1990) 913

[136] H. G. Petrow and R. J. Allen, "Catalytic Platinum Metal Particles on a Substrate and Method of Preparing the Catalyst", US Patent # 3,992,331, Nov. 16, 1976

[137] H. G. Petrow and R. J. Allen "Finely Particulated Colloidal Platinum Compound and Sol for Producing the Same, and Method of Preparation", US Patent # 3,992,512, Nov. 16, 1976

[138] P. Stonehart, Ber. Bunsenges. Phys. Chem. **94** (1990) 913–921

[139] R. Rosey, paper presented at the 1985 Fuel Cell Seminar, May 19–22, Tucson (AZ), Abstracts, 1985 p. 60

[140] A. J. Appleby, paper presented at the National FC Seminar, Newport Beach, CA, November 14–18, 1982

[141] V. M. Jalan, in Proceedings of the Workshop on The Electrochemistry of Carbon, S. Sarangapani, J. R. Akrdige, and B. Schumm, Eds., The Electrochemical Society, Pennington, NJ, 1984, p. 554

[142] Proceedings of the Workshop on "The Electrochemistry of Carbon", Aug. 17–19, 1983

[143] Proceedings of the Symposia on "Corrosion in Batteries and Fuel Cells and Corrosion in Solar Energy Systems", Proceedings Volume 83–1

[144] P. Stonehart and J. P. MacDonald, "Stability of Acid Fuel Cell Cathode Materials", EPRI, EM-1664, EPRI, Palo Alto, CA (1981)

[145] S. S. Penner, "Assessment of Research Needs for Advanced Fuel Cells by the DOE Advanced Fuel Cell Working Group (AFCWG), November 1985, Chapter 2: Phosphoric Acid Fuel Cells (PAFCs) by John A. Appleby, p. 41

[146] A. P Fickett, in Proceedings of the Symposium on Electrode Materials and Processes for Energy Conversion and Storage, ed.: J. D. E. Mcintyre, S. Srinivasan, F. G. Will, The Electrochemical Society, Inc., Pennington, NJ, p. 546, 1977

[147] A. J. Appleby, J. Electroanal. Chem., **118** (1981) 31

[148] A. J. Appleby, J. Electroanal. Chem., **118** (1981) 31

[149] J. Huff, "Status of Fuel Cell Technologies", 1986 – Fuel Cell Seminar, October 26–29, Tucson (AZ), 1986

[150] "Advanced Water-Cooled Phosphoic Acid Fuel Cell Development", Final Report, Report No. DE/MC/24221-3130, International Fuel Cells Corporation for US DOE unter Contract DE-AC21-88MC24221

[151] K. Harasawa, et al., "Fuel Cell R & D and Demonstration Programs at Electric Utilities in Japan", 1992 – Fuel Cell Seminar, Tucson (AZ), pp. 191–194

[152] J. Hirschenhofer, D. B. Stauffer, R. R. Engleman, "Fuel Cells A Handbook" (Revision 3), US-DOE under contract No. DE-AC01-88FE6184, pp. 3–13

[153] D. T. Shin, P. D. Howard, J. Electrochem. Society, **133** (1986) 2447

[154] S. T. Szymanski, G. A. Gruves, M. Katz, H. R. Kunz, J. Electrochem. Soc., **127** (1980) 1440

[155] K. Kordesch, "Brennstoffbatterien", Springer-Verlag, Wien 1984

[156] J. M. Feret, Journal of Power Sources **29** (1990) 59

[157] D. Newby, "Westinghouse Air-Cooled PAFC Program", CEC-Italian Fuel Cell Workshop, 4.–5. June, 1987, Taormina, Italy, p. 85

[158] K. V. Kordesch, J. C.T. Oliveira, "Fuel Cells", Ullmann's Encyclopedia of Industrial Chemistry, Vol. A 12, p. 55–83, 1989

[159] M. C. Williams, E. L. Parsons, T. J. George, "Status of Molten Carbonate Fuel Cell Technology Development", Proc. of the Third International Symposium on Carbonate Fuel Cell Technology", ed by: D. Shores, I. Uchida, H. Maru, J.R. Selman, Proc. Vol. 93-3, pp. 1–10

[160] A. Pigeaud, G. Wilemsky, "Effects of Coal-Derived Trace Species on the Performance of Carbonate Fuel Cells", Proc. of 4th Annual Fuel Cell Contractors Review Meeting, U.S. DOE/METC, July 1992, pp. 42–45

[161] ERC Topical Report for US DOE/METC, "Effects of Coal-Derived Trace Species on the Performance of MCFCs", US DOE/METC, DOE/MC/25009-T26, October 1991

[162] J. R. Selman, Energy **11** (1986) 153

[163] H. R. Kunz, J. Electrochem. Soc. **134** (1987) 105

[164] C. L. Bushnell, "Electrolyte Matrix for Molten Carbonate Fuel Cells" US Patent 4,322,482 (1982)

[165] Y. Yamamasu, et al., "Component Development and Durability Test of MCFC", Proceedings of the First IFCC, Makuhari, Japan, (1992), p. 161

[166] L. M. Paetsch, et al., "Review of Carbonate Fuel Cell Matrix and Electrolyte Developments", Proceedings of the Third International Symposium of Carbonate Fuel Cell Technology, Proc. Vol. 93-3, pp. 89–105

[167] Y. Yamamasu, et al., "Component Development and Durability Test of MCFC", Proceedings of the First IFCC, Makuhari, Japan, 1992, p. 161

[168] O. Hideyuki, et al. "Analysis of Matrix Degradation for MCFC", Proceedings of the First IFCC, Makuhari, Japan, 1992, p. 199

[169] H. Kasai, A. Suzuki, "Dissolution and Deposition of Nickel-Oxide Cathode in MCFC", Proc. of the Third International Symposium on Carbonate Fuel Cell Technology, Proc.-Vol. 93-3, pp. 240–251

[170] Energy Research Corporation, "NiO Dissolution in Carbonate Fuel Cells", Topical Report to US DOE, under Contract No. DE-AC21-90MC27168 May (1991)

[171] C. E. Baumgartner, J. Electrochem. Soc. **131** (8) (1984) 1850

[172] J. Doyon, T. Gilbert, G. Davies, L. Paetsch, J. Electrochem. Soc. **134** (2) (1987) 3035

[173] H. R. Kunz, J. Electrochem. Soc. **134** (1) (1987) 105

[174] Y. Miyazaki, et al., "Material studies on MCFC", 1994 – Fuel Cell Seminar, San Diego (CA), pp. 120–123

[175] K. Terada, et al., "Development of a new electrolyte matrix for MCFC", 1994 – Fuel Cell Seminar, San Diego (CA), pp. 128–131

[176] K. Tanimoto, et al., "Cell Performance of MCFC with Alkali and Alkaline Earth Carbonate Mixtures", Proceedings of the first IFCC, Makuhari, Feb. 1992, pp. 185–188

[177] A.J. Appleby, S.B. Nicholson, J. Electroanalyt. Chem. **53** (1974) 105, ibid **83** (1977) 309, ibid **112** (1980) 71

[178] J. R. Selman, "Technology Base Research for MCFC Development in the US", Proceedings of the Second Symposium on Molten Carbonate Fuel Cell Technology, Proc.-Vol. 90-16, pp. 187–205

[179] I. R. Selman et al., Proc. of the 3nd Int. Symposium on Carbonate Fuel Cell Technology, Proc. Vol. 93-3, pp. 309–324.

[180] S. W. Smith, W.M.Vogel, S. Kapelner, J. Electrochem. Soc. **129** (1982) 1668

[181] C. Y. Yuh, J.R.Selman, AIChem.E.J. **34** (1988), 1949

[182] D. A. Shores, J. R. Selman, S. Israni and E. T. Ong.; "Degradation of NiO Cathodes: Modeling and Experimental Studies"; Proceedings of the First Symposium of Molten Carbonate Fuel Cell Technology, Proc.-Vol. 90-16, pp. 290–317

[183] T. Nishina, I. Uchida, ibid, pp. 438–453

[184] R. C. Makkus, K. Hemmes, J.H.W. de Wit, ibid, pp. 454–467

[185] J. R. Selman, G. L. Lee, Proceedings of the Third International Symposium on Carbonate Fuel Cell Technology, Proc.-Vol. 93-3, pp. 309–324

[186] E. Fontes, M. Fontes, ibid, p. 454

[187] G. Chiodelli, "Synthesis and Electronic Conductivity of LiCoO$_2$ Materials for MCFC Alternative Cathodes", 1994 – Fuel Cell Seminar, San Diego (CA), pp. 176–179

[188] L. Giogi, et al., "Investigation on LiCoO$_2$ as a Novel Cathode Material for MCFC by Means of Electrochemical Impedance Spectroscopy", 1994 – Fuel Cell Seminar, San Diego (CA), pp. 172–175

[189] R. C. Makkus, K. Hemmes, J. H. W. de Wit, J. Electrochem. Soc., **141** (12) (1994) 3429

[190] L. Plomp, J.B.J. Veldhuis, E.F. Sitters, "Status of MCFC Materials Development at ECN", Proceedings of the first IFCC, Makuhari, 1992, p. 157

[191] L. Plomb, et al., Proceedings of the first IFCC, Makuhari, Japan, 1992, p. 157

[192] J. P. T. Vossen, L. Plomp, "Corrosion of Ni in Molten Carbonate", Proceedings of the Third International Symposium of Carbonate Fuel Cell Technology, Proc.-Vol. 93-3, pp. 278–291

[193] T. Yoshii, et al., "New Metallic Material Technologies for MCFC", 1994 – Fuel Cell Seminar, San Diego (CA), pp. 140–143

[194] R. D. Pierce, J. L. Smith, R. B. Poepell, Proc. Electrochem. Soc. Symp. on Molten Carbonate Fuel Cells, Montreal, May 9–14, 1982

[195] P. K. Adanuvor, R. E. White, A. J. Appleby, J. Electrochem. Soc. **137** (1990) 2095

[196] P. G. P. Ang, A. F. Sammells, J. Electrochem. Soc. **127** (1980) 1287

[197] A. J. Appleby, S. B. Nicholson, J. Electroanal. Chem. **112** (1980) 71

[198] T. Tanaka, T. Murashi and Nishiyama, Proc. 23rd Intersociety Energy Conversion Engineering Conference, 1988, p. 245

[199] T. Murashi, "Internal Reforming – An Overview", "MCFC", Proceedings of the First International Symposium of Molten Carbonate Fuel Cell Technology, Proc.-Vol. 90-16, p. 100

[200] J. Ohtsuki, T. Seki, S. Takeuchi, A. Kusunoki, "Development of Indirect Internal Reforming MCFC", Proceedings of the Third International Symposium of Carbonate Fuel Cell Technology, Proc.-Vol. 93-3, p. 48

[201] L. J. Bregoli, H. R. Kunz, J. Electrochem. Soc. **129** (1982) 2711

[202] T. G. Benjamin, et al.,"Status of MCFC Technology at M-C Power – 1992", 1992 – Fuel Cell Seminar, Tucson (AZ), pp. 21–24

[203] G. L. Anderson, P. C. Garrigan, Proceedings of the Symposium on Molten Carbonate Fuel Cell Technology, ed. by.: R. J. Selman, T. D. Claar, The Electrochemical Society, Inc., Pennington, NJ, p. 297, 1984

[204] H. C. Maru, M. Raroque, L. Paetsch, C.Y. Yuh, P. Patel, "Direct Fuel Cell Stack Technology Status", Proceedings of the First IFCC, Makuhari, 1992, p. 145

[205] H. C. Maru, M. Faroque, L. Paetsch, C. Y. Yuh, "Development of Direct Carbonate Fuel Cell Stacks", 1992 – Fuel Cell Seminar, Tucson (AZ), p. 29

[206] H. C. Healy, W. H. Johnson, C. A. Reiser, "IFC MCFC Stack Research Status", Proceedings of the First IFCC, Makuhari, Japan, Feb. 1992, p. 261

[207] J. Ohtsuki et al. "Development of indirect internal reforming MCFC", Proceedings of The First International Fuel Cell Conference, 1992, p. 251

[208] "Development of Internal Reforming MCFC", Fuel Cell RD&D, p. 43

[209] J. Ohtsuki et al "Development of indirect internal reforming MCFC", 1992 – Fuel Cell Seminar, Tucson (AZ), p. 17

[210] J. Ohtsuki et al "Development of indirect internal reforming MCFC", Carbonate Fuel Cell Technology, Proc.-Vol. 93-3, p. 48

[211] K. Ohtsuka, et al., "Status of Large Sized MCFC Cell & Stack Technology Development at Hitachi", 1990 – Fuel Cell Seminar, pp. 95–98

[212] "Development of Molten Carbonate Fuel Cells" in Fuel Cell R & D in Japan, August 1992

[213] S. Takashima, K. Ohtsuka, T. Kahara, "MCFC Stack Technology at Hitachi", Proceedings of the first IFCC, Makuhari 1992, p. 265

[214] K. Ohtsuka, et al., "Status of MCFC Stack Technology at Hitachi", 1992 – Fuel Cell Seminar, Tucson (AZ), p. 13

[215] A. Suzuki, M. Tool, M. Hosaka, T. Matsuyama, "Recent Development of Large Size MCFC Stacks at IHI", Proceedings of the first IFCC, 1992, Makuhari, Japan, p. 273

[216] A. Suzuki, et al., "Recent status of IHI MCFC Stack Development", 1992 – Fuel Cell Seminar, Tucson (AZ), p. 25–28

[217] "Development of MCFC", IHI, Fuel Cell RD&D, 1992

[218] Y. Izaki et al., "Experimental Results of 10 kW class MCFC", IECEC-89, Washington, D.C., August 1989, pp. 1523–1528

[219] Y. Mugikura et al., "6 kW Class MCFC with Gas Recycling", 1990 – Fuel Cell Seminar, pp. 103–106

[220] Y. Izaki et al. "The basic performance of large scale MCFC", Proceedings of the Third International Symposium of Carbonate Fuel Cell Technology, Proc.-Vol. 93-3, pp. 119–130

[221] N. Ueno, "Outline of Japan's Fuel Cell Projects", Sunshine Project, AIST, MITI in Fuel Cell News, Vol. X, Nos. 4, Page 10

[222] T. Yoshi, et al., "New Metallic Material Technologies for MCFC", 1994 – Fuel Cell Seminar, San Diego (CA), Dec. 1994, p. 140

[223] Y. Izahi, et al., "Study on Wettability and Oxidation of Materials with Molten Carbonate", 1994 – Fuel Cell Seminar, San Diego (CA), Dec. 1994, p. 168

[224] T. Kahara, et al., "Status of MCFC Cell and Stack Technology at Hitachi", 1994 – Fuel Cell Seminar, San Diego (CA), Dec. 1994, p. 222

[225] T. Murashashi, "Development of Internal Reforming MCFC", 1994 – Fuel Cell Seminar, San Diego (CA), Dec. 1994, p. 230

[226] M. Iijima, et al., "Development of DIR-MCFC Co-Generation System", 1994 – Fuel Cell Seminar, San Diego (CA), Dec. 1994, p. 226

[227] K. Uematsu, et al., "Demonstration of Fuel Processing Unit for MCFC Power Generation System", 1994 – Fuel Cell Seminar, San Diego (CA), Dec. 1994, p. 269

[228] T. Kakihara, et al., "Development of a 100 kW MCFC Stack Component", 1994 – Fuel Cell Seminar, San Diego (CA), Dec. 1994, p. 450

[229] T. Watanabe, et al., "100 kW Class MCFC Stack Performance", 1994 – Fuel Cell Seminar, San Diego (CA), Dec. 1994, p. 446

[230] A. Nakaoka, et al., "1000 kW Class MCFC Pilot Plant", 1994 – Fuel Cell Seminar, San Diego (CA), Dec. 1994, p. 458

[231] Peter C. v.d. Laag, T. N. Verbruggen, "Conceptual Designs of First-Generation 250 kW MCFC Demonstration Plants", 1994 – Fuel Cell Seminar, San Diego (CA), pp. 152–155

[232] F. H. Heuveln, J.P.P. Huijsmans, "Long Term Stability for Solid Oxide Fuel Cells Product & Systems Evaluation", SOFC Materials, Process Engineering and Electrochemistry, 5th IEA Workshop, Jülich, Germany, 2–4 March 1993, pp. 41–50

[233] M. Watanabe, H. Uchida, M. Shibata. N. Mochizuki, K. Amikura, J. Electrochem. Soc. **141** (2) (1994) 342

[234] W. Winkler, "Grundlagen der Konzeption und Auslegung von Kraftwerken mit Hochtemperaturbrennstoffzellen", Oral Lecture at the University of Technology, Graz, 15th Nov. 1993

[235] C. Milken, S. Elanovan, A.C. Khandkar, in Proc. 2nd Int. Symp. Ionic and Mixed Conducting Ceramics, Ed. By T. A. Ramanarayanan, W.L. Worell, H. L. Tuller, The Electrochem. Soc. Proc. Vol. 94-12, pp. 466–477, 1994

[236] M. Gödichermeir, K. Sasaki, P. Bohac, L.J. Gauckler, in Proceedings 6th-Workshop Advanced SOFCs, Rome, Italy, pp. 225–34, 1994

[237] K. Kinoshita, F. R. Mc Larnon, E. J. Cairns, Fuel Cell Handbook, DOE/METC-88/6096, Morgantown Energy Technology Center, U.S. Department of Energy, Morgantown, MV (May 1988)

[238] S.C. Singhal, H. Iwahara, Proceedings of the Third International Symposium on Solid Oxide Fuel Cells, Volume 93-4

[239] SOFC, Product information from Westinghouse Electric Corporation, December 1992

[240] N. Q. Minh, pp. 652–63, in Science and Technology of Zirconia V, Ed. By SPS Badwal, M.J. Bannister, R. H. J. Hannink, Technomic Publishing, Pennsylvannia (1993)

[241] N. Q. Minh, J. Am. Ceram. Soc. **78** (3) (1993) 563

[242] N. Q. Minh, T. R. Armstrong, J. R. Exopa, J. V. Guiheen, C.R. Home and J.J. van Ackeren, in Proc.Third International Symposium on Solid Oxide Fuel Cells, Proc.-Vol . 93-4, pp. 801–808, 1993

[243] D. T. Hooie, "Status of SOFC Development in USA", Proc. of the Third International Symposium on Solid Oxide Fuel Cells, Vol.-Proc. 93-4, pp. 3–5

[244] Fuel Cell News, Vol. X, Nos. 4, Winter 1994

[245] H. Tagawa, "Status of SOFC Development in Japan", Proc. of the Third International Symposium on Solid Oxide Fuel Cells, Volume 93-4, pp. 6–15

[246] P. Zegers, "Status of SOFC Development in Europe", Proc. of the Third International Symposium on Solid Oxide Fuel Cells, Volume 93-4, pp. 16–20

[247] Fuel Cell News, Vol. XI, No. 1, Spring 1994

[248] S. P. S. Badwal, K. Foger, "Status of SOFC Development in Australia", Proc. of the Third International Symposium on Solid Oxide Fuel Cells, Vol. 93-4, pp. 21–30

[249] U. G. Bossel, Proc. of the Third International Symposium of Solid Oxide Fuel Cells, Proc.-Vol. 93-4, pp. 833–840

[250] U. G. Bossel, "Performance Potentials of SOFC Configurations", Electric Power Research Institute, EPRI, Palo Alto, California, EPRI-Report 1992

[251] P. Zapp, "Environmental Issues of Solid Oxide Fuel Cell-Production", pre-publication of PhD-work, Sept. 1994, Programmgruppe Systemforschung und Technologische Entwicklung (STE), Forschungszentrum Jülich GmbH (KFA)

[252] K. Nakagawa, H. Yoshioka, C. Duroda, M Ishida, Solide State Ionics **35** (1989) 249–55

[253] K. Sasaki, P. Bohac, L.J. Gauckler, Extended Abstract, 2nd Expert Meeting of IEA-SOFC Task, Swiss Federal Office of Energy, Bern, Switzerland (1990)

[254] R. Metkemeijer, P. Achard, Int. J. Hydrogen Energy **19** (6) (1994) 535

[255] A. J. Bard, R. Parsons, J. Jordan, "Standard Potentials in Aqueous Solution", Marcel Dekker Inc., New York, 1985

[256] S. Birkle, R. Kircher, C. Nölscher, H. Voigt, B. Ganser, B. Höhlein, Energiewirtschaftliche Tagesfragen **44** (7) (1994) 441

[257] R. Parsons, T. V. d. Noot, J. Electroanal. Chem. **257** (1988) 9

[258] Dedicated Conference on "Electric, Hybrid and Alternative Fuel Vehicles: Progress in Technology & Infrastructure", 26th International Symposium on Automotive Technology and Automation, Aachen, Germany 13th–17th September 1993

[259] M. Waidhas, et al., "Research and Development of Low Temperature Fuel Cells at Siemens", 1994 – Fuel Cell Seminar, San Diego (CA), Nov./Dec. 1994, p. 477

[260] J.-M. Leger, C. Lamy, Ber. Bunsenges. Phys. Chem. **94** (1990) 1021

[261] K. Kunimatsu, Ber. Bunsenges. Phys. Chem. **94** (1990) 1025

[262] B. Beden. C. Lamy, A. Bewick, and K. Kunimatsu, J. Electroanal. Chem. **121** (1981) 343

[263] A. Hammett, et al., Ber. Bunsenges. Phys. Chem. **94** (1990) 1014

[264] W. Vielstich, P. A. Christensen, S. A. Weeks, A. Hamnet, J. Electroanal. Chem. **242** (1988) 327

[265] P. A. Christensen, A. Hamnett, S. A. Weeks, J. Electroanal. Chem **250** (1988) 127

[266] I. Iwasita, W. Vielstich, J. Electroanal. Chem. **250** (1988) 451

[267] J. Willsau, J. Heitbaum, J. Electroanal. Chem. **161** (1984) 93

[268] S. Wilhelm, T. Iwasita, W. Vielstich, J. Electroanal. Chem. **238** (1987) 383

[269] S. Wilhelm, W. Vielstich, H. S. Buschmann, T. Iwasita, J. Electroanal. Chem. **229** (1987) 377

[270] S. Wilhelm, T. Iwasita, W. Vielstich, J. Electroanal. Chem. **238** (1987) 383

[271] T. Iwasity, W. Vielstich, E. Santos, J. Electroanal. Chem. **229** (1987) 367

[272] H. W. Buschmann, S. Wilhelm, W. Vielstich, Electrochim. Acta **31** (1986) 939

[273] J. Willsau, J. Heitbaum, J. Electroanal. Chem. **161** (1984) 93

[274] J. Willsau, J. Heitbaum, Electrochimica Acta **31** (1986) 943

[275] P. N. Ross, H. A. Gasteiger, "Methanol Electrocatalysis by Pt Alloys and Adatom-Modified Pt", Proceedings of the First International Symposium on New Materials for Fuel Cell Systems, Montreal, Quebec, Canada, July 9–13, 1995, p. 286

[276] A. Aramata, I. Toyoshima, M. Enyo, Electrochimica Acta **37** (1992) 1317

[277] F. Kadirgan, B. Beden, J.-M. Leger, C. Lamy, J. Electroanal. Chem. **142** (1982) 171

[278] A. Hamnett, et al., Ber. Bunsenges. Phys. Chem. **94** (1990) 1014

[279] S. Swathirajan, Y. M. Mikhail, J. Electrochem. Soc. **138** (5) (1991) 1321

[280] A. Hamnett, B. J. Kennedy, Electrochim. Acta **33** (1988) 1613

[281] J. B. Goodenough, A. Hamnett, B. J. Kennedy, R. Manoharahn, S. A. Weeks, J. Electroanal. Chem. **240** (1988) 133

[282] A. S. Aricò, et al., J. Power Sources **50** (1994) 295

[283] H. A. Gasteiger, N. Markovic, P. N. Ross, Jr., E. J. Cairns, "Temperature-Dependent Methanol Electro-Oxidation on Well-Characterized Pt-Ru Alloys", ECS Spring Meeting, Volume 94-1, San Francisco (CA), May 22-27, 1994, Abstract No.: 623, p. 990

[284] W. Vielstich. A. Küver, M. Krausa, A. C. Ferreira, K. Petrov, S. Srinivasan, "Proton Exchange Membrane Fuel Cells Using Gas-Fed Methanol", Batteries and Fuel Cells for Stationary and Electric Vehicle Applications, edited by: A. R. Landgrebe, Z.-I. Takehara, Proceedings Volume 93-8, The Electrochemical Society, Inc., p. 269

[285] A. Aramata, M. Masuda, J. Electrochem. Soc. **138** (7) (1991) 1949

[286] M. M. P. Janssen, J. Moolhuysen, Electrochim. Acta **21** (1986) 869

[287] M. Shibata, S. Motoo, J. Electroanal. Chem. **229** (1987) 385

[288] Jet Propulsion Laboratory, Pasadena (CA), USA, 91109

[289] G. Halpert, Subbarao Surampudi, "The JPL Direct Methanol Liquid-Feed PEM Fuel Cell", International Conference on Fuel Cells, February 23–24, 1994, Long Beach (CA), pp. 148–156

[290] G. Halpert, Subbarao Surampudi, "The JPL Direct Methanol Liquid-Feed PEM Fuel Cell", International Conference on Fuel Cells, February 23–24, 1994, Long Beach (CA), pp. 148–156

[291] D. L. Maricle, B. L. Murach, L. L. van Dine, "Direct Methanol Fuel Cell Stack Development", Proc. of the 36th Power Sources Conference, 6–9 June, 1994, p. 99

[292] D. L. Maricle, B. L. Murach, "DMFC Stack Test Results", Extended Abstracts, ECS – Volume 95-1, Spring Meeting, Reno, Nevada, May 21–26, 1995

[293] R. P. Hamless, Handbook of Batteries, Chapter 38, 2nd Edition, David Linden, ed., McGraw Hill, 1995

[294] D. M. Drazic et al., "Neutral Electrolyte Aluminium-Air Battery", in Power Sources **7**, Proc. of the 11th International Symposium held at Brighton, Sept. 1978, edited by J. Thompson; pp. 353–363

[295] R. S. M. Patnaik et al., Journal of Power Sources **50** (1994) 331

[296] Advanced Battery Technology, Volume 31 Number 4, April 1995, "Around The Industry", p. 2

[297] S. Müller, F. Holzer, Chr. Schlatter, C. Comninellis and Otto Haas, New Generation of Rechargeable Zinc-Air Batteries, Paul Scherrer Institute, Switzerland, Abstract 32, Vol. 95-2, Fall Meeting Chicago 1995, The Electrochemical Soc., Inc., Pennington, NJ, USA

[298] E. S. Buzelli, "Power and Performance Characteristics of Rechargeable Zinc-Air Batteries for Long Run-Time Portabel Power Applications", Abstract 33, Vol. 95-2, Fall Meeting Chicago 1995, The Electrochemical Soc., Inc., Pennington, NJ, USA

[299] Y. Shimizu, H. Matsuda, A. Nemoto, N. Miura and N. Yamazoe, "Bifunctional Oxygen Electrode for Rechargeable Metal-Air Batteries", Progress in Batteries and Battery Materials, Vol. 12, (1993) p. 108–114, IBA Tianjin (China) Meeting, ITE-JEC Press Inc.

[300] K. Kordesch and A. Marko, "Rechargeable Air-Depolarized Cell with Third (Charge) Electrode", Austrian Patent Nr. 176889 (1953)

[301] S. Hattori et al., "A new design for the rechargeable Zinc-Air Battery", Power Sources **4**, p. 361

[302] S. M. Chodosh, M. G. Rosansky and B. E. Jagid, Leesona Corporation. 21st Power Sources Conference, Atlantic City, 1967 pp. 103. Proceeding Copies: The Electrochemical Society, Inc., Pennington, NJ

[303] R. E. Biddick, J. Cummins, R. Rubischko, Gould, Inc., 24th Power Sources Conference, Atlantic City, 1970 pp. 92, Proceeding copies: The Electrochemical Society, Inc., Pennington, NJ

[304] S. J. Bartosh, U.S. Army Electronics Technology and Devices Lab. (ECOM), Fort Monmouth, NJ., 25th Power Sources Symp., 1972, pp. 15, Proceedings Copies: The Electrochemical Soc., Inc., Pennington, NJ

[305] J. Goldstein, B. Koretz, "Ongoing In-Vehicle Testing of the Electric Fuel Zinc-Air Battery System", The 11th International Seminar on Primary and Secondary Battery Technology and Application, February 28 – March 3, 1994, Deerfield Beach, Florida

[306] J. R. Goldstein, B. Koretz, "Tests of a Full-Sized Mechanically Rechargeable Zinc-Air Battery in an Elctric Vehicle", IEEE, AES Systems Magazine, November 1993, p. 34

[307] M. Givon, B. Koretz, "Energy Storage Benefits of Electric Fuel ™ Regeneration to the Utility Network", Fourth Internation Conference Batteries for Energy Storage, Volume II, September 27–October 1, 1993, Berlin, Germany

[308] M. A. Klochko, E. J. Casey, Journal of Power Sources **2** (1977/78) 201

[309] F. G. Will in Power Sources for Electric Vehicles, B. D. McNicol and D. A. J. Rand (eds.), Elsevier, Amsterdam/New York, 1984

[310] J. Cerný, J. Jindra, K. Micka, Journal of Power Sources **45** (1993) 267

[311] K. Vijayamohanan et al., Journal of Power Sources **32** (1990) 329

[312] R. J. Brodd, "Iron Electrode Batteries", in "Handbook of Batteries and Fuel Cells", edited by David Linden, McGraw-Hill Book Company, New York, 1984, pp. 20-1–20-22

[313] K. Vijayamohanan et al., Journal of Power Sources **34** (1991) 269

[314] B. Andersson, Lars Öjefors, "Optimization of the Iron-Air and Nickel Oxide – Iron Traction Batteries", in Power Sources 7, Proc. of the 11th International Symposium held at Brighton, Sept. 1978, edited by J. Thompson, pp. 329–343

[315] S. U. Falk, A. J. Salkind, "Alkaline Storage Batteries", Wiley, New York, 1964, pp. 89–104

[316] H. S. Lim, G. R. Zelter, Journal of Power Sources **45** (1993) 195

[317] J. J. Smithrick, "Nickel – Hydrogen Batteries", in Modern Battery Technology, Ellis Horwood, New York, 1991, ed. by C. D. S. Tuck, pp. 472–487

[318] G. L. Holleck et al., Development of Silver-Hydrogen Cells", in Power Sources 7, Proc. of the 11th International Symposium held at Brighton, Sept. 1978, ed. by J. Thompson, pp. 271–284

[319] P. Antoine, P. Fougere, "Development of Silver-Hydrogen Cells", in Power Sources 7, Proc. of the 11th International Symposium held at Brighton, Sept. 1978, ed. by J. Thompson, pp. 285–300

[320] N. Furukawa, "Development and commercialization of nickel(oxide)-metal hydride secondary batteries" Sanyo Electric Co., Ltd., Journal of Power Sources, **51**, (1994) pp. 45–59

[321] Sealed nickel(oxide) – Metal hydride batteries Chapter 33 (author David Linden) in the Handbook of Batteries, 2nd Edition, ed. by D. Linden, MacGraw-Hill, Inc., 1995

5. Fuel Cells and Their Applications in Dispersed Energy Systems (Utility Use)

5.1. Utility Interest in Fuel Cells

Profound changes are taking place in the electric utility industry in the United States, and this trend appears to be followed by companies in Europe. These changes involve fundamental aspects of utility regulations, utility decision-making and will especially affect decisions to add power generation capacity to meet increasing demand for energy in the future.

Firstly, it should be noted that utility power plants account for roughly 25% of all CO_2 and 40% of the NO_x, SO_x emissions and other airborne pollutants. Environmental awareness is increasing worldwide, and utilities in the United States are implementing plans to comply with the Clean Air Act Amendments of 1990. As a result, utilities are increasingly forced to account more fully for the trade-offs between generation of power and the associated environmental consequences. The regulatory process is institutionalizing such considerations with increasing frequency, and forces utilities to evaluate total social cost as a criterion in planning for new investments.

Secondly, the electric utility industry is the largest single absorber of capital resources in the world. Major expansion of the utility resource base in the 1970s, at a time of rapidly escalating prices, led to dramatic overruns on power plant capacity – especially with nuclear units. Because of concerns regarding the interests of the rate-payer (the consumer) and about the prudence of some of these capital investments, regulators are putting increasing pressure on utilities to consider all possible alternatives. The thrust of this pressure is to identify those options with the lowest cost structure, taking into account all technical, institutional and regulatory uncertainties.

Thirdly, new regulatory structures for utility planning in the United States require that all resources be considered equally. All costs and power generation options should be considered in a structured way with the aim of being able to provide electricity at the lowest possible cost. The potential impact of major additions to power generation capacity on the rate base and on the environment should be minimized.

Utilities responded to the energy crisis of the 1970s by accelerating investment in nuclear and coal-fired, base-load power plant capacity. By the 1980s, growing concerns regarding the environmental and economic consequences of these large, capital-intensive power plants caused a change in the utility planning, and the eventual adoption of new planning criteria.

In North America, a long tradition allowed the private sector to create and to manage the monopolistic generation/distribution capability of the utility industry and the regulatory commissions were created as a counter-balance to the economic power of the private sector utilities. These commissions have introduced new and tougher regulations. There are already strong indications that the underlying approaches of these new planning methodologies are also emerging in Europe.

Against the backdrop of these profound changes in the utility planning process, major changes in the ranking of various power generation strategies and technologies will take place. The changes involve not only the preferred size of capacity additions, but also the options for power generation technology.

Fuel cell technology is unique in that it offers the opportunity to respond to the new planning for environmental consciousness, energy efficiency and smaller power addition projects. Fuel cells have been known for over 150 years, but they have only recently become available in the commercial marketplace.

5.2. Integrated Resource Planning (IRP)

The financial focus of utility expansion in recent years has been based on planning processes that have evolved from Least Cost Utility Planning (LCP), including a variety of Demand-Side Management (DSM) programs and perhaps independent power production and cogeneration. The trend to incorporate LCP and DSM into planning for additions to the power generation base has increased in recent years, and is known as Integrated Resource Planning (IRP). The IRP process is aimed at providing electricity at the lowest possible cost, thereby minimizing the impact of additions to power generation capacity on the consumer rate base.

The Clean Air Act Amendments (CAAA), and the practice of public utility commissions of incorporating the cost of environmental compatibility into the evaluation process for new power generation capacity, have led to a new accounting system for utility planning. It requires the calculation of the total social cost for meeting energy needs by factoring emissions and siting issues into the evaluation process.

The impact of expanded use of IRP is that alternative power generation technologies such as the fuel cell might emerge as a superior choice to achieve simultaneously the energy efficiency, environmental-friendliness and economic efficiency that are all required in today's utility planning.

The precise definition of Integrated Resource Planning is still the subject of active discussion involving regulators, utilities and environmental groups. However, IRP generally uses a number of planning methodologies already in widespread use by utilities – for example, Demand Side Management (DSM) and Least Cost Planning (LCP).

'Demand Side management' (DSM) is the planning, implementation and monitoring of utility activities, which influence customer-use of electricity, to optimize production of the utility's load. Utility programs falling under the umbrella of DSM include: load management, new uses, strategic conservation, electrification, on-site power and adjustments in market share.

Utilities have traditionally made studies to identify the lowest cost options, like new generating capacity or purchases of energy, where available, to meet expected growth in customer loads. Least cost planning (LCP) evolved from this process as a method to determine the best mix of new power generating capacity to meet society's energy requirements at lower cost.

Integrated Resource Planning (IRP) started in the late 1980s as a way to integrate DSM options into an LCP approach to utility planning. The result of the marriage of DSM and LCP put supply and demand resources on an equal footing in an LCP process. In addition, some regulatory commissions have also required that social and environmental costs be incorporated into the estimates, and that the roles of independent power producers and cogeneration be considered. It is also likely that the impacts of supply and demand strategies on transmission and distribution (T&D) costs will be factored into overall integrated analyses.

The implementation of Integrated Resource Planning (IRP) gives planners the ability to seek optimum combinations of demand and supply options. The principal objective is to provide energy at maximum value to the customers and at minimum cost to society. The establishment of a formal methodology for these considerations is critical. The determination of least-cost to society now includes a range of environmental effects.

The following sub-sections provide an overview of how fuel cell technology might be viewed within the LCP and DSM analyses.

5.2.1. Least Cost Planning (LCP)

Within the context of LCP, fuel cells will typically be compared to a variety of central station "power generation options". Figure 5-1 provides a comparison of central station power economics, including those attainable from fuel cell technology in the near-term. Increasingly, the gas turbine combined-cycle technology is a leading candidate for adoption when long-term gas contracts are available, because of the ready combination of high efficiency, low emissions and modest capital costs. Environmental factors are not incorporated in this analysis.

It is not expected that fuel cell technology will be competitive in central station applications until after the year 2000. As a result, most present commercialization efforts are directed toward its use in smaller capacities in a wide range of cogeneration and distributed power applications, where the inherently modular construction, the remarkably low emissions and the low noise/vibration situation can be used to advantage.

5.2.2. Demand-Side Management (DSM)

DSM is one of the cornerstones of the overall IRP process. DSM programs, such as high efficiency lighting programs, have usually been implemented to reduce load growth and/or to shift loads from peak to off-peak hours. The net impact of DSM programs is to reduce the load growth on the utility system and, thereby, reduce the need for new generation capacity. In an increasing number of states in the U.S., utilities must now consider DSM as an alternative to capacity additions, and are allowed to profit (through the rate base) from the DSM programs which they support. Similar policies are being actively considered in other countries and are also being inserted into donor policies (e.g. in World Bank programs).

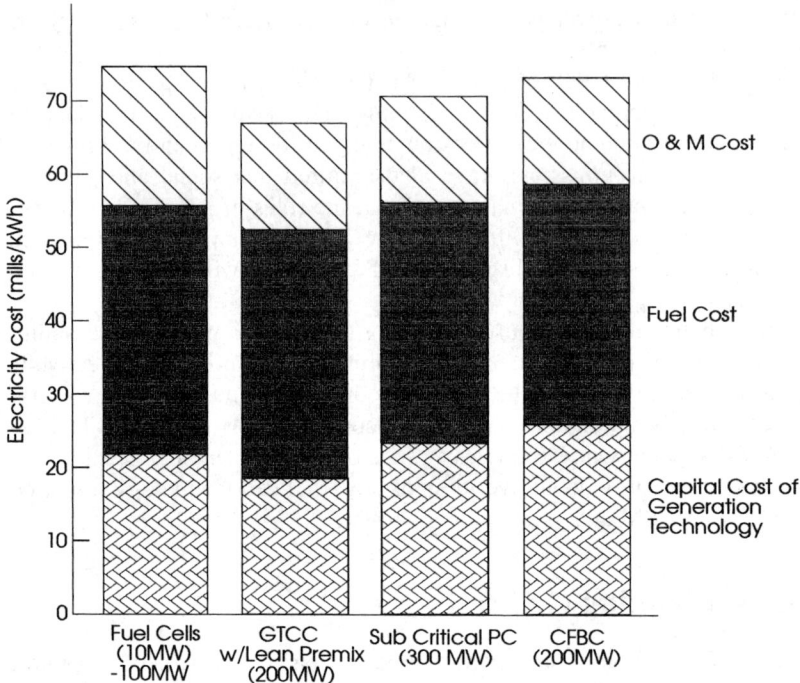

Figure 5-1. Comparison of Central Power Options. Environmental costs or other benefits for fuel cells are not included [1]

Fuel Cell technology would be particularly effective as an option within a DSM policy which supports on-site generation. It reduces utility load growth in the residential and commercial sectors (which account for over 60% of all electricity load). Two applications which show particular promise are:

- The use of fuel cell driven heat pumps, which could eliminate electric air conditioning and heating loads, and thereby reduce primary energy consumption and associated emissions
- The customer on-site generation of electricity, which would reduce load growth in general, decreasing the utilities' needs for power generation additions and T&D expansion.

In both applications, the fuel cell system would have major benefits within an IRP process by:

- Reducing the demand for utility-supplied power, during peak air conditioning (or heating) periods, which often require costly central capacity additions. In this role, the fuel cell systems would have a similar impact as the more "traditional" DSM options, such as efforts to improve the efficiency of electric air conditioners.
- Reducing the need for T&D upgrades in areas with T&D constraints.
- Reducing emissions of CO_2 by roughly 50% and over 75% for NO_x and other pollutants (referenced against the national average power generation mix).

Summarizing:

The key attributes of fuel cells in serving these functions are high efficiency over a broad range of power levels, low emissions, low noise levels and the potential for unattended operation. No other power generation technology offers this unusual combination of attributes.

5.3. Utility Phosphoric Acid Fuel Cell Plants

5.3.1. History

Electric utility interest in fuel cells dates back to the late 1960s and early 1970s. In 1971, the Edison Electric Institute (EEI), United Technologies Corporation (UTC) and a group of electric utilities began to study the benefits of fuel cells to the electric utility industry. Sufficient interest was generated to initiate the Fuel Cell Generator – 1 (FCG-1) program in 1972 between UTC and nine electric utilities. This ultimately led to the 1 MW demonstration power plant in the UTC Power Systems Division facility in South Windsor, Connecticut, in 1977.

After successful testing of the 1 MW unit, a scale-up was made to the 4.5 MW demonstration size for installation at a Consolidated Edison site at East 15th Street and the Franklin D. Roosevelt Drive in Manhattan. The original plan was that this unit should start to run before 1978, with testing to be completed by 1979. Its subsequent history is well known by now. Installation and testing were slowed down by licensing regulations and by fire-department uncertainties concerning untested aspects of the new technology, particularly in the fuel storage and processing area and for H$_2$ handling. This problem led to unforeseen pressure testing of parts of the system, including the heat exchangers, some of which were irreparably damaged by freezing of the water used for hydraulic testing during very cold weather. In consequence, installation of the stored power sections was not completed until the end of 1983, by which time the 1976 vintage technology cell stacks had "exceeded their electrolyte storage inventory lifetimes" (simply dried out). However, the plant did receive an operating license and became the only generating technology to be permitted in urban areas.

The second 4.5 MW demonstrator was ordered from UTC by the Tokyo Electric Power Company (TEPCO) in 1980. This power plant was essentially identical to the New York unit, but contained cell stacks with new ribbed-substrate technology (described in Section 4.4.). Since the cell stacks had more electrolyte inventory than the New York stacks, the shelf-life problems encountered in New York did not occur. Construction and licensing of the TEPCO unit, located at the Goi Power Plant on Tokyo Bay, proceeded close to schedule. It was modified in accord with requirements learned in New York, which included improvements to the burner and insulation of the reformer, and a better water-treatment and water cooling system. The Goi unit was started up in April 1983 and it was operated satisfactorily until December 1985.

Despite of the success at Goi, proposals to retrofit the New York unit with new stacks were abandoned as not cost-effective [2]. All these activities led to the construction of the next generation phosphoric acid fuel cell with 11 MW (AC) power for regional supply.

5.3.2. 11 MW PAFC Demonstration (TEPCO) [3–6]

5.3.2.1. Introduction

The 11 MW PAFC power plant located at the Goi Thermal Power Station Site is at present the largest fuel cell power plant in the world. The basic design of the 11 MW power plant, and the development and verification of its major components and subsystems, are the results of a joint venture by Toshiba and International Fuel Cells which began in 1985. The actual Goi power plant is a site-specific configuration, which was developed from 1986 onwards in a collaborative effort by the Tokyo Electric Power Co. (TEPCO), Toshiba and IFC. Based on the experience of TEPCO's 4.5 MW PAFC power plant, which was operated from 1980 to 1985 (construction period: '80–'83; operation period: '83–'85), many technological improvements were included in the design of the Goi 11 MW power plant. Toshiba, as the prime contractor, was responsible for the manufacture,

Figure 5-2. Goi 11 MW Fuel Cell Power Plant (Courtesy of the Tokyo Electric Power Co.)

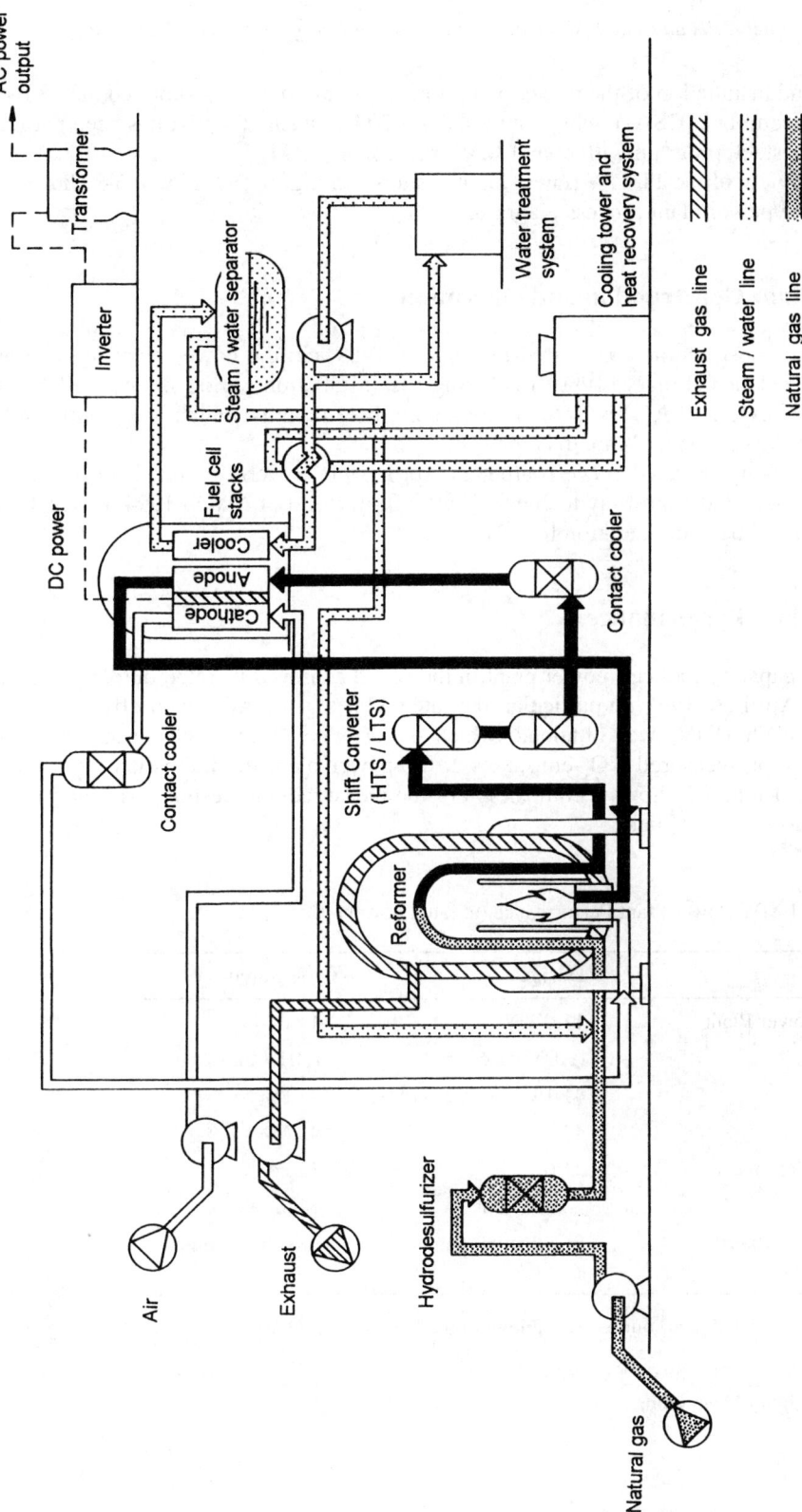

Figure 5-3. 11 MW Power Plant Flow Diagram
(HTS/LTS = High Temperature and Low Temperature Shift Converter)

assembly and installation of the power plant. International Fuel Cells supplied the 18 fuel cell stack assemblies (CSAs), which are rated at 670 kW each. Toshiba has been participating in the stack program with their 1 MW types, since 1994.

An aerial view of the 11 MW power plant is shown in Figure 5-2. Figure 5-3 shows an outline of the power plant process diagram.

5.3.2.2. Plant Construction and Operation

The Goi site construction was started in January, 1989 and the power plant installation was completed in February, 1990. Excluding the installation period of the Cell Stack Assembly (CSA), the PAC test actually took only three months, which was significantly shorter than the 4,5 MW 12 months long test.

After achieving its rated power output in April, 1991, several partial load operation tests were conducted from May to June of 1991. By the end of March 1994, the 11 MW power plant had been in operation for 9272 hours.

5.3.2.3. Plant Performance

The largest capacity fuel cell power plant in the world achieved its rated output power of 11 MW on April 26, 1991. In particular, the rated 43.6% gross AC power efficiency (i.e. 7826 BTU/kWh HHV) was obtained, which is 0.7% better than the design value of 42.9%. Also the measured NO_x-emissions at the power plant turbine exhaust were less than 3 ppm. Table 5-1 shows a comparison of the measured and design performances of the Goi 11 MW power plant.

Table 5-1. 11 MW Power Plant Performance (at rated power)

Item	Design	Measured
AC Gross Power Plant	11.0 MW	11.0 MW
Efficiency	(HHV base)	(HHV base)
AC Net	41.1%	41.8%
AC Gross	42.9%	43.6%
Waste Heat Recovery	31.6%	32.2*
Fuel	Natural Gas	Natural Gas
NO_x at Plant Exhaust (7 % O_2 base)	less than 10 ppm	less than 3 ppm

(*) Measured rejected heat, but recoverable with the heat recovery system.

1 HHV = Higher Heat Value

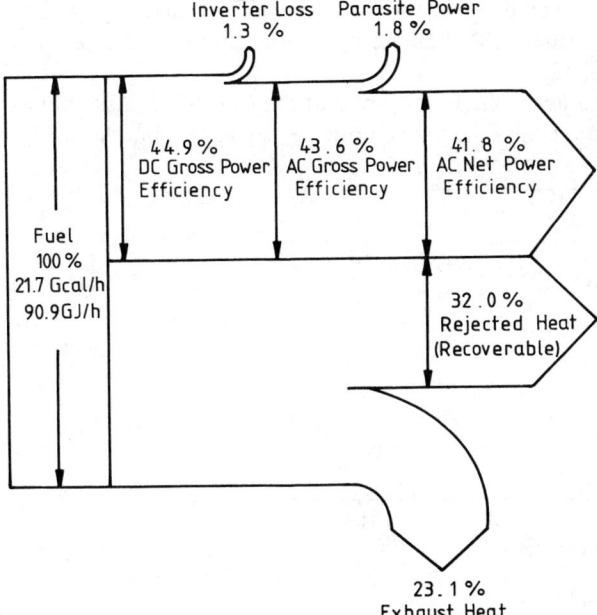

Figure 5-4. 11 MW Power Plant Energy Balance (HHV Base)

Figure 5-4 shows a measured energy balance at the rated 11 MW output of the power plant. These performance results measured, which are better than the design values, demonstrate the high efficiency and low emission characteristics of the 11 MW power plant.

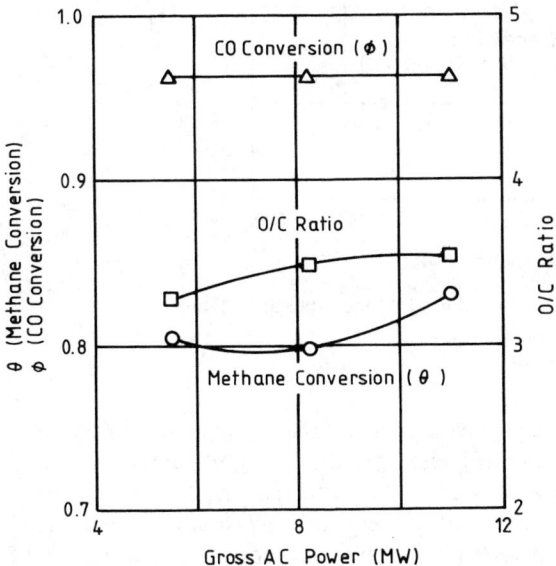

Figure 5-5. Fuel Processing Subsystem Performance

Partial load performance of the power plant is shown in Figure 5-5 and Figure 5-6. The power plant showed good performance and stable operation, throughout rated and partial power-operating conditions.

As is shown in Figure 5-7, the average voltage drop rate of the Cell Stack Assemblies (CSA) over the power generation time is less than 10% projected to 40 000 hours.

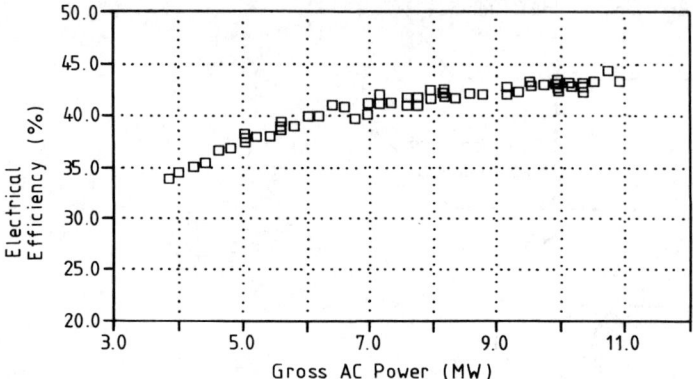

Figure 5-6. 11 MW Power Plant Efficiency

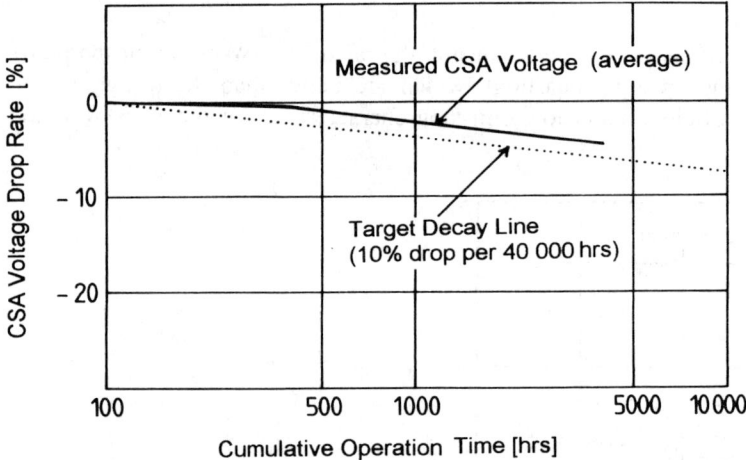

Figure 5-7. CSA ("Cell Stack Assembly") Voltage Trend Along Operation Time

Summarizing:

The GOI 11 MW fuel cell power plant was developed successfully and attained its rated power on schedule. The Process and Control test period was significantly shortened. High plant performance was obtained as designed. In particular, the highest electrical efficiency of a PAFC power plant was achieved. A very low emission characteristic of the power plant was verified and stable operation under automatic conditions was realized.

The demonstration test program will be continued to verify the total operation of the plant including waste-heat recovery-system operation. The final goal of this program is to establish the feasibility of the PAFC power plant as a dispersed power and heat combined-supply system to be utilized in urban and suburban areas.

5.3.3. 5 MW PAFC Demonstration (FE)

5.3.3.1. General

The basic technology for development of dispersed fuel cell power plants by Fuji Electric was completed with the successful trial operation from 1987 to 1988 of two 1 MW power plants planned by the New Energy and Industrial Technology Development Organization (NEDO) in the Moonlight Project which started in 1981.

Fundamental research and the investigation of development problems were conducted in co-operation with the Kansai Electric Power Co., Inc. (KEPCO). An order for design of a 5 MW class trial plant to be run by NEDO and the Technology Research Association for PAFC Power Generation Systems was received, and design was initiated.

5.3.3.2. Basic Specifications and Configuration of Plant

The basic specification of the 5 MW fuel cell power plant are shown in Table 5-2. Its system flow diagram is shown in Figure 5-8.

Table 5-2. Basic Specification of 5 MW Fuel Cell Power Plant

Rated output	5 MW AC
Net efficiency	42.2% HHV basis
Type	Phosphoric acid fuel cell
Fuel cell operating temperature	200 °C
Fuel cell operating pressure	6 kg/cm^{2}G
Fuel cell coolant	Water
Output voltage	6600 V
Output range	30–100%
Emissions at rated operation: No$_x$	max. 10 ppm
Acoustic	max. 55 dB (A)
Start-up time (warm)	within 3 hours
Load change rate	20% min^{-1}
Fuel	City gas (natural gas)
Waste heat	Recover to utilize

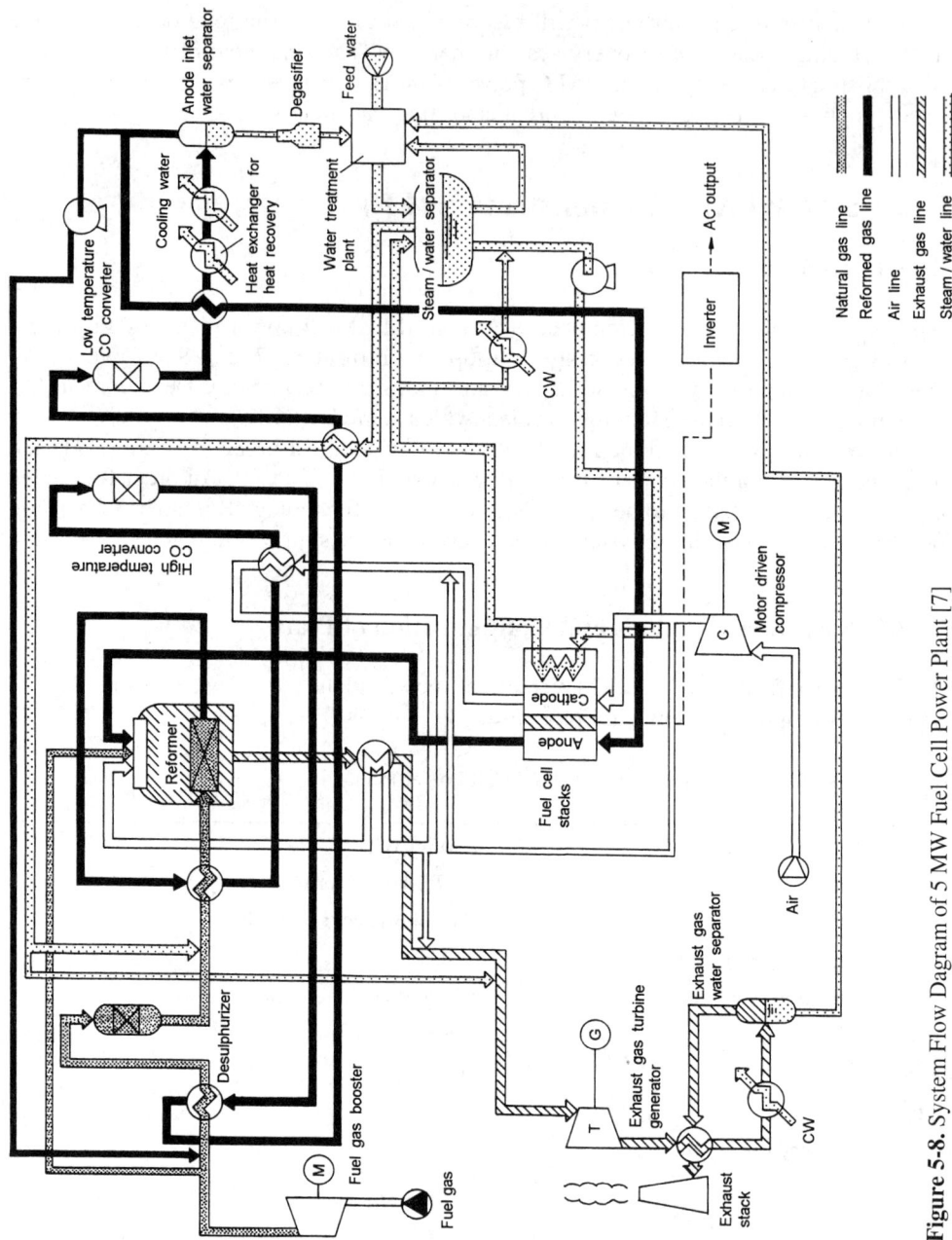

Figure 5-8. System Flow Diagram of 5 MW Fuel Cell Power Plant [7]

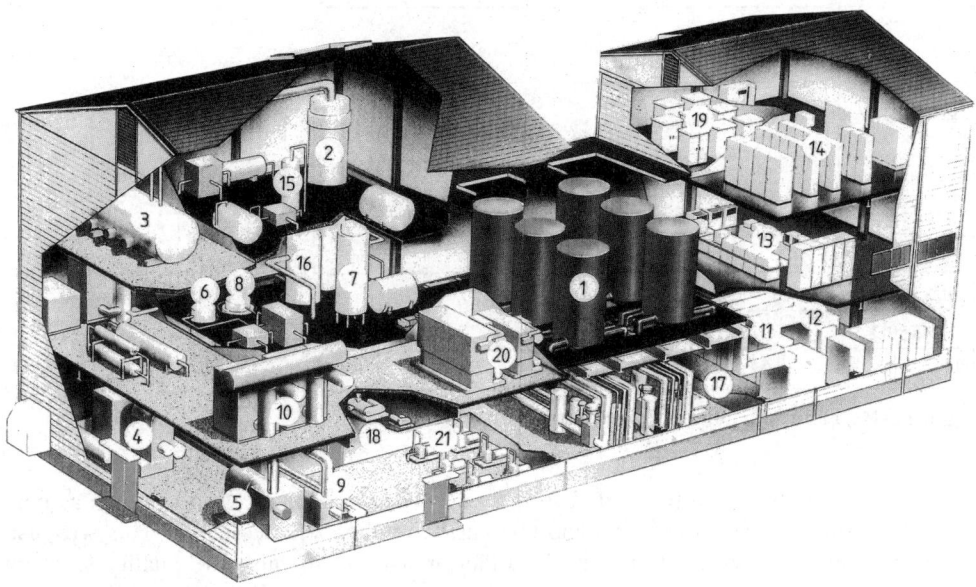

Figure 5-9. Perspective view of 5 MW plant (Courtesy of Fuji Electric Co., Ltd.)
1) Fuel Cell Stack, 2) Reformer, 3) Steam separator, 4) Air compressor, 5) Exhaust gas turbine generator, 6) Desulfurizer, 7) High temperature CO converter, 8) Low temperature CO converter, 9) Fuel cell coolant pump, 10) Start-up boiler, 11) Inverter, 12) Output transformer, 13) Control room, 14) Motor control center, 15) Anode inlet water separator, 16) Air preheater (for reformer burner), 17) Cathode recycle blower, 18) Reformed gas recycle blower, 19) Gas analyzer, 20) Refrigerator for heat recovery, 21) Pumps for heat recovery

The rated output is 5 MW, and the output range 30 to 100%. Net efficiency is 42.2% on an HHV basis. A motor-driven compressor is used for air pressurization. Operation and control of the entire system is stabilized, the start-up time is shortened and its load-following ability is improved by using a system which recovers power with an exhaust gas turbine.

The perspective view of the 5 MW plant is shown in Figure 5-9. The basic policy of the dispersed fuel cell power plant installation plan was the incorporation with existing equipment in urban areas. Therefore, efforts were made to make the plant compact enough to be installed at former substation sites and in the limited area and space under buildings. This aspect is very important for the siting problems in Japanese cities.

Further aspects are exhaust, waste water, noise, vibration, appearance and other environmental conditions. Good compatibility with the surrounding environment is necessary. Fortunately, a fuel cell power plant discharges few atmospheric pollutants and is compatible with the environment. Here, noise and appearance were taken into account and an indoor type was required.

The time schedule for construction of the 5 MW power plant is shown in Figure 5-10.

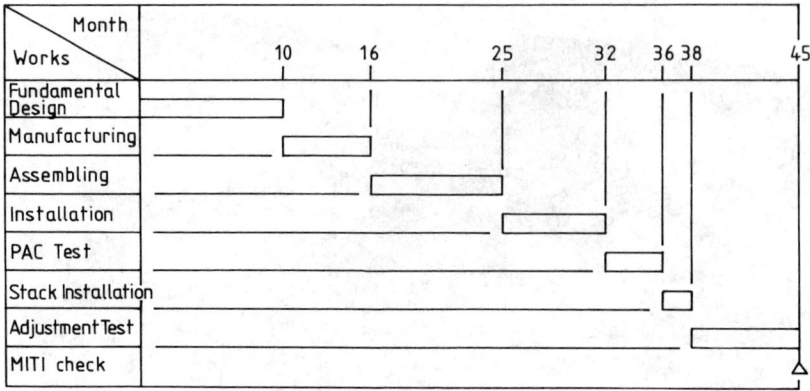

Figure 5-10. Time Schedule for Power Plant Construction

The planning shortened the period from design to pre-use inspection as much as possible. In particular, the installation and test period on-site was decreased. This was also done to shorten the installation time in urban areas, and to raise the quality level by manufacturing and testing the equipment in the factory where facilities are available.

5.3.3.3. Main Components

Fuel cell

The development of a large high-performance cell permits the generation of 5 MW of electric power with six stacks. From the standpoint of equal distribution of gas, the stacks are divided into two trains of three stacks and connected in a 3 in-series, 2 in-parallel configuration. The development target specifications of the fuel cell stacks are shown in Table 5-3. The configuration of a fuel cell stack is shown in Figure 5-11.

Table 5-3. Specification of One of the Six Stacks (480 Cells) of the 5 MW Plant

Output per stack	859 kW
Cell voltage	0.746 V
Current density	300 mA/cm^2
Operating pressure	6 kg/cm^2 G
Mean operating temperature	200 °C
Area of electrode	8000 cm^2
Configuration	Ribbed substrate
Coolant	Water

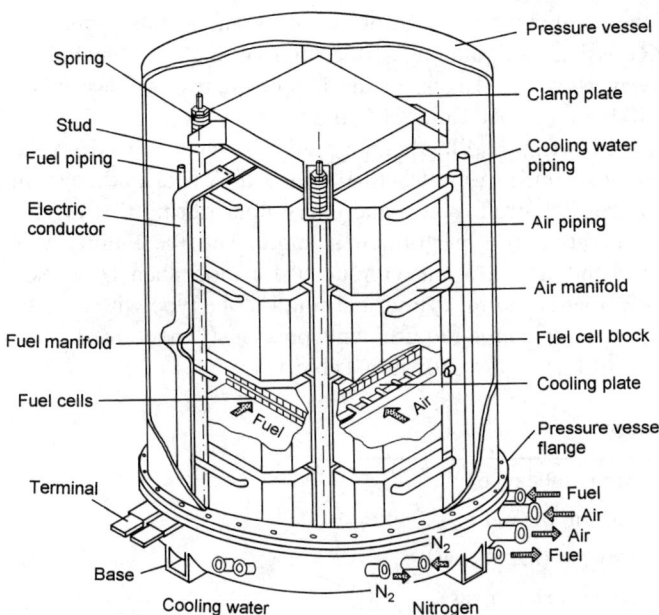

Figure 5-11. Configuration of Fuel Cell Stack

Reformer

The performance and functions required of a fuel reformer for a fuel cell are discussed below.

Compactness. For dispersed installation, site conditions require compactness of a power plant. The shape and dimensions of the reformer must also be chosen accordingly.

Load change speed. A reformer for the chemical industry is operated at almost constant load, but the electric output variations of a fuel cell are large. Therefore, the reformer must be able to change its hydrogen production rate to follow this change.

Stable combustion and efficiency improvement. The burner must be capable of lean burning by pressurizing the fuel cell anode and cathode exhaust gases. Minimization of the amount of heat applied from the outside for reformation use in the reformer itself also contributes directly to the thermal efficiency of the plant.

Low NO_x. With fuel cell exhaust gas, 10 to 20 ppm can be achieved. This is worthy of mentioning as a measure of the environmental-friedliness of the fuel cell power plant.

The following three main reformer types were trial-operated and test reports were issued. Pressurized reformers, suitable for dispersed fuel cell power plants, are of the following designs:

(1) Mono-tube type
(2) Multi-tube type
(3) Fluidized-bed combustion type

As yet there is no record of the combination of a reformer of the mono-tube type with a fuel cell. However, with EPRI assistance, Haldor Topsoe International A/S has built a 1.25 MW reformer test facility in Houston, Texas, in the United States and has ended trial operation after more than 4000 hours (See figure 8.5 in Section 8.3).

According to the test report, the start-up and load change time is 3 h 15 min from the cold condition to 25% load, and 42 seconds from 25% to 100% load. A heat cycle test, in which 100% and 25% load changes occurred, was made every hour during the day for one week. 25% constant load operation was performed at night, and the facility was completely stopped over the weekend. The test was conducted for more than 14 weeks. Concerning compactness, the reformer is a slim, cylindrical stand-alone type with a small footprint. The mono-tube type best satisfied the requirements previously mentioned.

The design specifications of the reformer are shown in Table 5-4.

Table 5-4. Specifications of Reformer

Type	Mono-tube type
Hydrogen production rate	3652 m^3/h (STP)
Fuel	City gas (natural gas)
Fuel for burner	Anode exhaust gas

5.4. Utility Molten Carbonate Fuel Cell (MCFC) Plants

5.4.1. Introduction

The original incentives for starting development of MCFCs derive from their ability to operate (at least indirectly) on coal. First generation PAFCs are usually considered to be more appropriate for higher quality fuels such as reformed natural gas, naphtha or methanol, since it was known that PAFCs could only consume hydrogen. Carbon monoxide is a poison for platinum anode catalysts in the PAFC at concentrations that depend on the operating temperature. Under normal utility fuel cell conditions, CO must be removed to a level of less than about 1–1.5% by using a water-gas shift converter (see Chapter 8). This extra fuel processing step results in a net cost and efficiency penalty.

At the high operating temperatures (~650°C) of molten carbonate fuel cells, anode kinetics are rapid and do not require noble metal catalysts, which are easily poisoned at lower temperatures. Furthermore, carbon monoxide in coal-derived fuel gas is readily converted to hydrogen and carbon dioxide under the enforced conditions of the MCFC, where one of the reaction products (hydrogen) is electrochemically consumed in situ, and one of its reaction products (water vapour) serves to drive the shift process. In principle, this permits the use of a wide range of fuels such as coal, distillates, and residuals in a suitable fuel processor and, in particular, it allows the integration of the fuel cell with a coal gasifier.

5.4.2. System Description
of Energy Research Corporation's 2-MW MCFC [8, 9]

5.4.2.1. Introduction

The Notice of Market Opportunity for Fuel Cells (NOMO) was released in Oct. 1988 by the American Public Power Association (APPA). Its goal was to identify a manufacturer for commercializing a multi-megawatt fuel cell power plant with attractive cost and perform-ance characteristics, supported by a realistic, yet aggressive commercialization plan leading to mid-1990s applications. Energy Research Corporation's program to commercialize its 2 MW internal reforming molten carbonate fuel cell was selected. The program was refined in the development of the principles and framework for commercializing Direct Fuel Cell Power Plants, which defines buyer responsibilities for promotion and coordination of in-formation development, for meeting certain milestones, for sharing the results of success in a royalty agreement and risk management features. Twenty-three electric and gas utilities in the US and Canada have joined the Fuel Cell Commercialization Group to support the buyers' obligations in this program. The City of Santa Clara (CA), the Electric Power Re-search Institute, Los Angeles Department of Water and Power, Southern California Gas Company, Southern California Edison, National Rural Electric Cooperative Association and Pacific Gas & Electric have formed the Santa Clara Demonstration Group to build the first 2 MW power plant. The preliminary design for this demonstrator is nearly complete. Integrated testing of a 20 kW stack with complete balance-of-plant has been successfully accomplished by Pacific Gas & Electric at its test facility in San Ramon, CA [10].

5.4.2.2. General Plant Overview

The fuel cell power plant will be a nominal 2 MW gross (1.8 MW net) facility using a molten carbonate fuel cell operating with natural gas. The plant will serve as a demon-stration site for molten carbonate fuel cell power plant technology, proving performance in terms of efficiency, power quality, operability, reliability and environmental impact. The plant will be situated in the City of Santa Clara. The system's flow diagram is shown in Figure 5-12.

The fuel cell power plant is based on the Energy Research Corporation's (ERC) sys-tems concept, which is as follows. Boiler feed water is raised to steam and mixed with desulfurized natural gas. The natural gas/steam mixture is then superheated and intro-duced to the fuel cell power block. The power block consists of two modules of carbon-ate fuel cell stacks, each module incorporating 10 stacks. A typical configuration will have modules connected in series to produce about 1 MW of power.

Unconverted fuel leaves the power block as anode exhaust, which is combusted in a burner to provide carbon dioxide for the cathode electrochemical reaction. The burner exhaust is cooled as needed by adding air, and enters the cathode side of the power block. The excess air in the cathode feed provides oxygen for the electrochemical pro-cess and cooling for the fuel cell stack assemblies. The waste heat in the cathode exhaust gases is used for preheating anode feed gas and raising steam. The entire system, includ-ing the fuel cell power blocks, operates at atmospheric pressure.

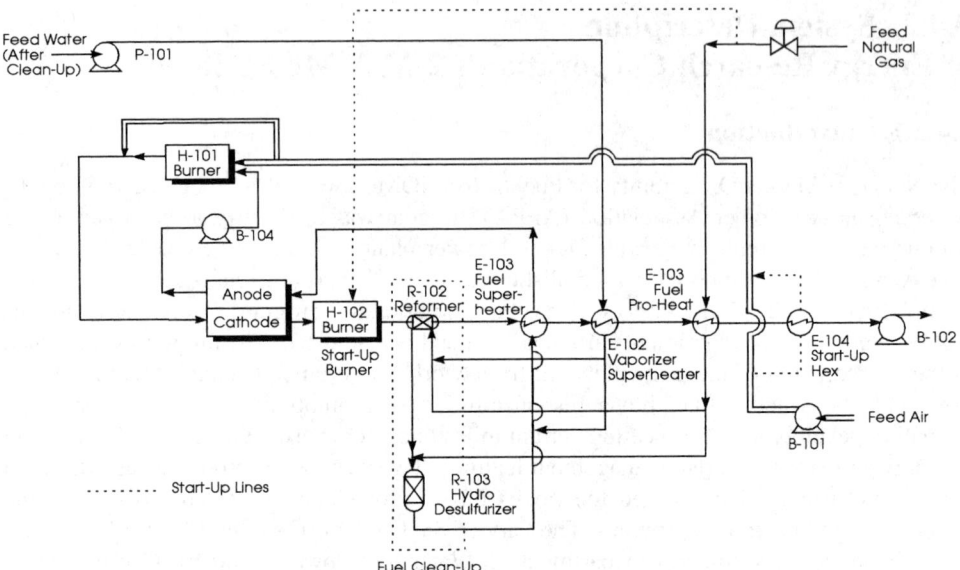

Figure 5-12. Two Megawatt Fuel Cell Power Plant System Flow Diagram [11]

The DC power block comprises a power conditioning system to convert direct current to alternating current. A transformer connected directly to the invertor boosts the AC output voltage to match the grid.

The plant control system is designed to provide completely unattended automatic operation during start-up, normal operation, standby and shutdown. The control system can accommodate operation from a remote dispatch station.

The balance of plant (under normal operation) will utilize the following:

Heat Recovery	PCU (Invertor)
Fuel Gas Supply	DC Switchgear
Oxidant Supply	Electrical Distribution
Auxiliary Fuel Supply	Digital Control
Water Treatment and Supply	Compressed Air

5.4.2.3. Main Components

1 MW Stack Module

The DC power source for the fuel cell power plant consists of two 1 MW stack module subsystems. Each 1 MW module is made up of ten 100 kW carbonate fuel cell stacks connected in a series electrical arrangement. Gas feeds and exit streams are fed to the 10 stacks in parallel.

Each 1 MW stack module is a self-contained subsystem in the plant with interfaces to the balance of plant for process streams, electrical power lines, instrumentation and control connections. The module will have a weather enclosure.

Heat Recovery Unit

The heat recovery system recovers heat energy from the fuel cell cathode side exhaust gases. Energy from the exhaust gases is utilized to:

- Preheat the natural gas supply.
- Generate superheated steam from a feedwater supply before mixing with the purified natural gas.
- Superheat the fuel supply before entering the anode side of the fuel cell.
- Preheat the fuel cell stacks during start-up operation.

The cathode exhaust gases leaving the heat recovery system are vented to the atmosphere through the exhaust blowers and exhaust stack.

The heat recovery system, with its start-up burner, provides make-up for additional heat energy to the heat exchangers whenever the temperature of the cathode exhaust gases is below a predetermined point.

The heat recovery system incorporates the reformer of the fuel gas supply system between the start-up burner and the fuel superheater.

Fuel Gas Treatment System

The fuel gas treatment system removes sulfur compounds from the natural gas feedstock and combines the sulfur-free natural gas with superheated steam. The sulfur compounds, present as odorants or impurities in the natural gas, must be removed to a concentration of less than 0.1 ppm to prevent contamination and degradation of the fuel cell internal reformer catalyst.

Oxidant Supply System

The oxidant supply system burns the anode gas and discharges the flue gas into the cathode as an oxidant. The equipment in this system is also used for start-up operation. In the start-up mode, the system burns methanol or desulfurized natural gas to supply hot gas at controlled temperatures to the cathode.

This system will operate in the flow path between the anode and the cathode. Anode and cathode requirements dictate a low pressure drop between these points, and pressure boost equipment is used to make up the pressure loss through the burner and piping. The system also includes a burner and air blowers with their associated auxiliary equipment. In addition, the system includes capabilities to discharge burner exhaust to the atmosphere to avoid violating the cathode requirements for gas compositions.

Auxiliary Fuel System

The auxiliary fuel system is the backup to the fuel gas supply system as an alternative source of fuel for the fuel cell power plant. The auxiliary fuel system is used if there is an interruption in the main fuel supply to the plant at any time during start-up, normal and shutdown operation.

The auxiliary fuel system is not intended to maintain prolonged operation after a loss of the main fuel supply. It provides up to 24 h of fuel supply to allow plant operators to locate the causes of main fuel interruption and to make a decision about whether to continue operation or to initiate an orderly shutdown of the plant.

Liquid methanol will be used as the auxiliary fuel.

Compressed Air Supply

The compressed air supply system is designed to supply clean, dry, low-temperature air for plant operations. The air will be used for operating fuel cell stack controls, purging of the fuel cells (blanketing) and operating pneumatic equipment.

Water Treatment and Supply

The water treatment system provides treated water used to generate the steam needed as a fuel cell feed gas component. The source of water is the Santa Clara municipal water supply system. The water treatment system, in conjunction with the heat recovery steam generator, provides the high quality steam needed to meet stringent fuel cell feed gas restrictions on contaminants, including sulfur, ammonia, chlorine and other dissolved solids.

Electrical Distribution

The standby generator subsystem of the electrical distribution system provides electrical power to the balance of plant systems through the critical services bus after a loss of the grid. The electrical power is used to operate equipment identified as essential for bringing the plant to a safe shutdown. After loss of the grid, the standby generator will switch on and start supplying power within 10 seconds.

Digital Controls

The control system for the fuel cell power plant will use a combination of distributed digital controls and conventional hardwired instrument and control loops. The distributed digital control and monitoring system includes microprocessor-based units located at strategic points within the station, and redundant data highways to interconnect the microprocessor units and an operator's control console located in the main control room. The DCS will include the necessary hardware, software and associated peripheral electronic equipment to monitor, control, display, alarm, record, log, and trend plant inputs to provide unattended operation and safe and orderly shutdown. Hardwire controls will be provided for electrical switchgear, protection, backup power (i.e. emergency generator) and other essential services required for safe shutdown.

Main Power

The main power system for the fuel cell power plant takes DC power generated by the fuel cell stacks, combines it in parallel and series combinations to optimize the DC voltage and current, and converts it to 12 kV AC power with the power conditioning units (PCUs) and standard electrical power system equipment. Included in this system will be all the required metering control, synchronizing systems and protective relaying requirements.

The DC outputs from the fuel cells will be protected by DC switchgear before input into the PCU.

The PCU will invert the DC to AC using proven state-of-the-art power conditioning units with transformation to 12 kV through a main step-up transformer.

Power will be delivered to the City of Santa Clara by an underground 12 kV feeder cable connecting into the City of Santa Clara's 12 kV system. A main breaker for the fuel cell output and a fused disconnect switch connection for the auxiliary power system are included in the main power 12 kV system.

5.4.3. Demonstration of MCFC Stacks and Plant System at Akagi Stack and System Square (3S)

Research and development on MCFC power generation systems have been carried out under a national program (Moonlight Project by MITI for efficient new energy systems) since 1987. Their final target is to demonstrate 1000 kW class power generation using a pilot plant. The project has been managed by the New Energy and Industrial Technology Development Organization (NEDO). The Technology Research Association for Molten Carbonate Fuel Cell Power Generation System (MCFC R.A.) was founded in 1988 as a contractor for NEDO to promote the MCFC project.

MCFC R. A. has made efforts to develop stack operation, fuel processing, heat recovery, system control technologies and equipment in cooporation with the project members. The test facilities for the technologies and equipments developed are under installation at Akagi Stack and System Square (Akagi 3S).

Table 5-5. Verification Test Schedule at Akagi 3S [13]

Test	1991	1992	1993
System and Control Test	Test		
100 kW class Stack Operation Test	Design, Manufacturing	Const.	Test
1000 kW class Fuel Processing Subsystem Test	Design, Manufacturing	Const.	Test
1000 kW class Fuel Processing Subsystem Test	Design, Manufacturing		Const.	Test
1000 kW class Heat Recovery Subsystem Test	Design, Manufacturing		Const.	Test

The demonstration program at Akagi 3S is summarized in Table 5-5. The system and control test was planned to obtain operation at characteristics data to construct and design the control system for the 1000 kW pilot plant. The test was finished in mid-July 1993. Two types of stack, multiple and rectangular, were used for the test. 250 kW stacks were fabricated according to the data obtained in the operation tests. Two types of reformer were also tested in the fuel processing subsystem test on a 1000 kW scale. One of them has been tested at a 100 kW scale at a domestic test stand [12].

5.4.4. Development Status of MCFC at the Central Research Institute of Electric Power Industry (CRIEPI)

In Japan, most of MCFC R&D is conducted under the Moonlight Project of MITI. The R&D under this project aims at the operation of a 1 MW pilot plant in 1997. This plant will be able to prove the consistency of MCFC power generation systems, but not be able to demonstrate their high performance. It will thus be necessary to operate demonstration plants before commercialization.

CRIEPI has been conducting a total system study of MCFC power generation system since 1987, and the investigations for a demonstration program have been running since 1991. LNG fueled, external reforming type MCFC power plants were investigated, as a first step.

Before commercialization of MCFC power plants, it will be necessary to demonstrate their high efficiency and to obtain an estimate of the provisional costs. A 1 MW pilot plant is too small for these purposes. Initially, in order to investigate what kind of plants are required and what items are to be demonstrated, a questionnaire was sent out to utilities and MCFC stack makers. According to their opinions, the introductory forms of MCFC power plants to be considered are as shown in table 5-6. The 100 MW system is a centralized power station proposed as a future substitute for current thermal power plants. The dispersed system will assume the role of an urban energy center and will be constructed in the re-built cities. One of the main objectives of this system is to maximize the total thermal efficiency, while keeping the electrical efficiency above 50%.

Table 5-6. Introductory Form of MCFC Power Plant

Type	Site	Operation	Fuel Supply	Cogeneration	Capacity
centralized	bay side	base/middle	LNG base	no	hundreds MW
dispersed	re-building	city gas	city gas	yes	tens MW

On the basis of the views of utilities and manufacturers, the basic specifications for two demonstration plants were decided upon as shown in Table 5-7. For a centralized power station, total plant capacity may be several hundred MW. The capacity of one discrete power zone of the plant for a centralized station must be at least 50 MW. Demonstrating this 50 MW system would assure fundings for the several hundred MW plant consisting of 50 MW systems. For dispersed power plants, the capacity has two limitations. To accomplish an efficiency of more than 50%, the capacity must be more than 20 MW. (This value was estimated from past studies on MCFC power plants.) On the other hand, the capacity is restricted by the heat demands of the area. Therefore, the capacity of the plant was chosen to be in the 20 MW class.

Table 5-7. Basic Specification of the Demonstration Plants

Characteristics	Importance placed on electrical efficiency	Importance placed on total thermal efficiency
Intention	Demonstration of Centralized power station	Demonstration of Dispersed power station
Capacity	50 MW	20 MW class
Net efficiency	about 52%	about 50%
Lines	1	1
Cogeneration	No	Yes
Site area	equal to existing power plants	no larger than 0.25 m^2/kW
Load response	10% min^{-1}	10% min^{-1}
Minimum load	less than 1/4 of full load	less than 1/4 of full load
Total efficiency	–	about 70%
Remarks	Combinations of this system result in a large capacity power plant	Design this system to supply as much low heat as possible

System flow diagrams of the two power plants are shown in Figure 5-13 and Figure 5-14. System A was designed to obtain high electrical efficiency using the simplest system. System B was designed to supply large amounts of low heat and has more heat exchangers than system A. The outlines of these two systems are almost the same.

The fuel-gas supply is pre-heated and mixed with steam, before being fed into the fuel reformer. The reformed gas is consumed at the MCFC anode, the anode exhaust gas is cooled (to separate water contained in the gas) and then utilized as the heat source for fuel reforming. The air is compressed, and part of the air is supplied to the reformer burner. Combustion exhaust gas is mixed with the remaining of the air, then utilized at the MCFC cathode as an oxidant gas. The cathode exhaust gas drives the expander turbine. Heat for steam generation and gas pre-heating is obtained from the anode exhaust and turbine exhaust gases.

Table 5-8 shows a summary of performance for each plant. The net efficiency, based on the high heat value of system A and B, is 54.4% and 50.1% respectively. The total thermal efficiency of system B is 72.6%. Although the efficiency of system B seems to be low compared to system A, this system is designed to obtain the highest total thermal efficiency and to keep electrical efficiency above 50%.

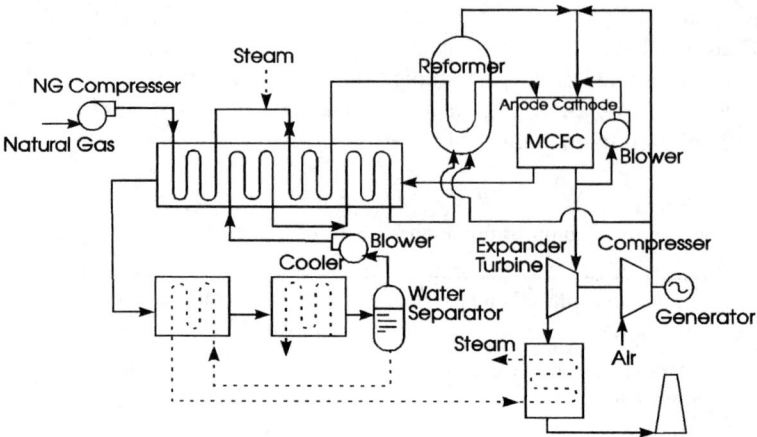

Figure 5-13. System Flow Diagram of System A [14]

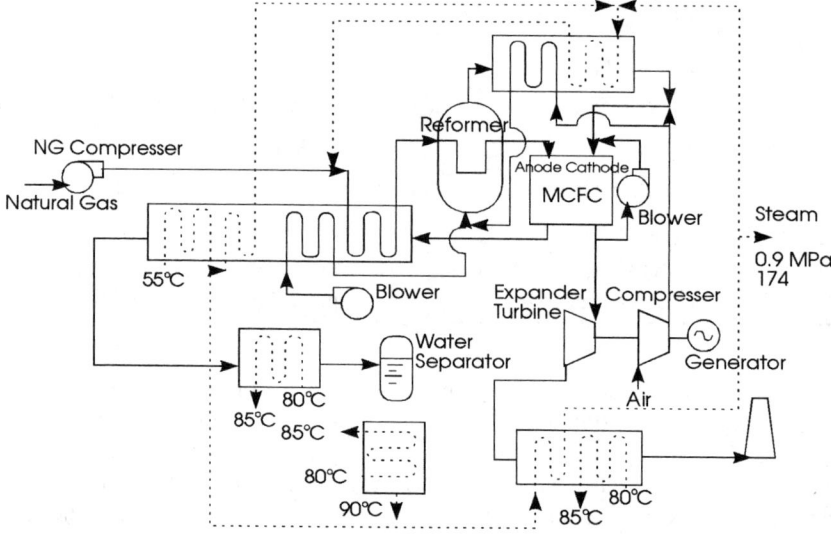

Figure 5-14. System Flow of System B [15]

The breakdown of the gross power output shows that about 11% of total power output are from the expander turbine. This is the advantage of high-temperature fuel cell plants. Without a bottoming cycle power generation, the efficiency of these plants would be about 45%. This value is the same as that of a conventional steam and gas-turbine combined system. It is therefore very important to make the best use of the waste heat from MCFC power plants.

The co-generation performance of system B is also listed in Table 5-8. Compared to conventional plants, this plant can supply very large quantities of low-temperature heat. The reason is that this type of plant can generate high-temperature heat and can be used in connection with a dual effect absorption-refrigerating machine.

Table 5-8. Plant performance summary

	Unit	System A	System B	Remarks
Fuel		LNG	town gas 13A	
Gross Power output	MW	51.3	26.0	
MCFC	MW	45.6	23.0	AC gross
Turbine	MW	5.7	3.0	generator side
Auxiliary power	MW	1.3	1.4	
Net power output	MW	50.0	24.6	
Gross electrical efficiency	%	56.0	52.9	HHV base
Net electrical efficiency	%	54.4	50.1	HHV base
Cogeneration heat	%	–	22.5	HHV base
Cold heat	%	–	21.1	HHV base
Warm heat	%	–	1.4	HHV base
Total efficiency	%	–	72.5	HHV base

5.5. Further Concepts Including SOFC

SOFCs are often labelled "Third Generation Fuel Cell Systems". It means that SOFC systems are being believed to enter the market after PAFC and MCFC power plants. NEDO has conducted a feasibility study of the Westinghouse SOFC technology for electric utility applications in Japan in cooperation with Westinghouse Electric Corporation and five electric utilities in Japan. EPRI was in charge of administration of this project.

In this project, conceptual designs are presented for three power plants using the Westinghouse tubular SOFC. Two of the power plants are rated at 300 MW for central station application, while the rating of the third plant is 20 MW intended for dispersed siting.

References for Chapter 5

[1] Arthur D. Little, "The New Role of Fuel Cell Technology in Utility Planning", prepared for the World Fuel Cell Council, Frankfurt (Germany), December 1992
[2] S. S. Penner, "Assessment of Research Needs for Advanced Fuel Cells by the DOE Advanced Fuel Cell Working Group", Nov. 1985, DOE, DE-AC01-84ER30060
[3] Kato, et al., "PAC Test Operation of 11 MW PAFC Power Plant at TEPCO Goi Thermal Power Station", 1990 Fuel Cell Seminar, Phoenix (1990)
[4] K. Yokota, et al., "Load operation characteristics of TEPCO 11 MW PAFC power plant", International Fuel Cell Conference (IFCC), Makuhari, Japan, p. 87 (1992)
[5] K. Shibata, "TEPCO fuel cell evaluation program", The 2nd Grove Fuel Cell Symposium, London, pp. 81–100 (1991)

[6] K. Yokota, et al., "GOI 11 MW FC plant operation interim report", 1992 – Fuel Cell Seminar, Tucson (AZ), p. 195

[7] Fuji Electric Review, Number 157 (2), Vol. 38 (1992)

[8] J. A. Serfass et al., Journal of Power Sources **37** (1992) 63

[9] J. A. Serfass et al., "Preparing for Commercialization: The Collaborative Matures"; 1992 – Fuel Cell Seminar, pp. 437–441

[10] P. H. Eichenberger, T. P. O'Shea, "Update on the World's first 2 MW carbonate fuel cell demonstration", 1992 – Fuel Cell Seminar, p. 183

[11] EPRI, "Availability Assessment of ERC's 2-MW Carbonat Fuel Cell Demonstration Power Plant", EPRI TR-101107, Project 3199-05, Final Report, Sept. 1992

[12] T. Mor et al., "Demonstration of MCFC stacks and plant system at Akagi Stack and System Square", Proceedings of the Third International Symposium on "Carbonate Fuel Cell Technology" (ECS), Vol. 93-3, p. 11

[13] N. Kajimoto, et al., "Demonstration program of MCFC power generation system at Akagi Stack and System Square", 1992 – Fuel Cell Seminar, Tucson (AZ), p. 179

[14] E. Koda, "The conceptual design of MCFC power plants – (LNG Fueled, External Reforming)", Proceedings of the Third International Symposium on "Carbonate Fuel Cell Technology" (ECS), p. 75

[15] "Development Status of MCFC at CRIEPI", Fuel Cell RD&D in Japan, August 1992, p. 39

6. Fuel Cells And Their Applications in On-Site Integrated Energy Systems and Industrial Co-generation

6.1. Introduction

On-site power plants co-generating useful heating power are especially suited to provide electric power in an environmentally-preferred fashion, contributing less to the climatic dangers of CO_2 production. Two aspects are important:

- Co-generating, on-site plants have a high overall efficiency because they produce electricity and usable heat (or steam). This combined (higher) efficiency in using the fuel materials results in the relative amount of emissions released into the air being reduced
- Natural gas has the additional advantage – beside higher efficiency of the co-generation – of being a pure fuel material. The flue gas after combustion or reformation of natural gas is practically free of sulfur dioxide, dust particles and heavy metal traces, and completely free of halogen compounds (see Figure 6-1).

Therefore, the most widely used fuel for on-site plants is natural gas. The use of other energy carriers, such as LPG, LNG, naphtha or methanol, would increase the spread of this technology. Figure 6-2 shows a 200 kW, CH_3OH plant from Fuji Electric on Toka-shiki Island.

Of the different kinds of fuel cell systems, only the phosphoric acid fuel and the molten carbonate and solid oxide systems are discussed for on-site use. Other concepts with alkaline and polymer electrolyte membrane fuel cells will only be mentioned for the sake of completeness.

6.2. Phosphoric Acid Fuel Cells for On-Site Use

6.2.1. Historical Review of Developments

The US gas industry became first interested in the concept of on-site fuel cell systems in the 1960s. This resulted in the TARGET program (Team to Advance Research for Gas Energy Transformation) over the period 1967–1976, during which 60 experimental, 12.5 kW total energy PAFC fuel cell systems were operated throughout the United States, Canada, and Japan. At that time, the equipment was short-lived and not cost-effective. Perhaps more important, institutional barriers existed, including very negative electric utility attitudes that severely hindered the introduction of natural gas-fired, on-

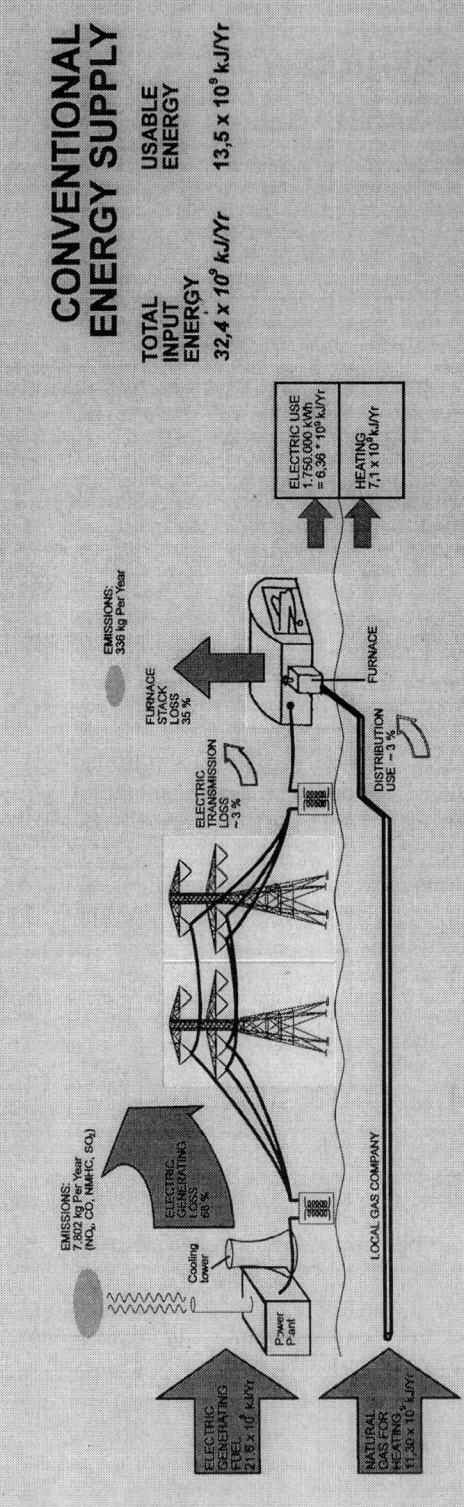

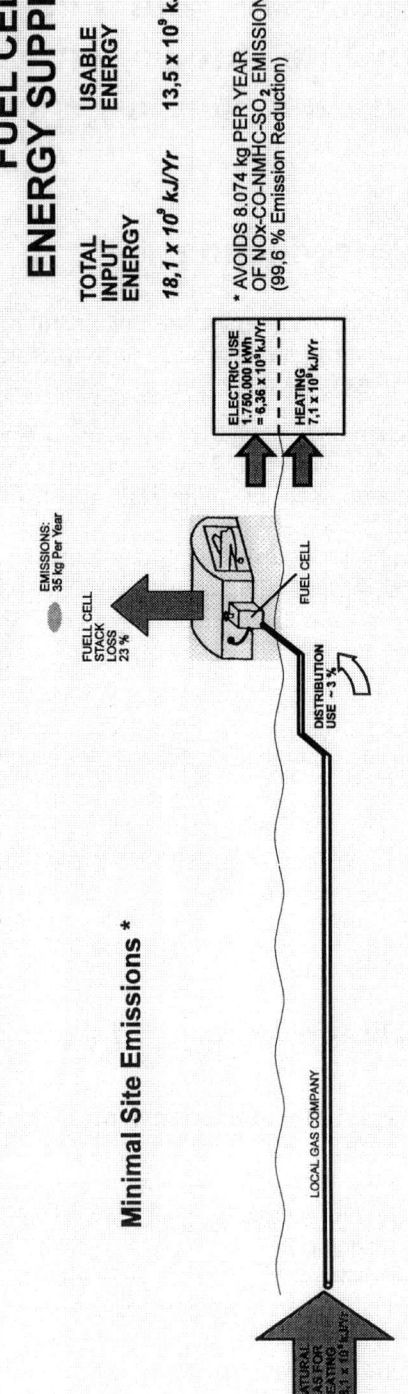

Figure 6-1. Estimate of the Total System Energy and Emissions Impact at Installation of a PAFC On-Site Power Plant at the Pittsburgh Airport [1].

Figure 6-2. The 200 kW Generating Plant for Offshore Islands (Okinawa Electric Power Co., Inc. [2] (Courtesy of Fuji Electric Co., Ltd.)

site fuel cell power plants. These factors, coupled with the energy crises of the mid 1970s and predictions of the unavailability of natural gas as a fuel, made it difficult to justify the on-site fuel cell at that time.

The outlook for the supply of natural gas is now much better, largely because of the economic incentives for exploration due to the deregulation of gas prices and the increase of BTU input per constant dollar of GNP. In the last years there has also been world-wide emphasis on higher energy end-use efficiency and lower environmental emissions, which has resulted in a shift towards dispersed power production and an increasing acceptance of co-generation by electric utilities and federal and state regulatory bodies. This shift has resulted in stricter national and international legislation.

For these reasons, about 30 major gas utilities have become determined to commercialize the on-site natural gas fuel cell in the 1980s, and have embarked on a program with the GRI (Gas Research Institute) and DOE (Department of Energy) to develop and commercialise natural gas-fuelled PAFCs. Forty-kilowatt units were initially selected as an optimum size to achieve further economics of scale compared with the 12.5 kW TARGET cell. The objective of the program was to investigate the practical potentials for use of on-site, natural gas fuel cell power plants in multi-family residential, commercial and light industrial buildings by installing and testing about fifty 40 kW units, designated PC-18 by UTC, at 20–30 selected commercial and light industrial locations in the United States, Canada and Japan [3].

As would be expected with any new technology, reliability and maintenance problems have occurred. The major problems showed up in the design and materials of the pressur

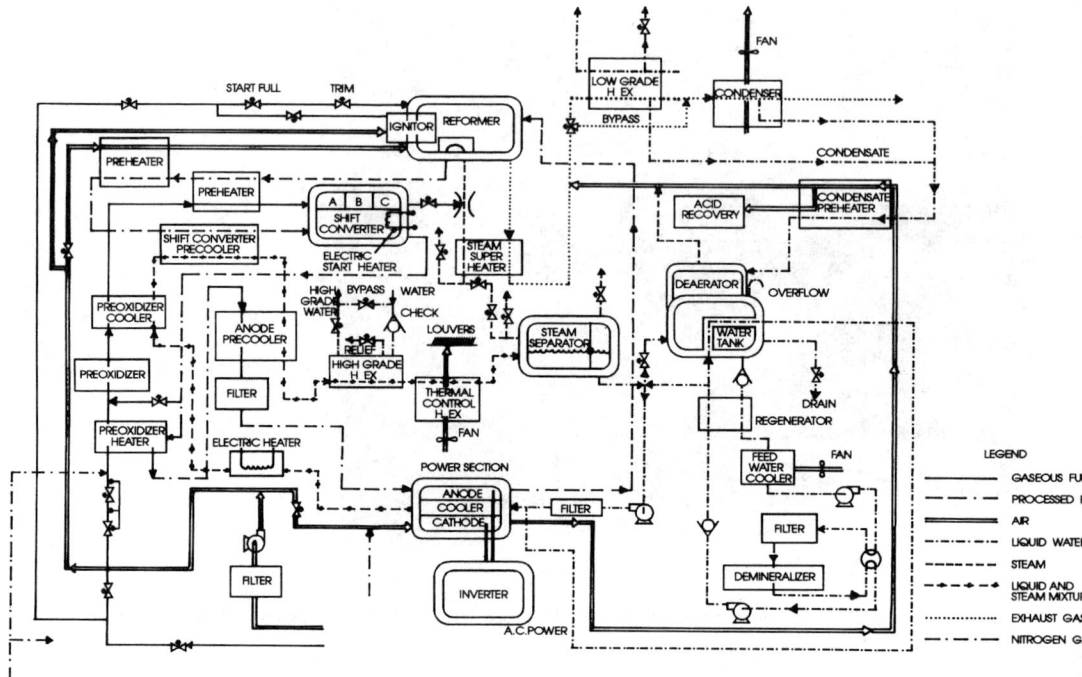

Figure 6-3. Flowsheet of UTC 40 kW system [4]

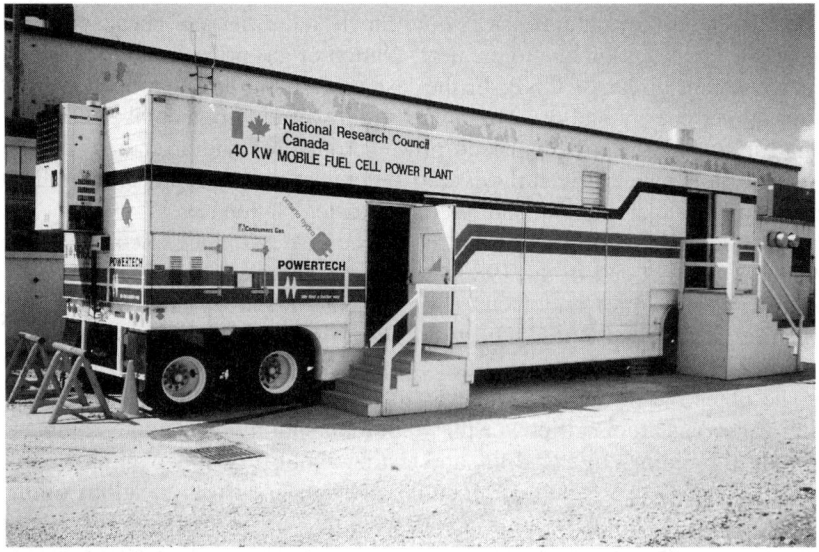

Figure 6-4. Photograph of the 40 kW System [5] (Courtesy of Ontario Hydro)

ized piping and manifolding of the fuel cell cooling plates. However, these problems have been progressively reduced or eliminated in later units. The field tests at different locations under varying conditions have been successful and the power plants met all the designated specifications. In the newer 200 kW-system program (PC-25), the objective is the continuous introduction and commercialization of phosphoric acid technology.

United Technologies (UTC) and Toshiba formed International Fuel Cells Corporation (IFC) in 1985.

6.2.2. International Fuel Cells Corp (IFC)

International Fuel Cells Corp. and its ONSI subsidiary are delivering phosphoric acid fuel cell power plants for commercial service. Initial manufacturing, installation and operating experience has been satisfactory. This experience and on-going product improvement activities provide evidence that phosphoric acid fuel cell power plants can provide economically competitive business growth opportunities for world-wide gas and electric utilities.

In January 1990, activities were initiated to establish a commercial fuel cell power plant model (a two hundred kilowatt unit designated the PC25) to meet a specification allowing this unit to be operated by any gas utility in the world. One requirement was to create a new facility with an initial capability to manufacture at a rate of 10 megawatts per year, growing to 40 megawatts per year. The first delivery was scheduled for December 1991. An initial production batch was completed by the summer of 1993 at a delivery rate of four units per month (achieved by the summer of 1992). An organisation was also to be set up to provide installation and service training to customers in the field. Figures 6-5 and 6-6 show the total fleet operating hours and the worldwide fleet availability of the PC25 from IFC/ONSI from 1992 to the first quarter of 1994.

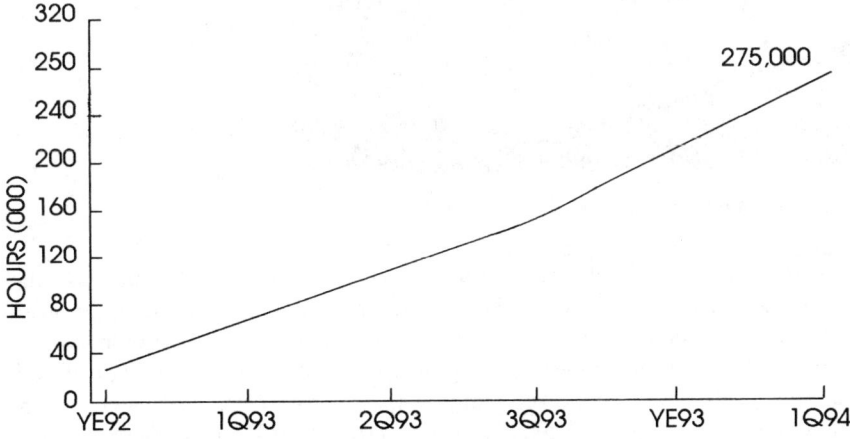

Figure 6-5. Total PC-25 Fleet Operating Hours (Cumulative) (Courtesy of IFC)

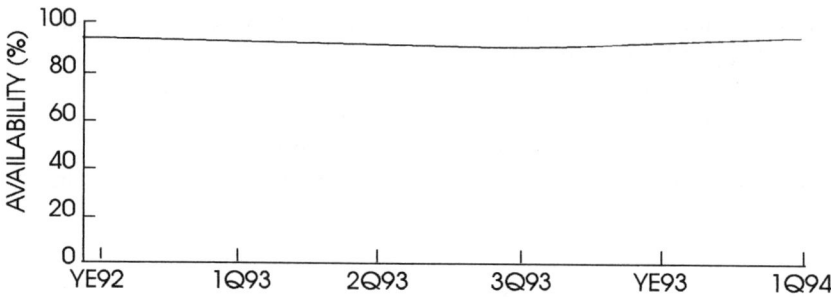

Figure 6-6. Worldwide PC-25 Fleet Availability (Cumulative Average) (Courtesy of IFC)

The PC25 Fuel Cell Co-Generation Power Plant (200 kW) [6–8]

General Description

The PC25 is a packaged, self-contained, fuel cell power plant with a continuous electrical rating of 200 kW/235 kVA. At full load, the PC25 provides 191 500 kcal/h of useful heat at temperatures between 40 °C and 80 °C.

Figure 6-7. PC25 – General Arrangement (Courtesy of IFC)

The ONSI power plant is available either as a plant powering loads independent of the utility grid, or as a grid-connected power plant operating in parallel with the utility grid. The grid-independent power plant has transient overload capability for motor starting and fault clearing and up to 6 grid-independent power plants can be paralleled to serve a common load. The grid-connected power plant includes protection features to ensure proper operation and can provide emergency power to critical loads in the event of a grid outage.

At its rated load, the PC25 consumes 54 m^3 (STP) of natural gas per hour and the electrical generation efficiency is 40 percent on a lower heat value (LHV) basis. Electrical efficiency remains at or above this value at loads between one quarter and full load. Exhaust emissions are lower than any current US Federal, CEE or local standard.

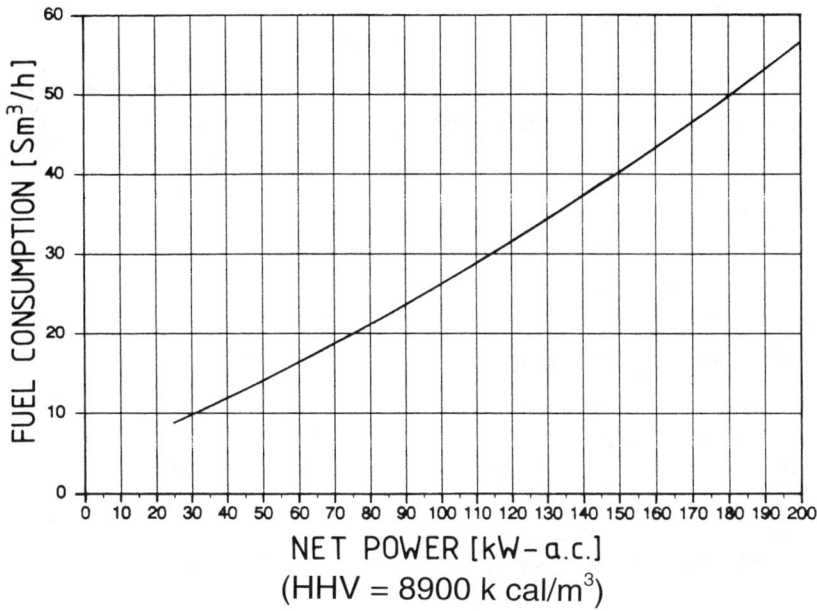

Figure 6-8. Fuel Consumption

Excess water produced by the power plant can be discharged to sanitation sewer facilities. Sound pressure level is below 60 dBA at 10 meters from the power plant.

The PC25 power plant is constructed and shipped in three modules designed for outdoor installation: a dc module that converts fuel to direct-current electricity, an electrical module that converts direct current to alternating current and a cooling module that rejects thermal energy in excess of heating/cooling requirements.

DC Module ($3 \times 7.3 \times 3.5$ m):   Fuel Processor
                                                           Cell Stack
                                                           Heat Recovery
                                                           Water Recovery

Electrical Module ($3 \times 3.3 \times 3$ m):   Inverter
                                                           Transformer
                                                           Controller
                                                           Diagnostics

Cooling Module ($1.8 \times 3.6 \times 2.2$ m)

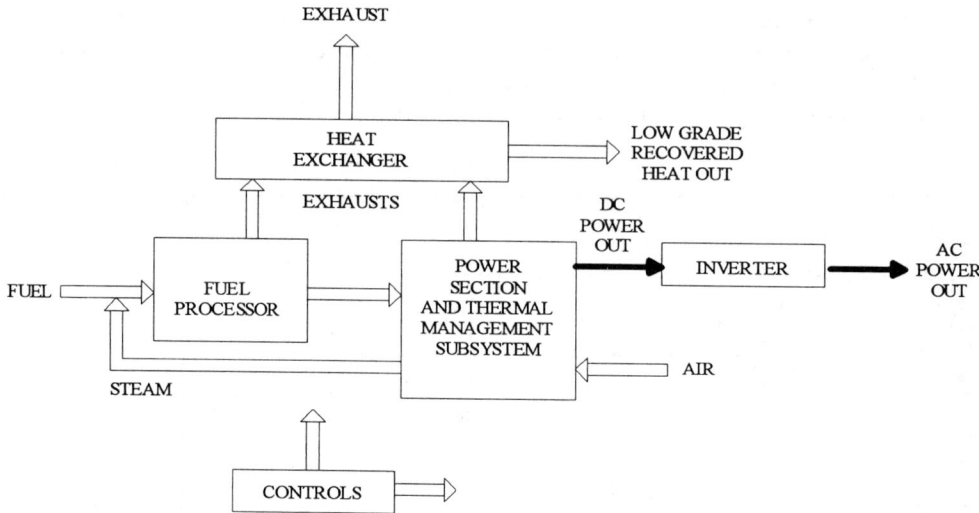

Figure 6-9. Power Plant Simplified Block Diagram

The power plant includes all components required to (1) convert natural gas into utility quality ac power, (2) provide useful heat to the customer and (3) reject excess heat to air. All ancillary power consumed in performing these functions has been deducted for the calculation of its output power. The power plant characteristics described in Table 6-1 are based on net useful power delivered.

The only additional equipment and material required at the site will be associated with the foundations for the equipment, electrical connections to the building and/or power grid, plumbing connections to the building and electrical and plumbing connections between the cooling and dc module.

Table 6-1. Characteristics of the 200 kW on-site power plant (200 kW)

Electrical Rating	200 kW/235 kVA continuous
Standard Electrical Output Selection	50 Hz 230 V – 3 phase, 3 wire 400 V – 3 phase, 4 wire 60 Hz 277/480 V – 3 phase, 4 wire 120/208 V – 3 phase, 4 wire
Electrical Efficiency	40% net (LHV) from 40% to 100% capacity
Thermal Rating	801 800 kJ/h hot water at 40 to 80°C at 100% capacity
Total Efficiency	84% at 100% capacity
Sound Pressure Level	below 60 dBA at 10 meters
Fault Current	grid independent: 475 amps line to line
	grid connected: not available
Transient Overload	grid independent: 120% rated kW/160% rated kVA
	grid connected: not available

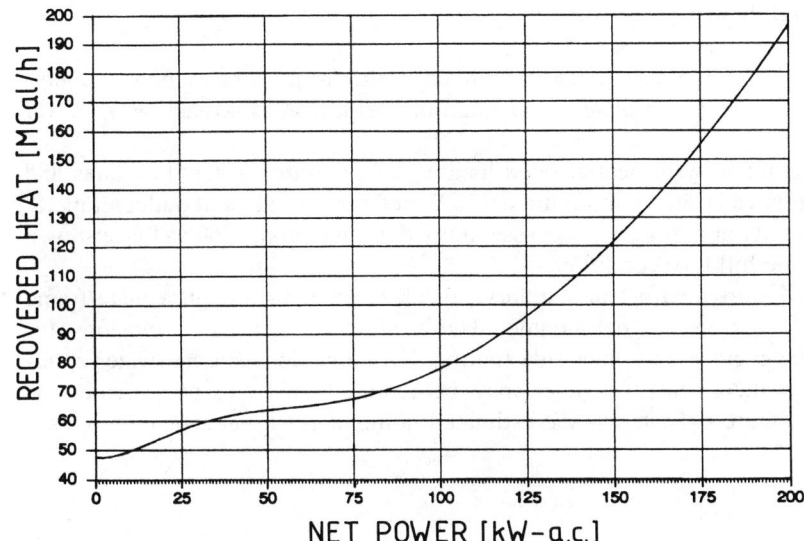

Figure 6-10. Estimated Heat Production

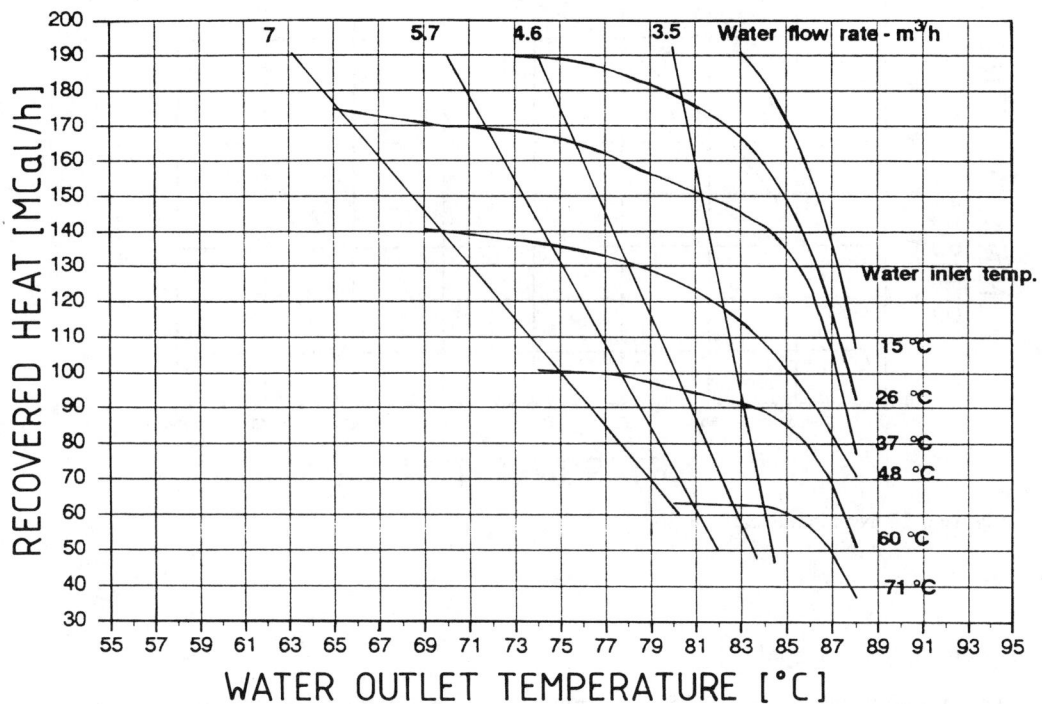

Figure 6-11. Recoverable Heat at Rated Electric Output

Useful Heat Availability

The PC25 power plant produces 801 800 kJ/h of useful heat at rated power conditions. Figure 6-10 shows the estimated production of useful heat as a function of power level.

The estimated availability of heat at rated load as a function of water flow rates and inlet/outlet temperatures is shown in Figure 6-11. The effects of inlet and outlet temperature on heat availability at part load are in essentially the same proportion as those shown in Figure 6-11 for the full load condition.

The useful heat is provided by an ancillary glycol-water loop through a plate-frame heat exchanger located in the dc module. The heat exchanger has provisions for core cleaning in the event of customer-side fouling. For some domestic hot water applications, a double-walled heat exchanger may be required, and can be provided on customer request. Figure 6-12 shows the estimated nominal electrical efficiency versus power.

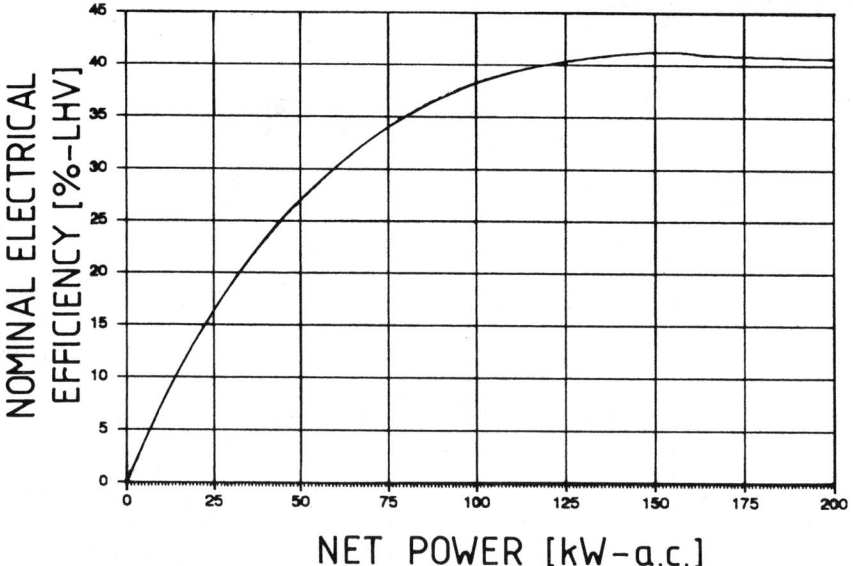

Figure 6-12. Nominal Electrical Efficiency

Environmental Capability

The emission level of the PC25 power plant is extremely low, considerably under present EEC, US or other national legislative requirements (Table 6-2).

Table 6-2. Emission Levels of PC25 Fuel Cell versus Existing Air Quality Standards

Substance	Emission level [g/kWh]	Emission level [mg/m³(STP)]	CEE Directive 609 – 24/11/88 [mg/m³(STP)]
NO_x	0.006–0.3	7–14	350
SO_x	none	none	35
CO	0.02–0.05	7–22	
Organics	0.03–0.07	12–17	
Particulates	none	none	50

6.2.3. Westinghouse/ERC

Parallel to their joint program to develop the PAFC system for electric utility use, Westinghouse Electric Corporation and the Energy Research Corporation (ERC) also have shown interest in the development of a pressurized module of four gas-cooled PAFC

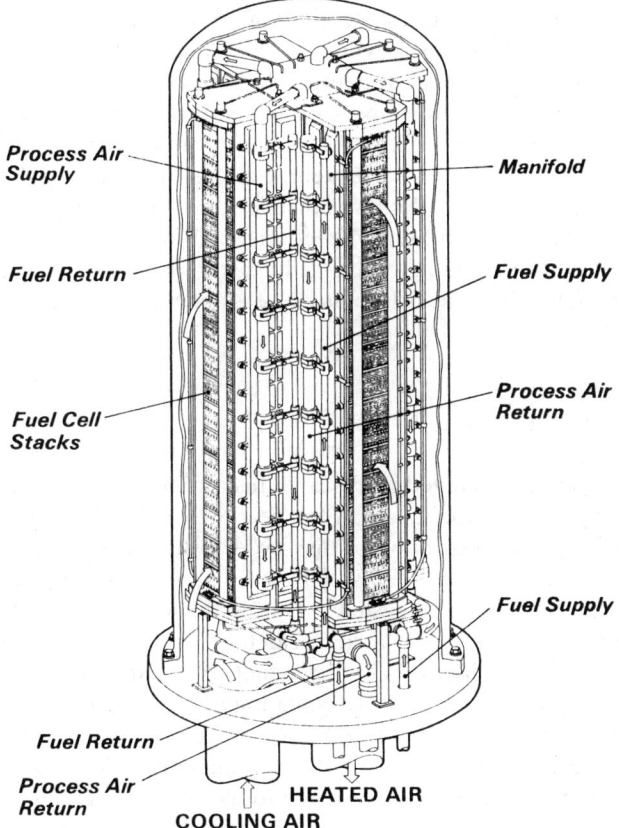

Figure 6-13. 375 kW Fuel Cell Module Design [9]

fuel cell stacks with a nominal output capacity of 400 kilowatts. ERC has indicated that its air-cooled, phosphoric acid fuel cell system might be especially attractive for use with on-site integrated energy applications, because air-cooled systems should inherently be simple, trouble-free, and inexpensive. Their coolant circuit does not introduce into the system any materials compatibility, materials joining or electrical isolation problems into the system, which can result in corrosion problems. It does not require high pressurization, which can cause a catastrophic failure if the cell stack is flooded with water.

Westinghouse was also planning standardized factory produced 3 and 13 MW power plants to be available commercially by the mid-1990s [10].

6.2.4. Engelhard On-Site Fuel Cell Program

The Engelhard Industries Division of Engelhard Minerals and Chemicals Corporation was engaged in a five-segment, DOE-funded program under contract to NASA Lewis Research Center to develop on-site integrated energy systems based on Engelhard's dielectric, liquid-cooled PAFCs.

The Engelhard cooling approach employed forced circulation of a dielectric liquid through the metal piping of the stack cooling plates, which were similar in principle to those in the UTC OS/IES (on-site/integrated energy systems) PAFC stacks. Liquid cooling is particularly attractive for OS/IES applications, because of the high quality of the heat that the cooling fluid can transfer from the stack to the heating system of the building. In addition to providing relatively high heat transfer temperatures, the use of a dielectric liquid coolant also permits the use of small heat exchangers, requires only minimal pumping power and offers good compatibility, with efficient absorption-type air conditioners. Finally, in the event of a total cooling-system failure, it may not lead to the total loss of a stack with all its peripherals by acid leaching and its attendant corrosion implications.

6.2.5. Developments in Japan

Research to develop fuel cells in Japan was initiated by research institutions beginning in 1955 and is backed by roughly 30 years of experience. In the years following 1965, the principal result of research was the commercialization of dissolved fuel-type cells using methanol and other substances as the fuel. The fuel cell should serve as a compact power source in energizing radio relay stations and other systems. During a short span of less than a decade, significant progress was achieved in related research.

The Moonlight Project is central to research efforts on these fuel cells and was implemented in 1981 by the Agency of Industrial Science and Technology (AIST) of the Ministry of International Trade and Industry (MITI). In addition, research conducted by governmental and private sector research institutions, and the introduction of American-made experimental plants, helped to promote fuel cell research to the present high level. Intensive research advanced three types of fuel cells: phosphoric acid fuel cells (PAFC),

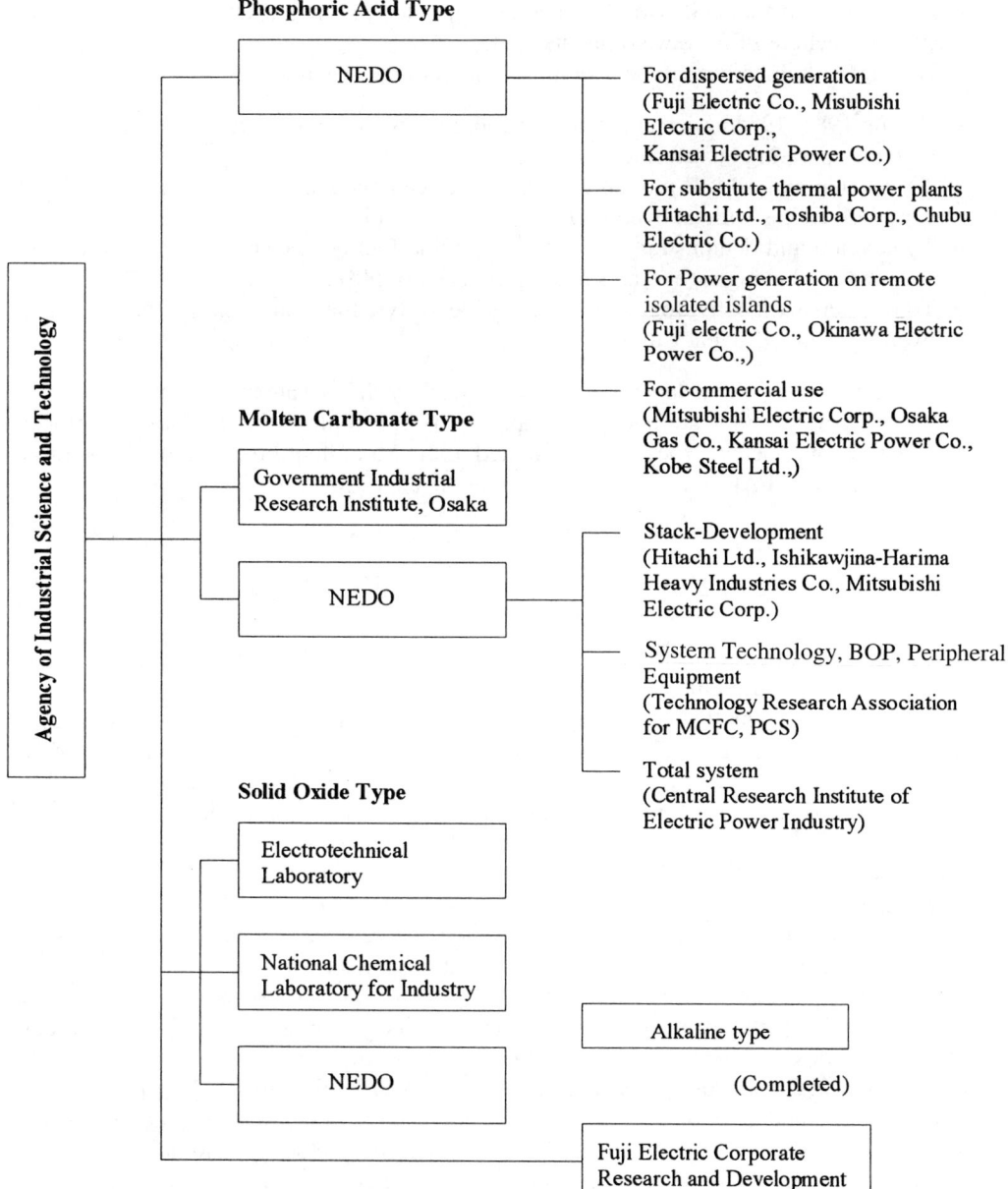

Figure 6-14. R&D Set-Up in the Moonlight Project [12]

molten carbonate fuel cells (MCFC) and solid oxide fuel cells (SOFC). Latest developments also include PEFC developments [11].

The fuel cell objectives of the Moonlight Project up to 1986 were:

- During 1984–1985 to design, construct and, in 1986, to demonstrate two 1 MW PAFC systems operating on reformed fuels.
- To develop a 5–10 kW molten carbonate fuel cell stack by 1986 that can operate on coal-derived gas with an overall efficiency of over 45%.
- To develop and demonstrate a 1 kW solid oxide fuel cell stack in 1986 operating on coal gas and having an overall system efficiency of 50%.
- To develop a 1 kW hydrogen-air, alkaline electrolyte fuel cell stack in 1984 having an overall system efficiency of 45%.

Research on PAFCs in Japan is being advanced by the private sector with the support of electric power companies and gas companies. Fuel cell plants from various developers will be presented below. They all participated in the Moonlight Project and will be offering a commercial PAFC on-site power plant in the next few years.

6.2.6. Work and Developments at Toshiba [13, 14]

Toshiba has dedicated itself to the commercialization of PAFC power plant for both large-capacity electric utility use (11 MW power plant) and small capacity on-site use.

Toshiba initiated its PAFC development in 1987 and completed a 35 kW cell stack and a 50 kW reformer in 1981. Toshiba's first power plant was constructed in 1982 to demonstrate the pressurized operation of an integrated system with a rated power of 50 kW. Based on these activities, Toshiba participated in the Moonlight Project and developed 100 kW and 260 kW cell stack in 1984 and 1986, respectively. Toshiba fabricated major components such as the fuel cell stacks, reformer and inverter for the Moonlight 1 MW power plant, which was completed in 1988.

Toshiba decided on international cooperation in order to accelerate fuel cell power plant commercialization. The first joint cooperation project was conducted from 1980 to 1985 with a 4.5 MW power plant for Tokyo Electric Power Company (TEPCO). The plant was manufactured by the United Technologies Corporation (UTC), and the construction work was done by Toshiba for the Goi Power Station.

In 1985, UTC and Toshiba established International Fuel Cells (IFC), a joint venture company. IFC and Toshiba cooperated in the engineering and development of an 11 MW PAFC plant (PC23) from 1985 to 1987. In this program, full-scale verification tests were conducted on major equipment such as a 670 kW cell stack, an 11 MW reformer and an inverter. As a result, TEPCO ordered an 11 MW plant from Toshiba in 1988. Although, IFC manufactured the cell stack for this plant in cooperation with Toshiba, most of the other parts of the system were manufactured by Toshiba.

For on-site fuel cell plants below 1000 kW, Toshiba made an investment in IFC to establish a subsidiary, ONSI, in January 1990. ONSI is now working on the commercialization of 200 kW power plant (PC-25) developed by IFC. Toshiba also has its own, in-house program for 200 kW and 50 kW class fuel cell development and commercialization in Japan, incorporating IFC/ONSI technology and experience.

The major components such as cell stack, reformer and inverter govern power-plant performance, reliability and cost competitiveness. In particular, the technology and cost level of the cell stack are the most critical items the PAFC power plant in both electric utility and on-site use. In 1990, Toshiba completed a 1 m²-class cell stack production facility.

To evaluate and verify the performance and life of a 670 kW-class cell stack (which is the same capacity as installed in the Goi 11 MW Demo Plant), the short-stack test and

Figure 6-15. 670 kW Cell Stack (Courtesy of Toshiba Corp.)

the full-stack test facility was installed in 1991. The long-life evaluation test of 670 kW class short stack was conducted with satisfactory results and a 670 kW full stack verification test was be carried out.

For on-site use, 200 kW cell stack manufacturing and testing has been started. In parallel with the present commercialization activity, the higher current density cell stack is now under development for further compactness and cost reduction. Toshiba has also manufactured the PC-25, 200 kW reformer and inverter jointly with ONSI. Furthermore, Toshiba is now making an effort to develop the low cost, more compact and low S/C (steam to carbon ratio) catalyst reformer. The characteristics of the TFC 200 are similiar to those of the PC25 (ONSI) power plant.

Table 6-3. Characteristics of the TFC200

Items	Design Specifications
Rated Ouput	200 kW
AC Output Voltage	210 V or 420 V, 3 phase
AC Output Frequency	50 Hz or 60 Hz
Electrical Efficiency	40% (LHV), 36% (HHV)
Fuel	natural gas, city gas 13A
Operating Pressure at CSA	atmospheric pressure
Load Response	instantaneous
Starting Time	1.5 hours from hot condition
Operating Mode	either grid connected or grid independent
Installation	indoor or outdoor installation
Emission	NO_x: less than 6–10 ppm, SO_x less than 0.01 ppm
Dimensions and Weight	7.3 x 3 x 3.5 (m); 27.3 tons

Figure 6-16. Toshiba's 200 kW on-site PAFC (Courtesy of Toshiba Corp.)

In addition, Toshiba is developing a 200 kW power plant with the option of obtaining steam at an efficiency of between 20 and 25% (max.). The level of temperature is about 170 °C, and total thermal energy output is constant at 40% (see Figure 6-16).

Toshiba has continued commercialization jointly with IFC/ONSI – in particular for ONSI's PC-25 200 kW initial batch of 53 units. Toshiba shared development and manufacture of the reformer and electric system assembly (consisting of inverter and controls), and their reliability has already been verified by prototype-unit operation at the ONSI factory. Toshiba is convinced that it will be possible to supply reliable, 200 kW-class on-site commercial plants by mass-production after 1994/1995.

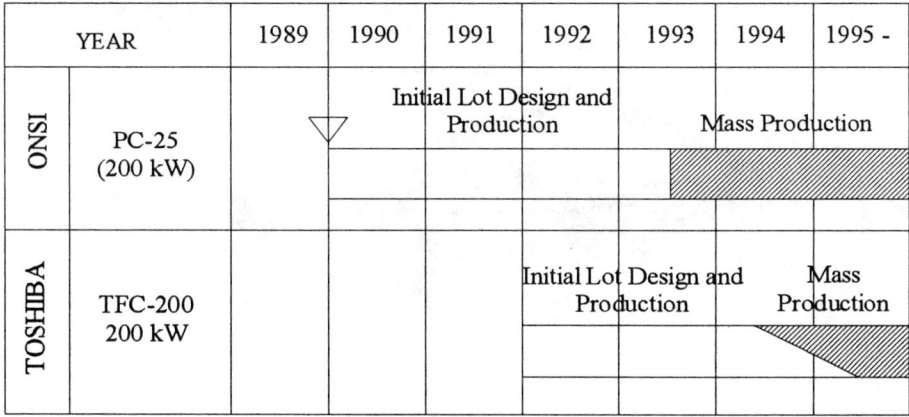

YEAR	1989	1990	1991	1992	1993	1994	1995 -
ONSI — PC-25 (200 kW)			Initial Lot Design and Production			Mass Production	
TOSHIBA — TFC-200 200 kW				Initial Lot Design and Production		Mass Production	

Figure 6-17. On-Site Fuel Cell Business Plan

Table 6-4. Basic Specifications of the 1000 kW On-Site PAFC Power Plant

	Basic Specificaiton
Rated Output	1000 kW (AC)
Electrical Efficiency [HHV]	$\geq 36\%$
Thermal Efficiency [HHV]	$\geq 35\%$ High Temperature Waste Heat: (170 °C Steam); 20–25% Low Temperature Waste Heat (65 °C Hot Water): 15–10%
Total Efficiency [HHV]	$\geq 71\%$

Fuel Cell Stack
   Operating Pressure
   Operating Temperature
   Number of Fuel Cell Stacks
	
	 Atmosphereric
	 205 °C
	 2 (Close Arrangement)

Reformer	1000 kW class, Multi-Tube Type
Performance Density	10 kW/m^2
Fuel	City Gas, 13A

Toshiba is also engaged in the development of a 1000 kW class PAFC plant (NEDO-Project) [15]. The target specifications are given in Table 6-4. The aim of this project is also typical for an on-site strategy: installation for large-scale commercial building, usage of cogeneration energy, effective waste heat utilization (steam) and compact construction of the power unit.

The Artist's concept of the 1000 kW-Class Plant is shown in Figure 6-18.

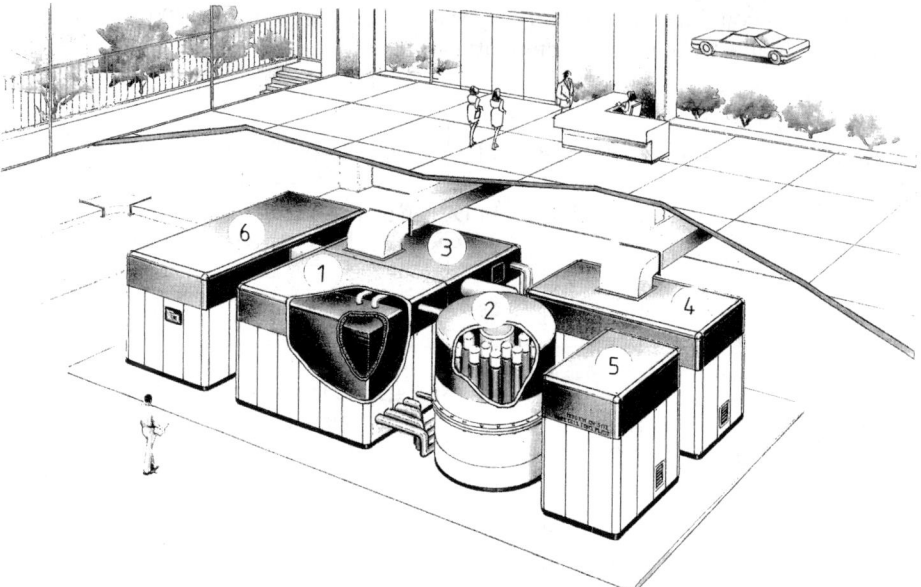

Figure 6-18. Artist's Concept of the Toshiba 1000 kW-Class Power Plant (Courtesy of Toshiba Corp.)
1) Cell Stack Assembly, 2) Reformer, 3) Fuel Processing System, 4) Thermal Management System, 5) Water Treatment System, 6) Electrical System Assembly

6.2.7. Fuji Electric

Fuji Electric commenced development of phosphoric acid fuel cells (PAFC) in 1973, and supplied 32 PAFC demonstration plants from 1982 to the end of 1991. Support from the Government (MITI), NEDO, Electric Utilities, Gas Utilities and others made this advance possible.

Key technologies, such as high performance cells and compact reformers have been developed for on-site applications. Working together with three gas utilities, Tokyo Gas, Osaka Gas and Toho Gas, reliable commercial power plants are expected to be completed by 1996.

Successful commercialization of the relatively high-cost early commercial units requires the close cooperation of government, customers (e. g. utilities) and manufacturers [16–20].

6.2.7.1. PAFC Plants Using Natural Gas (Town Gas 14A)

50 kW/100 kW On-Site PAFC Plant

The number of 50 kW and 100 kW packaged co-generation plants to be planned and demonstrated is more than 70, including four sets of 50 kW for European countries. The total capacity realizable for pre-commercial demonstration is about 5350 kW.

Development of the co-generation packaged plants was initiated and supported by the Tokyo Gas Company. Encouraged by the promising operating results with the first plant, there are currently ten such units being operated in gas and electric companies in Japan.

Figure 6-19 shows the 6 sets of 50 kW plants installed at the Rokko Testing Station, Kansai Electric Company, near Osaka. These plants are in operation under various test conditions, including several fault or transient operations (operating hours were 1500 h– 6000 h in 1993).

Figure 6-19. General view of 50 kW PAFC plants at Rokko Testing Station (near Osaka) [21] (Courtesy of the Fuel Cell Development Information Center)

Table 6-5 shows the key specifications of this 50 kW PAFC plant. The power density of 4.9 kW/m^2 is one of the highest attained in the world. Such a high power density is mainly achieved by the compact design of the fuel cell stack, reformer and inverter. The power density of the fuel cell stack alone was doubled compared to the first prototype delivered to Tokyo Gas in 1984. This was achieved by the improvement of a high performance electrode and separator, and by decreasing the number of cooling plates.

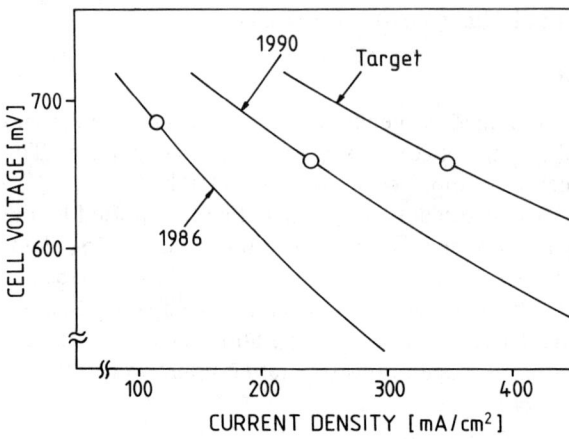

Figure 6-20. Increases in cell power density 1986–1990.

Table 6-5. Key Specifications of 50 kW PAFC On-Site Plant

Item	Specification
Purpose	on-site co-generation
Type	water-cooled, packaged type PAFC (indoor/outdoor)
Output [kW]	50 (net)
Electrical efficiency [%]	40 (LHV)
Total efficiency [%]	80 (LHV)
Heat recovery [°C]	65 outlet/ 40 inlet water
Fuel	town gas (13 A) (LNG) Inlet pressure 100 ~ 250 m/m Aq
Operation	unattended, fully-automatic operation independent and/or grid-connected
Load response [%]	± 50 (instantaneous)
Start-up time (cold) [h]	3 (with electrical heater) 1.5 (with start-up boiler)
Start-up power [kW]	30–35 (with electrical heater) 7 (with start-up boiler)
Output range [kW]	0–50
Water supply	not necessary
NOx [ppm]	< 10
Weight [ton]	5
Dimensions [m]	$2.9 \times 1.6 \times 2.2$
Footprint [m^2/kW]	0.09
Power density of plant [kW/m^2]	4.9

Performance (Normal Pressure)	Current Density [mA/cm^2]	Voltage [mV]	Power Density [mW/cm^2]	Ratio –
1986	115	683	80	1
1990	240	660	160	2
Target	350	660	230	3

Improvement of the reformer is an important factor in reducing plant size. Figure 6-21 shows a comparison of reformer size between the first model (1988 technology) and the second model (1990). The volume and weight of the second model was reduced to one-half and one-third, respectively. The comparison of key performance of these two models is shown in Table 6-6. An endurance test of this second model reformer combined with a fuel cell stack has been conducted under various operational conditions.

First model Reformer   Second model Reformer

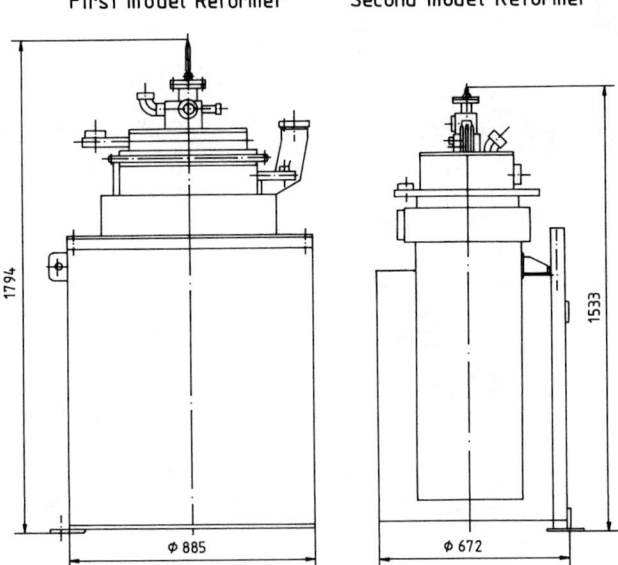

Figure 6-21. Comparison of Reformer Size (1988–1990)

Table 6-6. Comparison of Key Performance of Reformers

	First model (1988)	Second model (1990)
Fuel	town gas 13 A	town gas 13 A
Process	steam reforming	steam reforming
H$_2$ generation [m^3/h (STP)]	50	50
Reaction temperature [°C]	700	710
Working pressure	atmospheric	atmospheric
Steam/carbon ratio	3.25	2.5

Compact design of the inverter also offers a substantial reduction in plant size. The second generation inverter is using a high frequency dc-dc converter and IGBT. These modifications resulted in the volume and weight of the second generation inverter being reduced to one-third and one-fifth, compared the first model. This high-frequency inverter also contributes to a substantial noise reduction. Figure 6-22 shows a non-interrupting switch-over from independent operation to grid-connected operation, which is preferred with this inverter system.

In addition to the substantial size reduction of key components, system simplifications were realized through integral design of many other components such as heat exchangers.

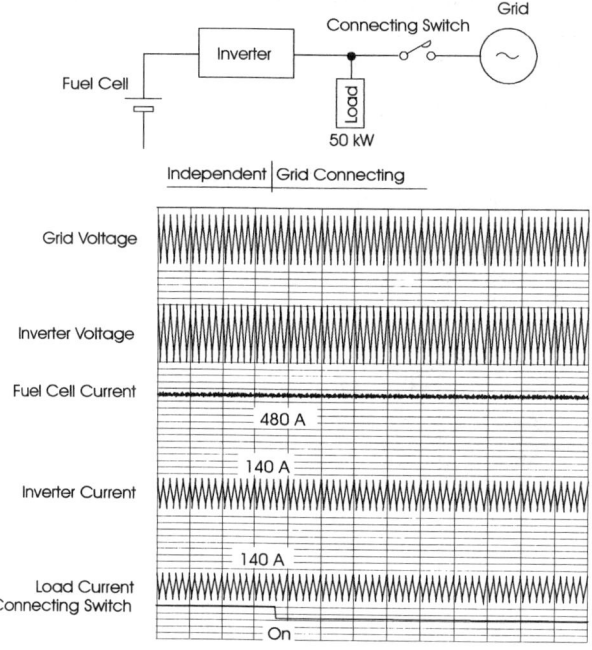

Figure 6-22. Non-Interrupting Switch-Over of Independent/-Grid Connected Operation

Cost Reduction

The cost-reduction aspect for the packaged type of PAFC is of prime importance. Substantial cost reduction, however, can be achieved via technical improvements, standardization and mass production. To reach the cost target goal, Fuji has standardized all cell sizes to 2000 and 8000 cm^2 (effective cell area), and plant capacities as shown in Table 6-7.

Modular design can be achieved generally at a slightly higher unit cost in comparison to large systems. This increase in cost can be offset by mass production of standard plants, short construction periods etc. The final step in cost reduction is mass production. Fuji Electric has installed a semi-mass production facility with an annual capacity of 15 MW per year for producing on-site power plants and stacks for dispersed power plants. Figure 6-23 shows an assembly line for on-line, packaged plants installed in this semi-mass production facility.

Table 6-7. Standardization of Plant Capacities

Application	Type	Output [kW]
On-site use	FP-50	50
(atmospheric pressure system)	FP-100	100
	FP-200	200
	FP-500	500
Dispersed station use	FPD-5	5000
(pressurized system)	FPD-20	20000

Figure 6-23. Assembly Line of On-Site Package PAFC Plants (Courtesy of Fuji Electric Co., Ltd.)

In cooperation with Osaka Gas, Fuji Electric is currently developing three 500 kW, on-site PAFC plants. The first prototype has been operated for 3000 hours [22]. (see Figure 6-24).

Figure 6-24. Osaka Gas 500 kW Power Plant [23] (Courtesy of Fuji Electric Co., Ltd.)

6.2.7.2. PAFC Plants Using Various Fuels

Apart from the natural gas-fuelled power plants described above, Fuji Electric has also been developing various types of PAFC plants using methanol, LPG and naphta. The following gives a very brief description of these various PAFCs.

Methanol Reforming PAFC Plants

Fuji Electric has supplied many methanol reforming PAFCs, as shown in Table 6-8. Methanol is liquid at ambient temperature and is appropriate for plants installed in isolated areas, remote islands or vehicular applications, where methanol is the fuel of choice. Since the reforming temperature of methanol (about 300 °C) is much lower than natural gas (about 800 °C), a methanol-reforming fuel cell plant has generally a better efficiency and a shorter start-up time. The 200 kW PAFC plant shown in Figure 6-2, installed on a detached island in the Okinawa Electric Power Company, was constructed and has operated very successfully over more than 8300 hours under the NEDO project.

Table 6-8. Methanol Reforming PAFCs

Capacity [kW]	Number of sets	Customer	Delivery	Remarks
4	2	Electric Utility Companies	1986–7	research
4	3	Oil Companies	1988	research
25	1	US DOE (Bus Project) (Phase I)	1989	oil cooling
50	3	US DOE (Bus Project) (Phase II)	1992–3	oil cooling
200	1	Okinawa Electric PC	1989	NEDO-Project

Table 6-9 shows the key performance of the 200 kW methanol reforming PAFC plant. The plant entered into operation in January 1990, and supplied electric power to the island grid in parallel operation with existing diesel generators of 1400 kW capacity. After successful operation under very severe weather conditions, including a typhoon, the plant stopped operation at the end of July 1991, due to project budget limitations.

Table 6-9. Key Performance of 200 kW Methanol-Reforming PAFC Plant for Detached Island

Items	Performance (test results)
Rated power	215 kW (generating end)
Efficiency [%]	39.7 (HHV) at rated load
	40.6 (HHV) at 75%
	36.7 (HHV) at 50%
Stack working conditions	190 °C, 1 atm
Cooling	water cooling
Cold start-up time [h]	3
Minimum load [%]	20
Response time [s]	25
NO_x [ppm]	< 2 (O_2: 7%)
Operation mode	full-automatic operation

A 25 kW brassboard, experimental PAFC plant, manufactured under the Phase I Bus Contract of the US-DOE, has also operated successfully in the USA. This is a typical example of a fuel cell suitable for vehicular application. The project expects support for the next phase, under which three sets of 50 kW PAFC plants will be installed in three demonstration buses. The test results of the 25 kW brassboard plant and the specification of the 50 kW on board plant are shown in Tables 6-10 and 6-11, respectively. Details of this project were reported to the Fuel Cell Seminar held in Phoenix during the Fall of 1990. One bus was presented at the Seminar in 1994 [24].

Table 6-10. Test Results of the 25 kW Brassboard PAFC Plant (Phase I)

Item	Results
Output [kW]	
at generating end	25 (dc. 52.8 V × 480 A
at chopper end	23.2
Efficiency [%]	38.0 (LHV, excluding auxiliary power)
Exhaust gas [ppm]	
CO	30
NO_x	0.5
HC	not detected

Table 6-11. Design Specifications of a 50 kW PAFC On-Board Plant (Phase II)

Item	Specifications
Plant output [kW]	50
Efficiency [%]	38 [LHV]
Fuel consumption [kg/h]	Methanol: 23.6 pure water: 26.0
Start-up time [min]	< 25
Dimension [m]	$1.3 \times 2.1 \times 1.45$
Weight [t]	1.2

LPG-Reforming PAFC Plants

Fuji has had trial operation experience with an LPG-reforming, 50 kW PAFC plant for Tohoku Electric Power Company, and is also developing an LNG/LPG-reforming, 50 kW PAFC plant (nine sets) for domestic gas companies. Furthermore, Fuji Electric will start development of a comprehensive, LPG-reforming, 100 kW PAFC plant under a government contract. This 100 kW project started in 1991 and should finish in 1994. The purpose of this project is to expand the fuel cell market, because the LPG-supplied area in Japan is much greater than the LNG-supplied area. As a result, the applicability of LPG-reforming plants might be larger than that of LNG.

Naphta-Reforming PAFC Plants

Fuji is developing naphta-reforming PAFC plants in close co-operation with PEC (Petroleum Energy Centre), and supplied two of these PAFC plant systems to the Idemitsu Oil Company in 1989 and 1990. The construction of the plants is basically similar to the 50 kW prototype plants, but the naphta-reformer catalyst was developed and supplied by Idemitsu. These plants are now being tested at the Idemitsu Oil Company Research Centre. These developments should contribute to expanding the marketability and applicability of PAFC plants.

6.2.8. Mitsubishi Electric Corporation [25–27]

The Mitsubishi Electric Corporation (MELCO) has been developing phosphoric acid fuel cell (PAFC) over the past decade, aiming at early commercial applications of the fuel cells. The three demonstration power plants, which were mainly supported by the Ministry of International Trade and Industry, operated successfully. In particular, the on-site 200 kW power systems installed at Hotel Plaza in Osaka demonstrated over 13000 hours of operation without acid addition (see Figures 6-25 and 6-26).

Figure 6-25. 200 kW PAFC-Plant at Hotel Plaza (Courtesy of Mitsubishi Electric Corp.)

MELCO was encouraged by these results, and put PAFC power plants into pilot production in the middle of the 1990s. MELCO is now developing the technologies and reducing the plant costs.

Test Results and Designs of Power Plants

Table 6-12 shows the test results and planning. Hotel Plaza's ambient pressure-type power plant obtained excellent results. The system demonstrated a capability of about 20000 hours total and 3000 hours continuous operation time. The power section needed no maintenance such as acid replenishment.

MELCO intends to overcome some remaining technical problems while continuing to supply the next plant and the commercial plant. Furthermore, MELCO is developing large size (8000 cm^2) electrode cells and an 1 MW class reformer.

The cost of the 200 kW standard plants is intended to be reduced to 200 ~ 250 thousand yen/kW (approx. US-$ 1800–2000/kW) in 2000 by means of technical developments, mass-production techniques and the effect of material superiority. This cost means that fuel cell power plants, taking account of their specific characteristics and other credits, could hold a dominant position among co-generation systems,.

Consequently, MELCO has accomplished an initial performance of about 650 mV at 300 mA/cm^2 with subscale (100 cm^2) cells (see Figure 6-27), and is preparing for a practical size cell (3600 cm^2 class) short stack using these improvements to verify its actual performance and durability.

Table 6-12. Test Results and Designs of Power Plants

	Subject	Hotel Plaza 200 kW 1989	ROKKO Island 200 kW 1991	Next plant 200 kW 1992-5	Commercial 200 kW 1996 ~
Power Section	1. PERFORMANCE Current density	150 mA/cm²	200 mA/cm²	250 mA/cm²	300 mA/cm²
	2. LIFE Running Time	13038 h	(40000 h)	(40000 h)	40000 h
	3. RELIABILITY & MAINTENANCE Acid addition	Non	Non	Non	Non
Reformer	1. SIZE REDUCTION Compact ratio	35 Nm³/h m³	35 Nm³/h m³	60 Nm³/h m³	70 Nm³/h m³
	2. RELIABILITY 3. PERFORMANCE S/C ratio	3.5	3.5	3.0	2.5
Total system	1. SIZE REDUCTION Compact ratio	0.16 m²/kW	0.15 m²/kW	0.12 m²/kW	0.08 m²/kW
	2. RELIABILITY: Continuous running time	2656 h	–	–	1 year
	3. CONTROLE	Full Auto (Package)	Full Auto (Package)	Full Auto (Cogen. Syst.)	Full Auto (Cogen. Syst.)
	4. HIGH TEMPERA-TURE Heat-recovery (recovery ratio)	170°C (18%)	–	170 °C Steam (21%)	170 °C Steam (25%)

Figure 6-26. Rokko Island 200 kW PAFC Power Plant (Courtesy of Mitsubishi Electric Corp.)

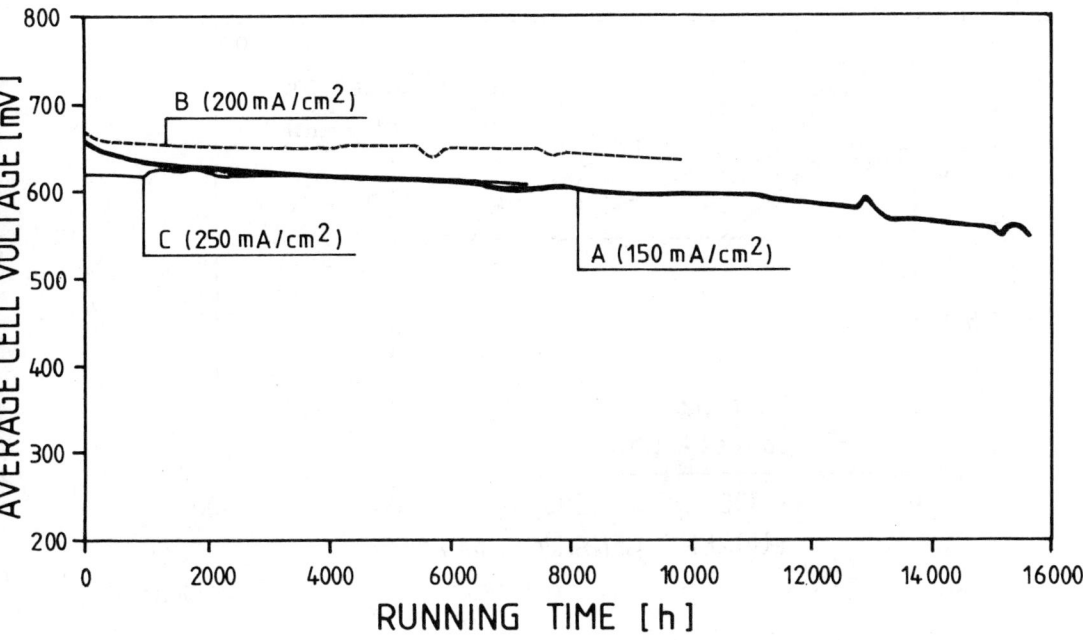

Figure 6-27. Performance History of Short Stacks (Increase of Current Density) (Courtesy of Mitsubishi Electric Corp.)

Three ideas have been applied to improve cell life:

- Series fuel gas flow in the stack to alleviate corrosive conditions for the electrodes.
- Well-balanced, micro-pore size reservoirs among cell components to avoid flooding of electrodes.
- High corrosion-resistant carbon support for the cathode catalyst.

These improvements have resulted in a smaller decay rate of the cell voltage, e.g. 20 mV/10000 hours at 200 or 250 mA/cm^2 in tests with practical size cells (3600 cm^2 class) in short stacks, (see Figure 6-28).

At the Kobe Works, a 10 kW short stack has been successfully operated for 16000 hours without acid refilling [28]. From chemical analysis of the stack, it was estimated that it would be able to operate for more than 40000 hours without addition of acid.

On the other hand, testing of a large size cell (8000 cm^2 class) short stack has been started in order to develop 500 kW-class stacks. Its performance was comparable to conventional short stacks. Another research topic was the development of a reformer. The development includes catalyst evaluation, a 1/6 size burner model of the 1 MW type and a 50 kW mono-tube reformer model. The burner model was tested under various conditions.

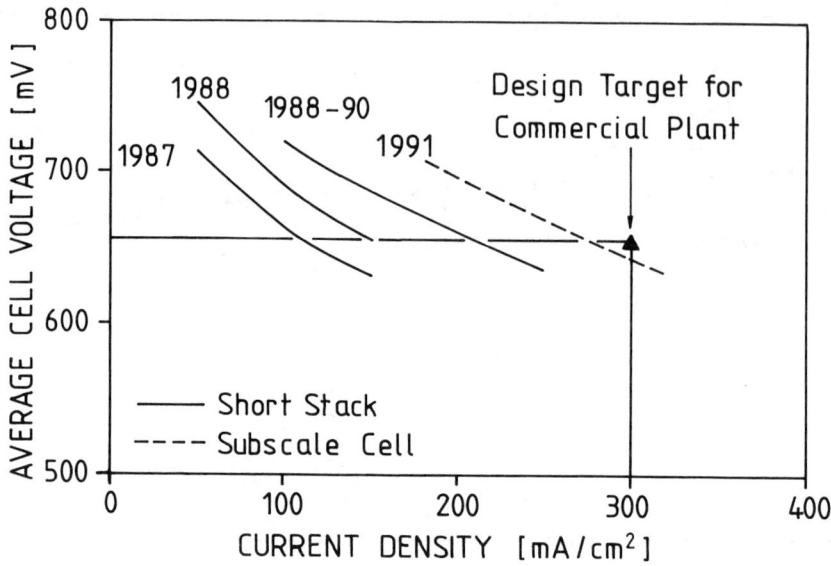

Figure 6-28. Cell Performance (Decrease of Decay Rate) (Courtesy of Mitsubishi Electric Corp.)

6.2.9. Technological, Ecological and Economic Comparison Between PAFC and Conventional On-Site Power Plants

In the past, producer and user always discussed the total energy efficiency of a power plant. However, the different forms of energy (electrical and thermal energy), and especially the different quality of the thermal energy, have to be evaluated accordingly. Thermodynamics offers the model of exergy. Exergy is defined as the maximum amount of work that is available in a gas, fluid or mass as a result of its nonequilibrium condition relative to an environmental reference point. More precisely, exergy is the maximum work potential of a system, stream or matter or a heat interaction in relation to a reference environment called the datum state [29]. It is characterized by a perfect state of equilibrium, i.e., absence of any gradients or differences involving pressure, temperature, chemical potential, kinetic energy and potential energy. The environment constitutes a natural reference medium with respect to which the exergy of different systems is evaluated. Unlike energy and mass, exergy is not subject to a conservation law. The exergy data can conveniently be presented by an equivalent method as the energy data (Sankey diagram) in a so-called Grassmann diagram. Figure 6-29 shows such a diagram for the first European fuel cell plant, namely the 25 kW fuel cell demonstration system of Kinetics Technology International (KTI), Zoetermeer in 1989.

The exergy of the incoming natural gas can be defined as 100% (in the same manner as in the case of Sankey diagrams). The produced electrical exergy W_{el} is 39.1% of the exergy available in the inlet feed + fuel stream. The amount of exergy produced as heat in the fuel cell is about 7% of the exergy in the system inlet stream. From the exergy

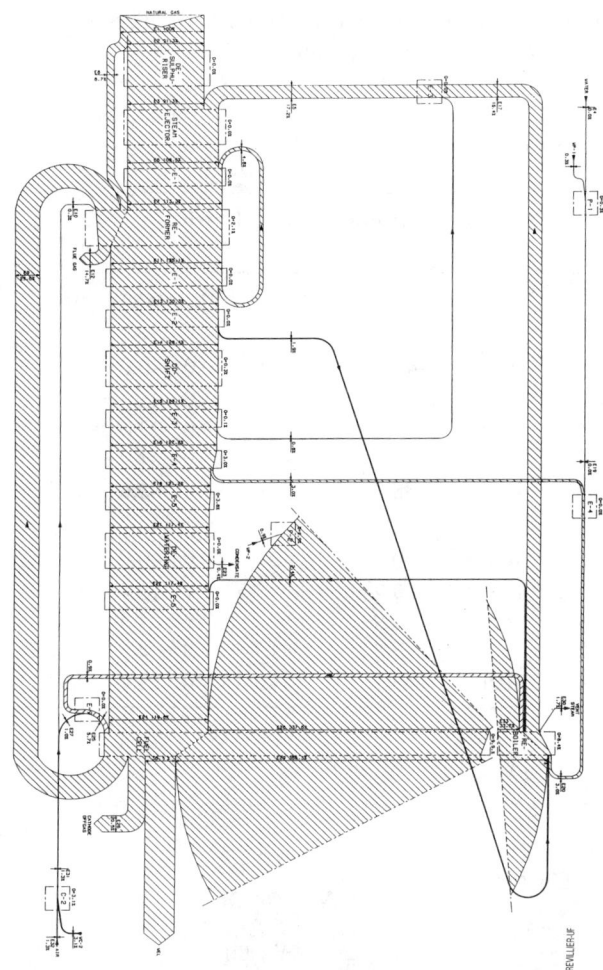

Figure 6-29. Grassmann diagram of the 25 kW PAFC fuel cell system (KTI, Fuji Electric) [30]

analysis it can be concluded that the largest irreversibilities occur in the reformer and the fuel cell and to a lesser extent in the reboiler and the steam ejector.

The comparisons in Figure 6-30 were made under special consideration of the Japanese developments [31]. The following phosphoric acid power plants were investigated: IFC/ONSI (PC25, 200 kW), Toshiba (TFC200, 200 kW), Mitsubishi (200 kW), Fuji Electric (50 kW, 100 kW, 500 kW) and a comparable gas-engine power plant (350 kW).

The comparable, total energetic efficiency of the gas engine and the phosphoric acid fuel cell changes with regard to the total exergetic efficiency. The more recent phosphoric acid fuel cell plants (thermal energy in the form of steam at 170 °C) have an exergetic efficiency of between 49 and 50%. The comparable, total exergetic efficiency of the gas engine is 43% (see Table 6-13).

Among all the fossil fuels, natural gas (CH_4) contributes the least to CO_2-emissions because of its high H/C ratio of 4. Figure 6-31 shows the calculated CO_2-emissions for the TFC-200 (200 kW, Toshiba) and the gas engine shown in Table 6-13. The basis of the calculation is an energetic, exergetic and electric consideration.

Table 6-13. Energetic and Exergetic Potential Estimation

Producer	IFC/ ONSI	Toshiba	Mitsubishi	Fuji Electric			Gas- engine
Electric power [kW]	200	200	200	50	100	500	350
Energetic efficiency							
Electric efficiency [%]	40.0	40.0	40.0	40.0	40.0	40.0	33.9
Thermal efficiency [%]	44.0	40.0	45.0	40.0	43.0	45.0	51.2
Steam	–	25.0	21.0	–	23.0	23.0	–
Hot Water:	44.0	15.0	24.0	40.0	20.0	22.0	51.2
Energetic total efficiency [%]	**84.0**	**80.0**	**85.0**	**80.0**	**83.0**	**85.0**	**85.1**
Exergetic Efficiency							
Electric Efficiency [%]	40.0	40.0	40.0	40.0	40.0	40.0	33.9
Thermal Efficiency [%]	7.0	10.0	9.7	4.7	9.0	10.4	9.2
Steam	–	8.2	6.9	–	7.2	7.5	–
Hot Water	7.0	1.8	2.8	4.7	1.8	2.9	9.2
Exergetic total efficiency [%]	**47.0**	**50.0**	**49.7**	**44.7**	**49.0**	**50.4**	**43.1**

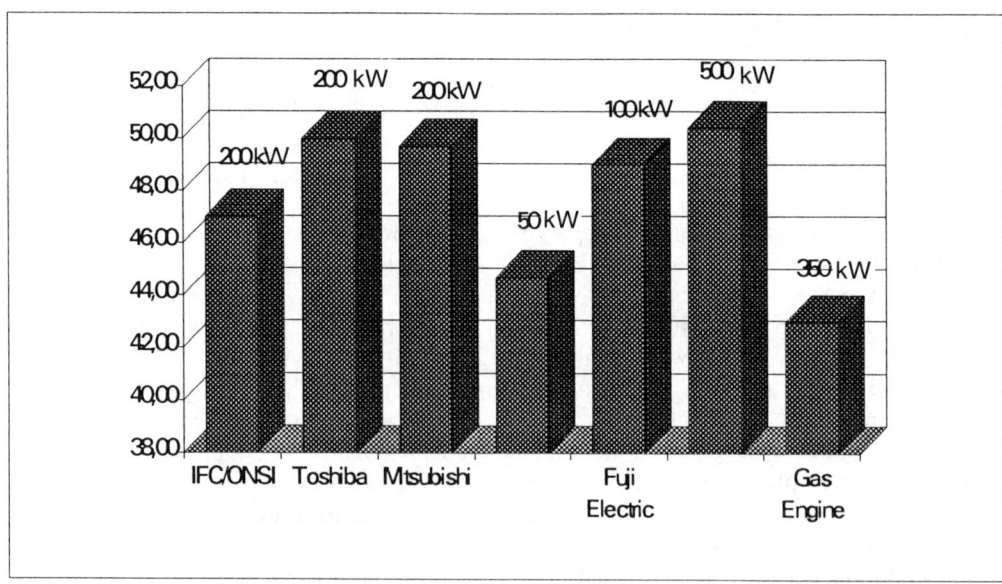

Figure 6-30. Total Exergetic Efficiency of the different Power Plants

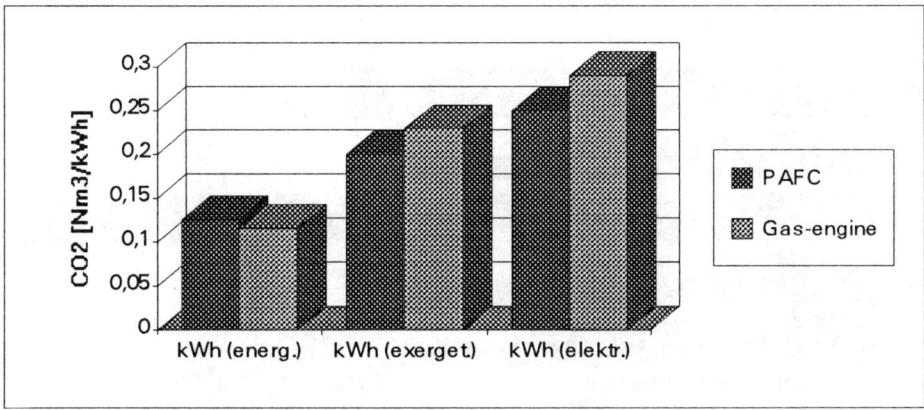

Figure 6-31. Comparison of CO_2 Emissions

The emissions of the classical pollutants NO_x, CO and HC from phosphoric acid fuel cell plants are negligible compared with the gas engine emissions. Thus national and international laws represent no problems or restrictions for the introduction of fuel cell technology. Table 6-14 shows the emissions from the PC25 (IFC/ONSI) and a comparable gas engine.

Table 6-14. Comparison of Emissions

Emissions[1]	PAFC-on-site plant [mg/Nm^3]	Gas engine [mg/Nm^3]	TA-Luft[2] [mg/Nm^3]
NO_x	< 10 (3.1)*	500	500 (250)
CO	< 20 (5.7)*	650	650 (325)
HC	< 15	150	150

* PC25: after 1700 operation hours

Gas and electrical utilities are mainly interested in the cost of energy delivered by the power plants. For the following calculations, data from the PC25 of Austrian Ferngas (Gas Utility), four data sets concerning gas engines from STEWEAG (Austrian Electrical Utility) and two published records (BRD85, VKW) from German studies have been considered. Conventional gas engine, on-site power plants are in operation only 3000 to 4000 hours per year. This is the reason why different PC25 operating life-times have been used (Figure 6-32).

1 O_2: 5 Vol.-%
2 German Emission Standard

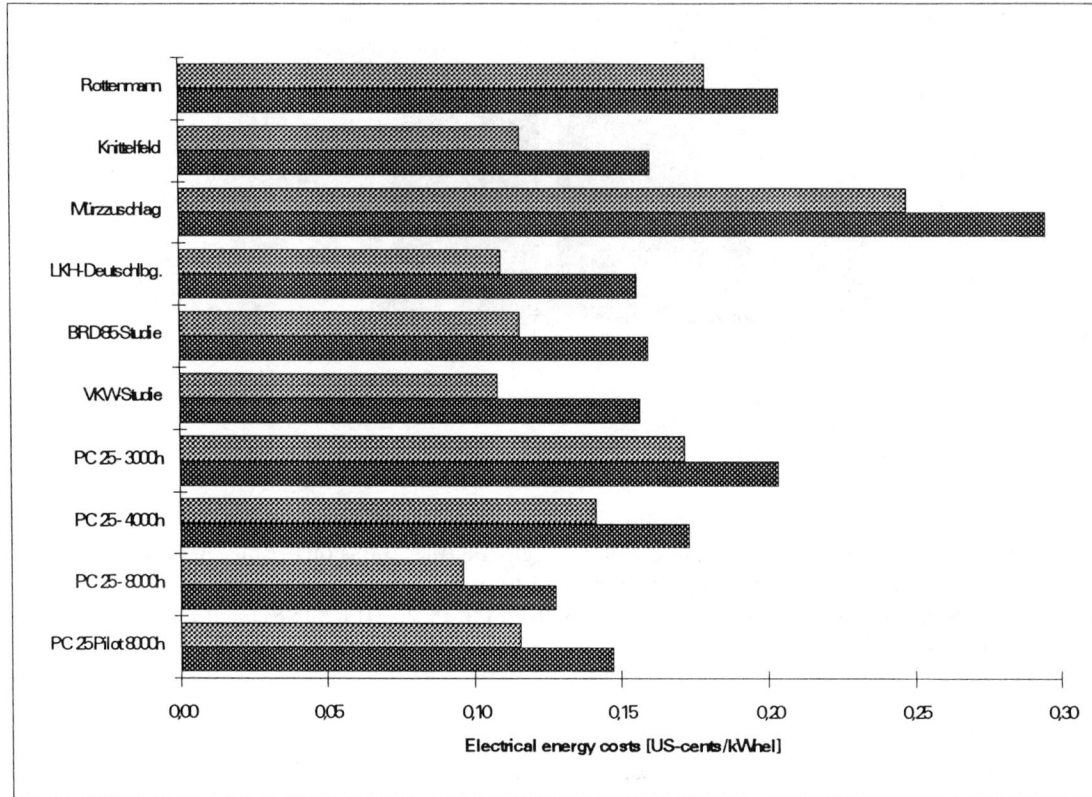

Figure 6-32. Electrical Energy Costs in US-cents/kWh (with and without credit of thermal energy at 2.75 US-cents/kWhel[3]

The thermal energy – as a co-generation product – is obtained at a calculated cost of 2.75 US-cents/kWh. The electrical energy costs from the different gas engine power plants can be summarized with US-cents 11,53/kWh. The comparison shows that the PC25 ("pre-commercial pilot plant") has electrical energy costs similar to those of conventional, on-site distributed power plants. This means that the phosphoric acid fuel cell plant is already competitive with presently-planned, conventional gas-engine power plants in Austria. After mass-production of the fuel cell plants has started, we can expect that fuel cells will have even more competitive advantages relative to gas engine plants in future dispersed applications for on-site power plants. In summary, the estimates of thermodynamic and ecological potential confirm the statement that the phosphoric acid fuel cell technology represents an important approach for a sustainable future development. These evaluations pertain to dispersed fuel cell power plants in the range of 10–100 MW.

---

3 Calculation basis considering Austrian economic data sets, September 1994

6.2.10. Troubles and Their Eradications Referring to PAFC

The number of operational troubles occurred in PAFC can be used as a reliability index. The New Energy Foundation in Japan carried out a survey on plant troubles and found that the number of occurrences was considerably reduced for units installed in 1992 when compared with those installed before 1989, i.e., the average number per one unit was 43.3 previously, while reduced to 3.3 in 1992. Also the number of troubles occurred per 1000 h of operation decreased from 5.6 to 2.2, indicating that plant reliability has improved. Also, it is reported that the equipment prone to troubles is changing from the cell stack and reformer at the initial stage to peripheral equipment and system related matters.

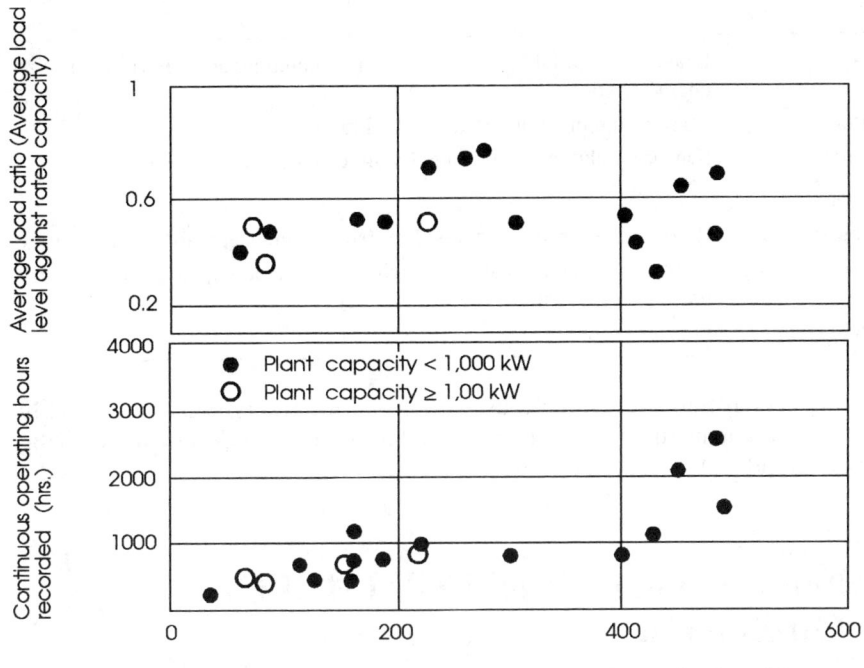

Figure 6-33. Continuous operating hours recorded and average operating load levels [32]

Table 6-15 shows the causes of breakdown and troubles occurring in fuel cell plants. These are all technically correctable, and no serious breakdown problem exists.

Many instances of breakdowns and troubles stem from instrumentation, steam leakages and electrical parts. Also, there occurred a decrease in the performance of cell stacks and the corrosion of heat exchangers, which are considered unique to PAFC plants. The occurrence of the same troubles being repeated tends to decrease due to the countermeasures taken, but new problems are beginning to arise as the hours of operation are further extended.

Of the breakdowns occurred at TEPCO's 11 MW plants, 25% are due to leakage of fluids from flanges, valves, main devices, etc., 19% are due to the vibration of rotating machineries, and 18% are due to faulty sensors in the control system and breakdowns related to control valves. Thus, there are few phenomena which are unique to fuel cell plants. As for main troubles relating to cell stacks, there was an incident in which a part of the corners of cell stack were damaged electrochemically. As this was attributed to the composition of combustion exhaust gas from the reformer, which was used as purge gas for the containment vessels of the cell stack. Plant operation resumed after improving the vessel purge system.

Table 6-15. Potential causes of breakdowns

Electrical:	
Fuel cell stack	Leakage, material breakage, operation, handling and deteriorated characteristics
Ac/dc converter	Cooling fan breakdown, dc ground fault
Control system	Sensor breakdown, control software changes, faulty parts
Process devices:	
Fuel processing system	Reformer durability, catalyst life, faulty flame detection, faulty valves
Peripheral devices/piping	Faulty valves, pipe leakage, machinery breakdown, clogged filter mesh, rotor vibration
Cooling water system	Poor water quality

By conducting demonstration tests, the causes of breakdowns and troubles are steadily removed. Also, as problems that were not considered at the time of design are showing up, they will be considered in future designs.

6.3. Molten Carbonate Fuel Cells (MCFC) for On-Site Use

M-C Power Commercialization Program for MCFC Power Plants [33, 34]

Based on successful stack development efforts to date, and confidence in the ultimate viability of the IMHEX® (internally manifolded heat exchanger) technology, M-C Power has established a comprehensive MCFC commercialization program. The goal of this commercialization program is to develop a market-responsive, natural gas-fuelled MCFC power plant. Figure 6-34 shows both the fuel cell stack mounted inside of the vessel base and a view of the assembled vessel of the 250 kW power plant.

Contingent orders for this unit are expected in 1995, with deliveries starting in 1997. Initial market entry units will be configured for the on-site and distributed generation markets, ranging in size from 500 kW to 3 MW. Subsequent IMHEX® products will serve dispersed generation, repowering and new central station applications.

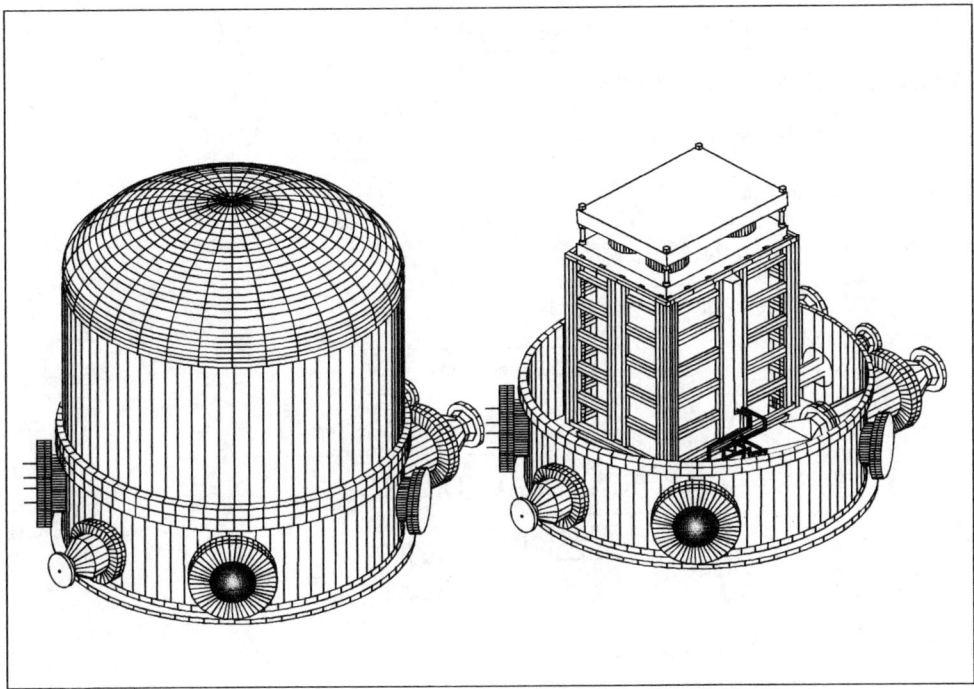

Figure 6-34. 250 kW IMHEX® Fuel Cell Power Module [35]

The IMHEX® Commercialization Program is an integrated effort which focuses on four major areas of interrelated activity.

- Technology development: component and sub-systems development, integrated system design and testing,
- Field verification: proof-of-concept, field experiment and commercial prototype unit testing.
- Manufacturing: development of manufacturing techniques and quality programs, design and construction of fabrication facilities.
- Market development: obtaining customer involvement, product definition, commercial business assessment.

Figure 6-35, the program schedule, shows the overall work efforts to be carried out under each major program activity.

The IMHEX® Commercialization Program includes a number of major technical milestones to allow tracking of overall progress versus plans. The milestones and projected completion dates are shown in Table 6-16.

Figure 6-36 shows the polarization and power results of a 20-kW, full-area (approximately 1 m²) stack. This stack included components that were closer to the "standard" active materials, and used assembly techniques that were simplified and less rigorous than its precursor model.

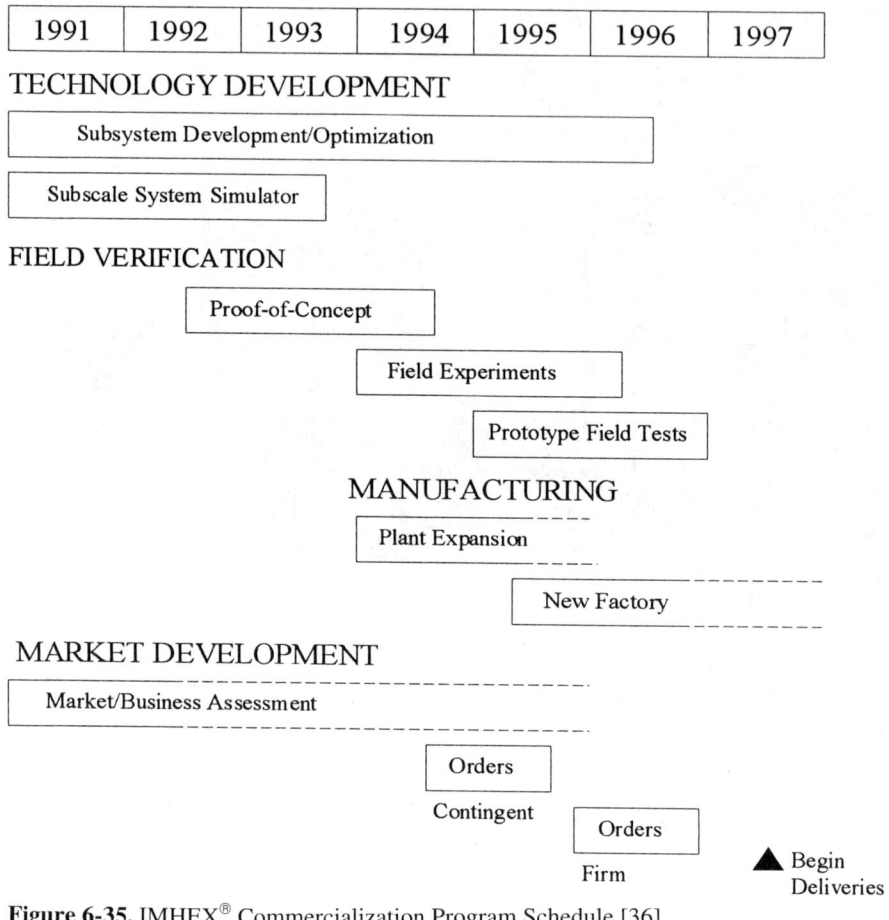

Figure 6-35. IMHEX® Commercialization Program Schedule [36]

Program support

The IMHEX® Commercialization Program enjoys wide technical and financial support from government, industry and research institutions. The following organizations are providing funding to the program:

Government:	US Department of Energy, State of Illinois, and South Coast Air Quality Management District.
Industry:	San Diego Gas & Electric, Southern California Gas Company, Union Oil of California, M-C Power and IGT
Institutions:	Electric Power Research Institute and Gas Research Institute

An industry group has been established to guide, support and stimulate commercial introduction of IMHEX® fuel cell power plants. Members of the Alliance to Commercialise Carbonate Technology (ACCT) include gas and electric utilities, pipeline companies, and other potential users of MCFC power plants. The major functions of ACCT include monitoring technology developments and providing market insight and intelligence. This input from ACCT will ensure that the IMHEX® products developed are responsive to market and customer needs.

Table 6-16. Major Technical Milestones

Tall stack – 70 cells Confirm stackability and no carbonate pumping	mid 1991
Full-area stack – 20 cells Verify fabrication and operation	late 1991
20 kW subscale system Operate reformer, stack and controls	late 1992
Dual-fueled 250 kW proof-of-concept Demonstrate stack in partially integrated system at UNOCAL	mid 1993
Natural gas-fueled 250 kW proof-of-concept Verify feasibility of integrated system at SDG&E	mid 1994
Natural gas-fueled field experiments Verify design and performance characteristics	mid 1995
Natural gas-fueled field tests Simulate commercial service, begin market entry	late 1996

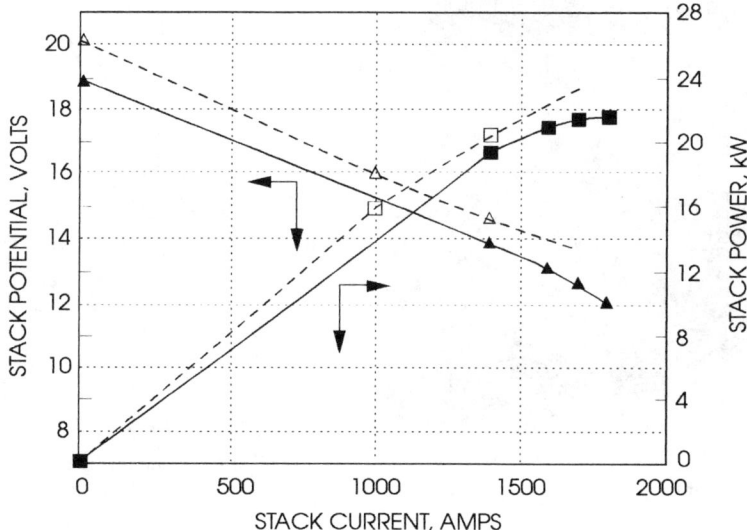

Figure 6-36. Polarization and Power Curves of a 20 kW Commercial m^2-Area Stack (MCP4) [37]
Diagnostic Gases – dashed lines
System Gases   – solid lines
(Courtesy of M-C Power)

6.4. Other Projects for On-Site Use

Solid Oxide Fuel Cell (SOFC) System Development for On-Site Use

SOFCs, with their high temperature exhaust, high efficiency, low emissions, fuel flexibility and modularity, can be used in a broad spectrum of applications. Electric utilities have identified applications for SOFC technology that include multi-megawatt size, distributed electric power plants (see Chapter 5) located near load centers, repowering of existing units and large central station, all-electric power plants. The gas utilities interest in SOFC technology is primarily for on-site power generation for commercial and small industrial applications.

Osaka Gas (Japan) has high expectations for the potential of natural gas-fueled SOFCs as a co-generation system, and they are therefore interested in a keen development effort. A 25 kW generator was developed by Westinghouse jointly with Osaka Gas and Tokyo Gas in March 1990. The system started operating at the Iwasaki test center of Osaka Gas in October 1992. Various problems arose, including the heat recovery steam system, and damaged cells caused operation to be stopped shortly after start-up. A modified co-generation system will be operated from December 1994 on.

Figure 6-37. 25 kW Class SOFC Co-Generation System, Front View (Courtesy of the Fuel Cell Development Information Centre).

In addition, three independent units of 20 kW-class SOFC generators (Westinghouse) were operated from February 1992 to March 1994. One unit achieved 7064 hours – the longest operation for an SOFC module. (See Table 6-17) This program was carried out by the Kansai Electric Power Company, Tokyo Gas Company and Osaka Gas Company.

Table 6-17. Operation Test Summary of SOFC Modules [38]

		Module A	Module B	Module B (after repair)
Test period	from:	25 Feb. 1992	27 Feb. 1992	31 March 1993
	to:	12 July 1992	29 April 1992	10 March 1994
Maximum output		18 kW	20 kW	18 kW
Operation time		2 061 hours	1 579 hours	7 064 hours
Cumulative output energy		43 125 kWh	28 029 kWh	108 253 kWh

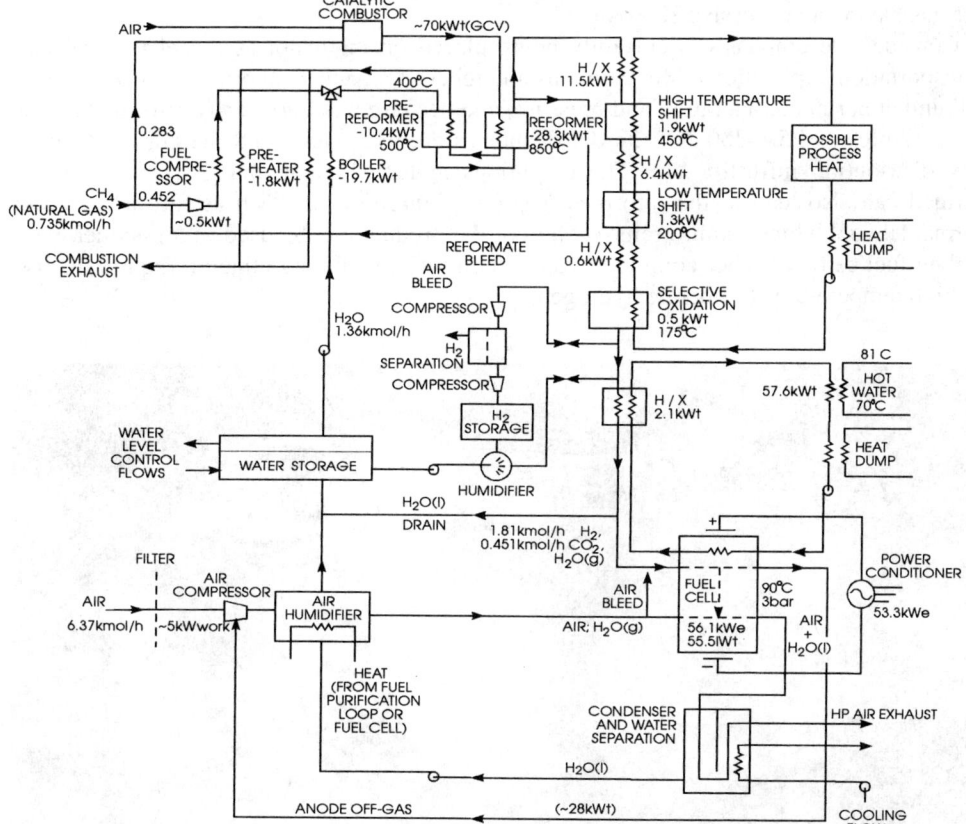

Figure 6-38. Example of Process Requirements for SPFC System for Co-Generation

Polymer Electrolyte Fuel Cell (PEFC) System Development for On-Site Use

The inherent qualities of SPFCs (high power density, high full and part-load efficiencies, low temperature, responsiveness, robustness) make it an attractive proposition for transportation, but it also has potential for power generation – particularly for stand-alone applications or co-generation. Figure 6-38 shows an example of a possible 200 kW PEFC co-generation plant [39, 40].

The choice of the best reforming technique for any particular PEFC application is not simple. What is certain is that steam reforming, which has become a fairly common choice for phosphoric acid fuel cells (PAFCs), is not necessarily the best choice for PEFCs. The bulky, slow starting steam reformer degrades the most desirable attributes of the PEFC. In addition, an PEFC (unlike a PAFC) does not provide a source of waste steam.

Autothermal reforming has some disadvantages compared with steam reforming, particularly the additional N_2 and CO_2 in the product stream. This results in a small increase in pumping power, and increases the size of the shift reactors. It is also important to ensure that the stack is designed to accommodate this nonreactant gas without the anode off-gas having an excessive H_2 content.

Considerable emphasis is currently being placed on methanol as a fuel for fuel cell transportation applications. Methanol has a number of advantages, primarily because it is a liquid at normal temperature and pressure, and because it is very easily reformed at low temperatures (~ 250–300 °C). Its disadvantages are its toxicity and the current lack of any distribution infrastructure. The conditions under which methanol is typically reformed can also lead to the reformate gas containing poisons such as formic acid and formaldehyde, which require removal before the product can be used in a polymer electrolyte fuel cell. All other common hydrocarbon fuels, including ethanol, require the use of high temperatures to liberate hydrogen.

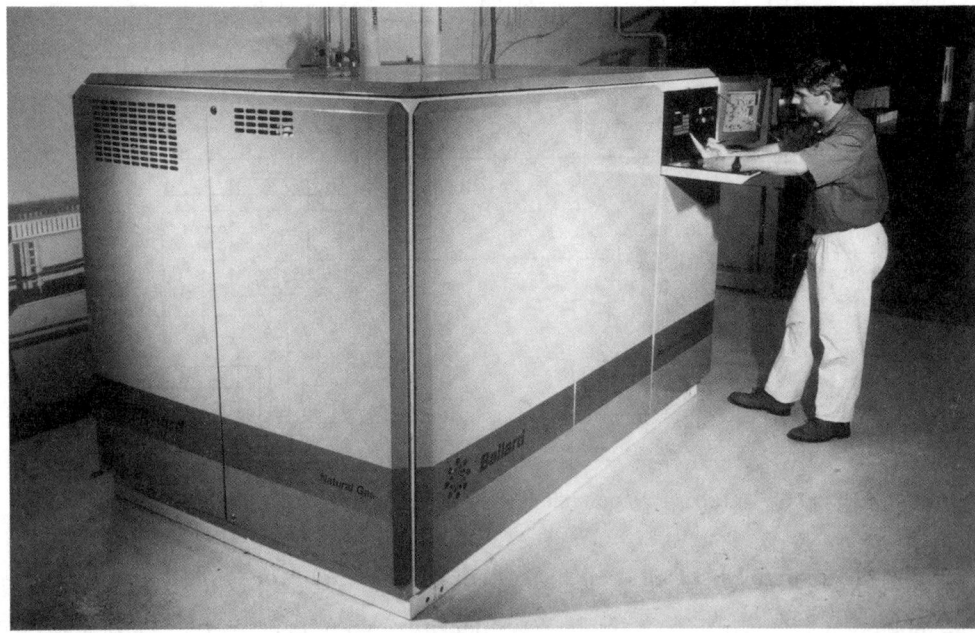

Figure 6-39. Ballard's 10 kW natural gas PEM on-site power plant (Courtesy of Ballard Power Systems)

Considerable R & D activities in this sector are mainly carried out by Ballard Power Systems in Vancouver (B.C., Canada) in cooperation with the Dow Chemical Company [40, 41]. Ballard started in 1991 with the Utilities Development Program (UDP), funded by Ballard and the governments of Canada and British Columbia. The Can-$ 14.3 million first phase was completed in mid-1994. It resulted in the successful demonstration at an industrial facility of a 30 kW fuel cell power plant operating on by-product hydrogen from an adjacent chemical plant. (see Figure 4-37 in section 4.3). A 10 kW natural gas power plant was also developed to demonstrate the key sub-systems for the commercial proto-type 250 kW natural gas fuel cell power plant, that is scheduled to be completed in 1996.

References for Chapter 6

[1] Fuel Cell Times, A Quaterly Publication of ONSI Corporation, Volume 1, Issue 3, October 1993
[2] Fuel Cell RD&D in Japan, Fuel Cell Development Information Center, June 1990, p. 31
[3] H. Wendt, BWK, **41,** 10 (1989) 463
[4] M. M. Papic, D. W. Casson, "Installation and Operation of a 40 kW Phosphoric Acid Fuel Cell Power Plant", Powertech Labs Inc., Canada, March 1989
[5] P. Kalal, Ontario Hydro, research division, 1992
[6] International Fuel Cells (IFC), "The PC 25 200 kW Fuel Cell Plant", CEC-Italian Fuel Cell Workshop, Proceedings (P. Zegers, ed.), Taormina, Italy 1987, p. 81
[7] Contract for the Purchase of the PC 25 Fuel Cell Power Plant with plant description, ONSI Corporation, South Windsor, Connecticut, USA, 1990
[8] Ansaldo "The PC 25 Fuel Cell Co-Generation Power Plant" CLC S.r.L., Genova, Italy, 1994
[9] D. Newby, "Westinghouse Air-Cooled Phosphoric Acid Fuel Cell Program",CEC-Italian Fuel Cell Workshop, 4th/5th June, 1987, pp. 85–90
[10] J. M. Feret, Journal of Power Sources **29** (1990) 59
[11] "Fuel Cell RD & D in Japan", 1994, Fuel Cell Development Information Center, pp. 81 - 85
[12] N. Itoh, Journal of Power Sources **29** (1990) 29
[13] T. Matsushita, et a., Commercialization Plan of Toshiba's PAFC", International Fuel Cell Conference (IFCC), Febr. 2.–6, 1992, Makuhari Japan, p. 137
[14] Technology and Business Development at Toshiba on Fuel Cell Power Plant, Personal Communication, June 1993
[15] T. Koshimizu et al., "Development of 5000 kW-Class & 1000 kW-Class Plants", 1992 – Fuel Cell Seminar, Tucson (AZ), pp. 501–504
[16] R. Anahara, Journal of Power Sources **37** (1992) 119
[17] R. Anahara, International Conference "Cooperative Clean Air Technology" at Santa Barbara, USA on March 30 -April 1, 1992
[18] K. Okano, "Fuel Cell Commercialization by Fuji Electric., Ltd.", IFCC, Makuhari (Japan), febr. 3–6, 1992, pp. 129–132
[19] Y. Sawada et al., "Development of 500 kW Class PAFC", 1992 – Fuel Cell Seminar, Tucson (AZ), pp. 366–369
[20] M. Sakurai, K. Okano, R. Anahara, "Phosphoric Acid Fuel Cell Activities in Fuji Electric for Commercialization", 1992 – Fuel Cell Seminar, Tucson (AZ), pp. 641–645
[21] Fuel Cell, Information broschure from the Fuel Cell Development Information Center (FCDIC), 1994

[22] Fuel Cell R & D in Japan, 1994, Fuel Cell Development Information Center, p. 35
[23] Fuel Cell Development Information Center (FCDIC), Brochure, 1994, Secretary General: Rijo Anahara
[24] J. C. T. Oliveira, Trip Report "The International Conference on Fuel Cells", Feb. 23–25, 1994, Long Beach (CA), Electrochemical Science and Technology Centre, University of Ottawa
[25] H. Yamamoto, "Field Test of a 200 kW On-Site Fuel Cell at a Hotel", IFCC, Febr. 3–6, 1992, Makuhari (Japan), pp. 91–94
[26] M. Matsumoto et al., "PAFC Commercialization and Recent Progress of Technology in Mitsubishi Electric", IFCC, Febr. 3–6, 1992, Makuhari (Japan), pp. 133–136
[27] K. Usumi et al., "PAFC Commercialization and Recent Progress of Technology in Mitsubishi Electric", 1992–Fuel Cell Seminar, Tuncson (AZ), pp. 366–369
[28] B. M. Barnett, W. P. Teagan, Journal of Power Sources **29** (1990) 29
[29] T. J. Kotas, "The Exergy Method of Thermal Plant Analysis", Butterworths, Guilford, UK, 1985
[30] P. F. van den Oosterkamp, A. A. Goorse, L.J.M.J. Blomen, Journal of Power Sources **41** (3) (1993) 239
[31] G. Simader et al., "Fuel Cell Developments in Japan", Final Report for the South Austrian Energy Cooperation, June 1993
[32] K. Shibata, K. Watanabe, Journal of Power Sources **49** (1994) 77
[33] R. A. Figuera et al., "IMHEX® Fuel Cell stack endurance testing and its relation to commercialization", 1992 – Fuel Cell Seminar, Tucson (AZ), p. 370
[34] E. H. Càmara et al., "IMHEX® Commercialization Program Status", IFCC, Makuhari, Japan, 1992, p. 149
[35] R. R. Woods et al.; "IMHEX® Fuel Cell Power Plant Demonstration Projects", International Conference of Fuel Cells, Proceedings, February 23–24, 1994, Long Beach (CA), pp. 281–289
[36] E. H. Camara et al., Journal of Power Sources **37** (1992) 168
[37] D. K. Fleming, T. T. Benjamin, R. J. Remick, "Performance of 20 kW IMHEC® full-area stack", Proceedings of the third international symposium on Carbonate Fuel Cell Technology, Volume 93-3, The Electrochemical Society,
[38] Fuel Cell RD&D in Japan, 1994, Fuel Cell Development Information Center, p. 61
[39] J. P. Shoesmith, R. d. Collings, M. J. Oakly, D. K. Stevenson, Journal of Power Sources **49** (1994) 129
[40] D. S Watkins, B. C. Peters, "Development of a PEM Power Plant for Stationary Power Applications", 1994 – Fuel Cell Seminar, San Diego (CA), pp. 250–252
[41] Ballard Power Systems, News Release, "Ballard Receives $ 3.3 Million in Orders from Dow for fuel Cells", February 1, 1995
[42] D. S. Watkins, B. C. Peters, "Development of a PEM Power Plant for Stationary Power Applications", 1994 – Fuel Cell Seminar, San Diego (CA), pp. 250–252

7. Fuel Cell Powered Electric Vehicles

7.1. Introduction

Due to the increasing environmental problems caused by the use of automobiles powered by combustion engines, the electrical vehicle has been considered as the alternative solution for keeping and even increasing the level of personal mobility and satisfying the transportation needs of the coming century.

Considerable efforts have been made to reach the goal of producing electric vehicles with satisfactory performance characteristics, but the results have been generally disappointing. The main reason for the slow progress in this direction can be found in the inability of available rechargeable battery systems to provide the energy and power densities needed to make electric propulsion systems competitive with the range and power output of heat engines operated on fossil fuels.

Electricity can be generated by several energy conversion methods, but it cannot be stored in a tank, like gasoline or diesel fuel. It must be stored in batteries. The disadvantage of batteries is generic: electrochemical reactions, which are the means of storing and subsequently generating the electric power on demand, occur within the confines of the masses of the electrodes. These are, however, limited in weight and volume by their specific chemical compounds.

Even the theoretically most energetic electrochemical systems could only provide a small fraction of the energy content of e.g. diesel fuel, which can be supplied on demand to a relatively small and light-weight combustion engine. The electrochemical reaction must also be reversed (the battery must be recharged after its capacity is exhausted), in order to establish the original condition. This requires far more time than filling a fuel tank, and needs special charging facilities.

Early electric vehicles (around 1910–1920) satisfied the needs of local personal mobility with a rechargeable lead-acid battery as the source of electric power. Today, special city vehicles (small personnel carriers, buses etc.) can still be operated with very well-defined advantages with this storage battery. However, the real challenge is the modern automobile operated on fossil fuels. An intermediate choice (before fuel cells are available) may be the gasoline-battery combination [1].

The only electrochemical systems which can successfully fulfill the task of providing mobility from renewable energy sources or non-fossil fuels are the fuel cell (or Fuel Cell-Hybrid-) Systems: "Electrochemical systems, acting as power generators, supplied with fuels on demand". They have a long research and development history, with great success stories in space technology, but they were generally considered as long-range future technology for "niche" applications on earth [2, 3]. This situation changed recently, as environmental problems reduced the desirability of combustion engines using fossil fuels.

7.2. The Environmental Benefits of Fuel Cell Vehicles

Fuel cell electrodes operate on hydrogen, no matter what fuel is actually supplied to the fuel cell system. In direct fuel cells, pure hydrogen is fed to the electrodes. In indirect fuel cells the hydrogen carrier (methanol, ammonia, hydrazine etc.) must first be converted at a certain efficiency, producing also side products (equivalent to amounts of CO_2 in the case of hydrocarbon conversion).

The use of a fuel cell system in a vehicle will, in any case, lower the accountable pollution emissions due to its higher conversion efficiency. Compared with the heat engine, this results in less carbon dioxide being produced per unit of energy originating from the fossil fuel. For the case of a road vehicle, where the heat of reaction cannot be of any use (except perhaps for heating in winter), a highly efficient, low-temperature fuel cell (like the PEM or the alkaline system) is especially suitable.

The advantages in reducing overall CO_2 production is also true for the case where coal, oil, or natural gas-burning power stations first produce electricity and then use it for electrolysis to make hydrogen. Of course, the specific circumstances must be carefully checked to arrive at valid comparisons. Only the exergy values are indicative of the true efficiency of the whole chain of conversions. No CO_2 pollution problem seems to exist with electric power production via water power, nuclear energy, wind or solar installations.

However, the energy consideration will also count the CO_2 which was produced during cement production for the water dam or nuclear containment, for melting of the silicon, etc. The danger is that such calculations are taken too far, and become self-defeating for any technology. A border-line case is, for example, the question of how much of the energy used for building highways and bridges should be counted as a CO_2 source due to the existence of the automobile or the railways as a means of needed transportation.

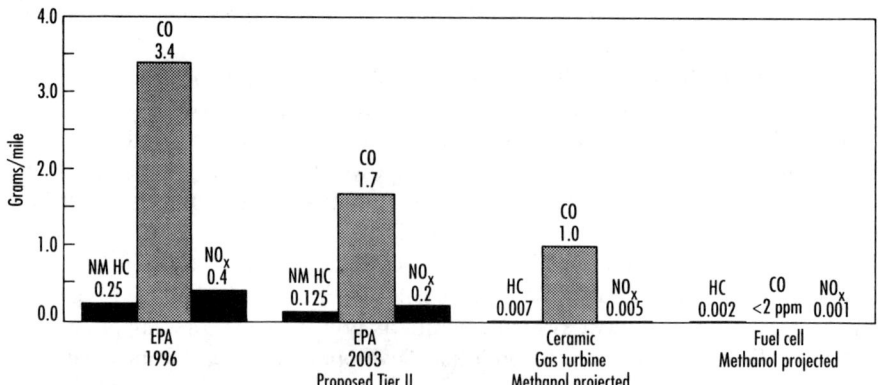

Figure 7-1. Emissions of Present and Future Cars Compared with the Fuel Cell Vehicle.

Estimates have been made on the basis that in the year 2030 fuel cells will be operating in 5% to 30% of the light-weight vehicles. The savings, expressed in barrels of oil saved as compared to 1990, are estimated to be in the order of 2 to 4 billion barrels per year. Total economic benefits from the methanol fuel cell scenario are estimated at US-$ 26 to 42 billion. US-$ 116 Billions could be saved by operation on hydrogen fuel. The amount of pollution avoided can also be estimated and air quality predictions can be made. Detailed data were published by the U.S. Department of Energy, Office of Propulsion Systems and the U.S. Environmental Protection Agency within the multi-year National Program Plan for Fuel Cells in Transportation [4].

7.3. Present Day Electric Vehicle Technology

7.3.1. The Performance of Rechargeable Batteries

The performance of electric vehicles, especially with lead-acid batteries, is generally poor if compared with that of gasoline or diesel engine powered automobiles. It must be reconized that this fact is mainly the result of the energy limitation of the battery, which determines the operational vehicle range possible. The power output is not the dominating problem. In order to be able to drive as far as possible, the weight of the batteries installed in some vehicles reaches nearly half of the total weight. In a vicious cycle this might decrease the acceleration and grade climbing capability. The safety of some high temperature batteries is also a question [5, 6].

Table 7-1. Presently (1993) Proposed Electric Vehicle Batteries

Systems	Pb/PbO_2	Ni/Cd	Na/S	Zn/Br	Ni/Fe	Ni/Zn	$Na/NiCl_2$
[Wh/kg]	35	50	120	70	60	80	100
[W/kg]	160	200	180	80	100	200	100
Cycles	400	500	500	500	500	200 ?	400 ?
[US-$/kWh]	250	1500	1200	400 ?	800 ?	600 ?	1000 ?

Notes: The Ag/Zn system is not considered due to its cost.
The bipolar Pb/PbO_2 system is only in the prototype stages.

Table 7-2. Proposed/achieved performance of "Advanced Batteries"

Systems	Li/FeS_2	Li-Polymer	Zn/MnO_2	Al/Air	Zn/Air
[Wh/kg]	180	200 ??	80 ?	200 ?	150 ?
[W/kg]	200	400 ??	100	100	100
Cycles	1000 ?	300 ?	200 ?	Al	Zn
[US-$/kg]	500 ?	500 ?	250 ?	500 ?	500 ?

Notes: Al and Zn-Air cells exchange the anodes and electrolyte.

The double logarithmic diagram of the following Figure 7-2 presents a comparison between different electric vehicle propulsion battery systems. It is the specific energy content versus specific power density ("Ragone Plot") for various battery systems, each weighing 300 kg (fuel cells have sufficient fuel in the tank, included in this weight). A vehicle weight of one ton is assumed. Such diagrams have been essentially unchanged for 15 years to a certain extent, indicating the lack of progress in the battery field as far as available systems are concerned.

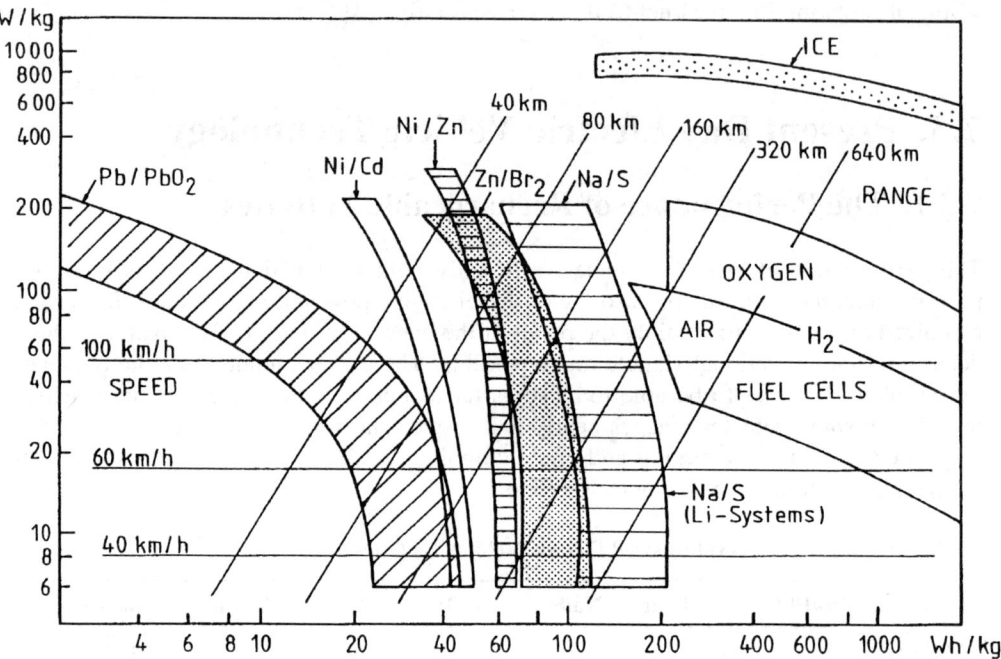

Figure 7-2. Working Range of different Battery Systems (Ragone Plot) showing Energy Density as a Function of the Power Density. (Future bipolar lead-acid batteries are not included.)

The production qualities and life-times have changed, but not the application possibilities for vehicles competing with combustion engines. Gasoline (IC) engine data occupy the right upper corner of the diagram, the 500 to 600 km range at very high specific power values (over 400 W/kg). This indicates immediately that a Ni-Cd battery in parallel with a H_2-air fuel cell could match that performance in a hybrid configuration.

A serious disadvantage of the combustion engine is that fuel consumption rises quickly on intermittent operation, while the fuel cell system functions even better on part-time service. Also, at high peak power the fuel cell hybrid system is superior, in addition to the principal twofold thermodynamic efficiency of the fuel cell over the heat engine. The life-times of combustion engines are in the 2000 operating hour range over a 10 year calendar life. Fuel cells can also match that. As will be shown later; calendar time can be very much prolonged if intermittent use is considered. In this respect, fuel cells for vehicles can be of lower cost per kW and will be designed differently from fuel cell systems for power plants, which require 40 000 hours of activated operational life.

2000 operating hours correspond to approximately 200 000 km driving distance of an average automobile. Only trucks and buses reach millions of km driving distances during their life time.

7.3.2. Electric Motors

Electric motors built today (e.g. for fork-lift trucks) are still 50 to 100% heavier (on a kW/kg basis) than modern combustion engines. However, this situation is expected to change with demands for electric road vehicle motors in the future. Induction-motor electronics and permanent-magnet motors have made considerable progress. For example; the 75 kW induction motor used in the General Motors "Electrovan" (1967) weighed only 60 kg, but the necessary electronic circuits increased the weight of the electrical system by 200 kg. The early, high price of light-weight permanent magnet materials has now decreased due to mass-production, and larger motors (in the multi 10 kW-ranges) are becoming available.

7.4. Historical Review of Fuel Cell Vehicles

7.4.1. The Early Developments

The technology of fuel cell systems for electric vehicles arose to a large extent from the history of the Alkaline Fuel Cell (AFC) after WW II [7–9].

Recollecting those efforts, it seems that technology trends often make full cycles over a period of 30 years. During the 1960s and 1970s alkaline systems with circulating KOH electrolyte were favored and this technology was definitely a spin-off from space technology. NASA achieved the journey to the moon during the "Apollo" project with an improved version of Francis T. Bacon's alkaline fuel cell system, using porous metal electrodes [10, 11].

One of the first fuel cell operated vehicles used "on earth" was the Allis Chalmers fuel cell powered tractor, built by W. Mitchell at the Allis Chalmers Mfg. Co in Milwaukee, Wisconsin, in 1959. The tractor had a 750 V, 15 kW hydrogen/oxygen fuel cell battery with a weight of 917 kg. It was the prototype of high-voltage, bipolar stacks with porous metal electrodes, catalyzed with platinum and using an asbestos matrix for immobilization of the liquid KOH-electrolyte. Gas pressure kept the pores free, but the microporosity of the asbestos did not allow cross-leakage of the gases. A new method of water removal by vapour diffusion through a third electrode – later applied in many other porous metal electrode systems and space fuel cells – was used [12].

In Europe, the developers of alkaline fuel cells with porous metal electrodes chose the designs of E. Justi and A. Winsel to fixate the interfaces of gas and liquid KOH. The electrodes had a dual porosity structure for better pressure balance [13, 14].

ASEA in Sweden constructed a 200 kW submarine battery following these principles. Unfortunately, an accident prevented the ultimate realization of this project [15–19]. Varta AG and Siemens AG built a motor boat for demonstration purposes with dual-porosity nickel electrodes. Later, a 20 kW hydrogen-oxygen system was built [20, 21]. Thornton Ltd. – Shell operated a truck with a hydrocarbon-converter and a shift reformer to clean the H_2-gas for the fuel cell system [22]. The first all-carbon (3 mm thick) electrodes were produced by the Union Carbide Corp. for the US-Navy, and for an experimental, stationary 90 kW H_2/O_2 prototype battery for the Ford Motor Co.

7.4.2. The General Motors Corp. "Electrovan"

After introduction of the thin composite carbon-metal electrodes by Union Carbide, General Motors decided to equip a six-passenger "Electrovan" in 1967 with a UCC Fuel Cell System operating on hydrogen and oxygen, planning at the same time also to test liquified gas-storage possibilities. For safety reasons, the vehicle was only used on company property, but it performed like the "GM-Gasoline-Van" version [23, 24].

The electric motor was a specially developed, 125 HP three-phase induction motor (13 700 rpm) with full electronic control. The fuel cell system was built by Union Carbide, and contained 32 modules with a continuous rating of 2 to 3 kW and a top output of 5 kW. The electrodes were 0.2 mm thick, of the so-called "thin-carbon/fixed-zone" type, with increasingly wetproofed working and gas diffusion layers on hydrophobic porous nickel sheets. The layers were sprayed-on sequentially [25].

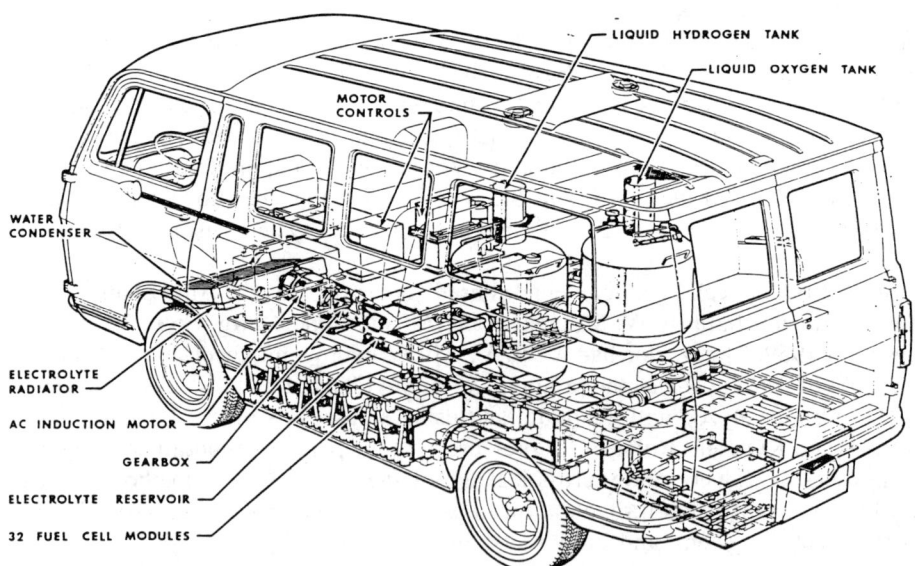

Figure 7-3. A Phantom View of the General Motors Co. "Electrovan". Rated Fuel Cell Output: 200 A/400 V. Peak Power 160 kW.

The use of jet-pumps for the automatic pressure-differential and consumption-dependent recirculation was a special novelty. The vehicle had a driving range of over 200 km, and a top speed of 105 km/h in spite of its weight of 3400 kg. However, there were also serious shortcomings:

- The fuel cell and drive system had excessive weight.
- Fuel cell life after activation was only 1000 hours.
- The cost of the Pt-catalyst per kW-output was too high. At that time, both electrodes, used 1 mg of Pt per cm^2 electrode.
- The starting time after shut-down was sometimes prolonged to several hours due to internal cell reversals, which had to be corrected before a heavy load could be put on the system. The reason was that cells in the series connected stacks received the active gases at different times, and the flow of parasitic currents caused the wrong polarization of cells.
- The high voltage of 400 V dc was dangerous if short-circuits occured.
- Operation on liquefied hydrogen and oxygen is objectionable.

7.4.3. The Kordesch Fuel Cell-Battery Hybrid Passenger Car

7.4.3.1. Description of the Vehicle

In 1970, learning from the experience with the "Electrovan", Kordesch privately built the first, fuel cell-operated city car to be driven on public roads. The electrodes were the same as those used in the "Electrovan" and were donated by Union Carbide for this project of building and testing a 90 V / 6 kW hydrogen-air fuel cell system, which was connected in parallel with a nominally-rated 84 V, 7 kWh (later 96 V, 8 kWh) SLI-lead acid battery [26–28] (see Figures 7-4 to 7-9).

No power converter was used between the fuel cells and the secondary battery. The different load characteristics of the two systems provided a smooth transfer for various driving conditions.

This hybrid system was fitted into a four-passenger car, an Austin A 40, weighing 730 kg before and 950 kg after the conversion. The lead-acid battery provided sufficient acceleration and short hill-climbing power for city driving needs. The light-weight, pressurized hydrogen tanks provided sufficient fuel for about 300 km, and could be refilled within two minutes at an H_2-storage tank farm equipped with "quick-connectors".

The motor was a 7.5 kW (continuous load), or 20 kW at peak power, dc series motor with a 4000 rpm max. speed limit, corresponding to a car speed of 75 to 80 km/h. The four-speed gear-box and flywheel-coupling were retained in order to obtain optimal climbing and smooth starting properties. "Weakening of the field" (winding – switching) was used as an overdrive feature to save energy under continuous and level driving conditions.

A regenerative braking system was installed initially, but was later removed because it did not prove to be efficient enough due to the relatively low weight of the vehicle. There was also no intention to drive in mountainous areas. The principal disadvantages of the "Electrovan" could be remedied to a large extent.

Figure 7-4. Photograph of the converted Austin A 40. The lead-acid batteries are in front, the fuel cells in the rear of the car.

- The life expectancy of alkaline fuel cells with circulating electrolyte was increased by deactivating the battery between operating periods. Shutting off the hydrogen supply and emptying the electrolyte into the tank below removed the hydrostatic pressure, exposed the hydrogen electrodes to air (which has a catalyst-regenerative effect) and eliminated all parasitic currents (and H_2-consumption) during standing.
- Operating on air was found to ensure a long life for the cathodes, if the CO_2 was removed to at least half of the normal content (0.03%) by a soda-lime scrubber with a colored indicator. The KOH-electrolyte absorbed a part of the CO_2, but could be changed easily when the pH dropped. An operating life of 2000 hours would correspond to a driving distance of 100 000 km at a speed of 50 km/h, although this limit was never even approached during the 3 years of test operation.
- The vehicle could be driven immediately after turning the key, with the lead-acid batteries delivering initial power until the electrolyte is pumped into the stacks and circulation starts. After safety controls have determined the readiness of the system, hydrogen is allowed to enter the fuel cell battery.
- The correct polarity of all cells in series was ensured by a very small current flow supplied by the lead-acid battery via a resistor-diode connection. The diode prevented

Figure 7-5. Front view of the car, showing the lead acid batteries mounted on top of the electric drive motor.

Figure 7-6. The 6 kW fuel cell system in the rear of the car.

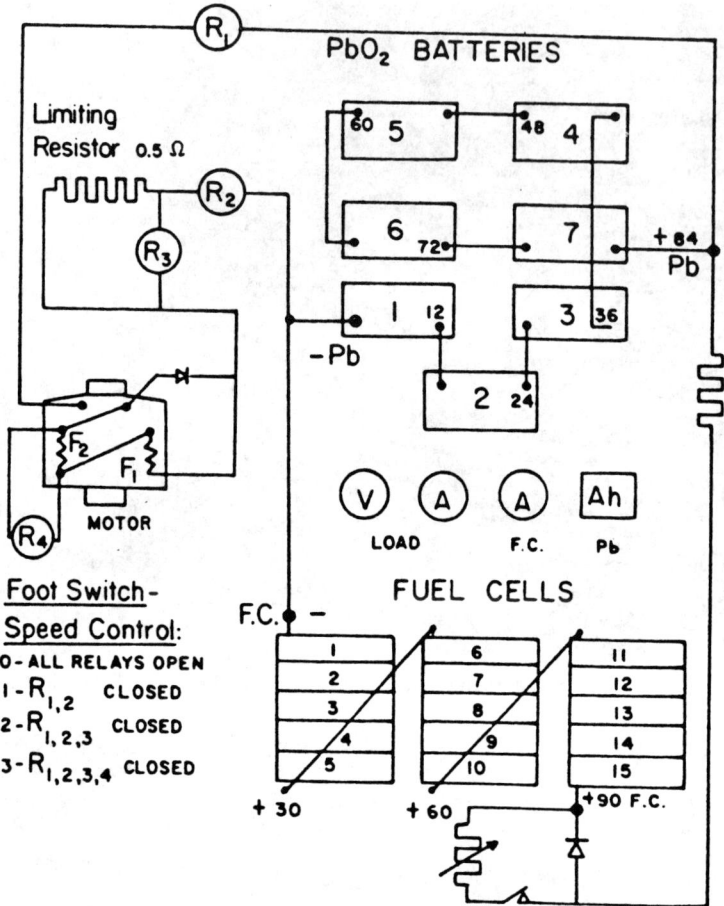

Figure 7-7. Complete diagram of the vehicle power system with the battery arrangement and motor controls.

any electrolysis when the fuel cell system was off but, when bridged during the programmed starting cycle, counteracts all parasitic currents which would appear during delayed or uneven H_2-gas supply to the individual cells.

- The incoming hydrogen always mixed first with nitrogen, which accumulated during longer non-operating periods after the oxygen from the air is catalytically recombined. Nitrogen from a small tank could also used be during intentional or fully-automatic flushing of the system in critical or accident conditions, such as leakage or short circuit.
- The optimum operating temperature of the fuel cell system was between 60 and 80 °C. After a few minutes of driving on the lead-acid batteries, the temperature had risen to the proper level and the radiator cooling loop started only when needed.
- Safety was a prime concern. A voltage in the range of 80 to 100 V was considered sufficient. Liquefied hydrogen was definitely not suitable for a city vehicle with

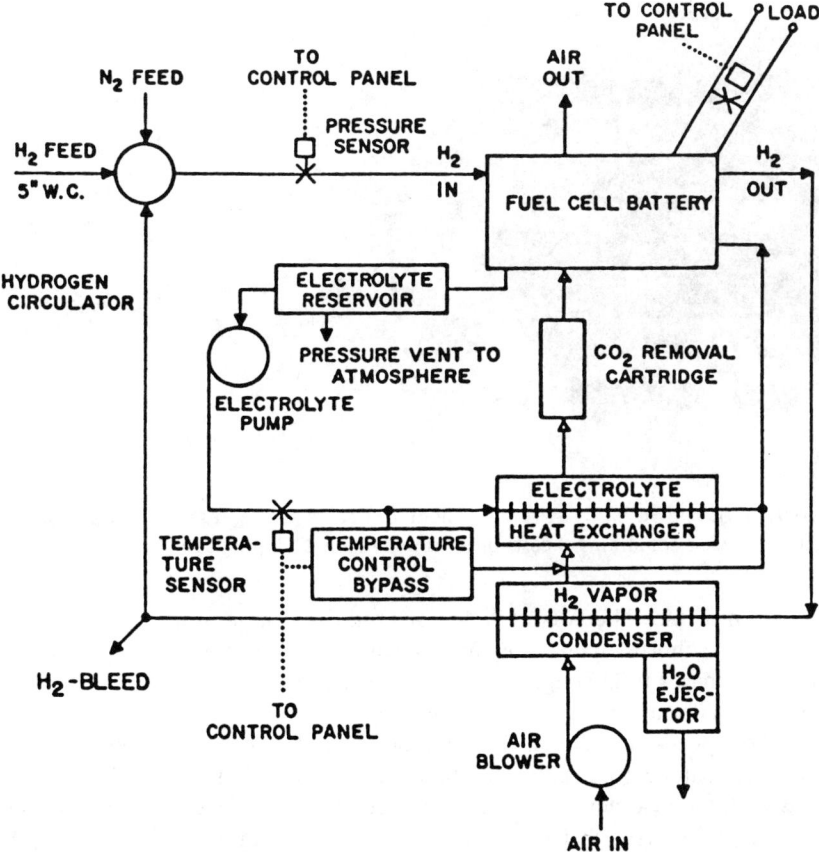

Figure 7-8. Layout of the hydrogen-air fuel cell system. In practice, the KOH tank was on the low-est level. The 90 V, 6 kW air-H$_2$ System consisted of three 30 V, 65 A (2 kW) stacks electrically connected in series, but KOH and gases circulated through the stacks in parallel manifolds.

- Extended periods of standing in a garage. The hydrogen tanks (a scuba-type design) and all pressure regulators were mounted on the outside of the car and all hydrogen lines inside the passenger compartment had a gas pressure below 20 cm water-column. The vehicle was admitted to public traffic under Ohio license no. KK 359. The required warnings in red letters; "Pressurized Gas" and "Hydrogen", had to be clearly visible. Smoking in the car was not allowed.

7.4.3.2. The Fuel Cell System

The six hydrogen tanks (weight: 13 kg each) were mounted on the roof and contained a total of 20 to 25 m^3 of hydrogen if filled to a pressure of 130 to 150 bar. This amount of hydrogen is equivalent to 60000 Ah or 45 kWh (at 0.7 V per cell). After system losses, about 33 kWh can be supplied to the motor. This means that a 5-hour operation at the

Figure 7-9. Photograph of one 30 V, 65 A (2 kW) module. Each stack represents an assembly of five 6 V, 65 A batteries, each containing 8 cells built from so-called "duplex cells".

6 kW output level is possible. The Austin A 40 car used 6 KWh at an average speed of 45 km/h in the city and suburbs, with frequent stops at lights. The energy density of the fuel cell system, which had a total weight (including gas cylinders) of 250 kg, was therefore about 140 Wh/kg. The calculated efficiency was 58%.

Operating at higher speeds or loads at the three-hour rate takes more power from the lead-acid batteries than can be replaced by the fuel cell unless some periods of idling are included, during which the lead-acid cells are recharged. This deficiency was only partly remedied by using eight 12-volt SLI-Batteries instead of the original seven. However, this measure did not increase the performance of the vehicle significantly. The conclusion was that the fuel cell system capability should be increased to a continuous 8 to 10 kW, with a peak 15 kW power output. Considering that the current density of the fuel cell electrodes produced in the 70s was limited to a maximum of 100 mA/cm^2, it can be expected that fuel cells of the 90s should fulfill the need for an increase of the specific power density.

7.4.3.3. The Hybrid System: Fuel Cell and Rechargeable Battery

The use of a rechargeable battery connected in parallel to provide starting power and peak power is a necessity. The balancing of the two systems is a matter of economy and performance requirements. The fuel cell system is the more expensive partner due to the catalyst requirement, even at a loading of 0.2 mg Pt/cm^2 and the cost of accessories is high. The continuous load rating is important. 150 mA/cm^2 can be accepted as being quite sufficient for air operation, considering that the life of the cells is also a function of the loading and control circuits, and accessories like pumps, blowers and condensers become more expensive, and their operation more critical, at higher outputs.

The lead-acid batteries used in the Kordesch fuel cell vehicle were standard 12 V SLI lead-acid batteries with a 90 Ah (20 h) 1 kWh-rating. The rating had to be downgraded to 0.5 kWh per battery due to the acceleration requirements demanded by city traffic (40 km/h reached in 10 seconds). With a total battery weight of 150 kg, specific energy density was only 22 Wh/kg. However, the short-time and peak power output was 200 and 400 A, respectively, which was quite sufficient for steep grades and acceleration maneuvers (passing at high speed). The calculated specific power of 200 W/kg was typical for SLI batteries. Unfortunately, the life time was not satisfactory. The lead-acid system was completely replaced three times during the fuel cell hybrid operation period (about every year), but there was no other choice – industrial type lead-acid batteries could not be used at all. A Ni-Cd battery would have performed much better, but could not be afforded.

Nevertheless important conclusions could be reached regarding street performance of a parallel hybrid system. The combined weight of fuel cells and lead-acid batteries was 400 kg, or about 1/3 of the curb weight of the car. The permitted load (after increasing spring tension, tire size and fitting larger brake drums) was 320 kg or four passengers. There was no trunk space left. The average energy usage per km was 0.2 kW. By including all battery capacities available for driving the car, a value of 40 kWh was reached, which corresponds to an overall energy density of 100 Wh/kg. A regenerative braking system and special radial tires were not used after experimental road tests. The savings were minimal.

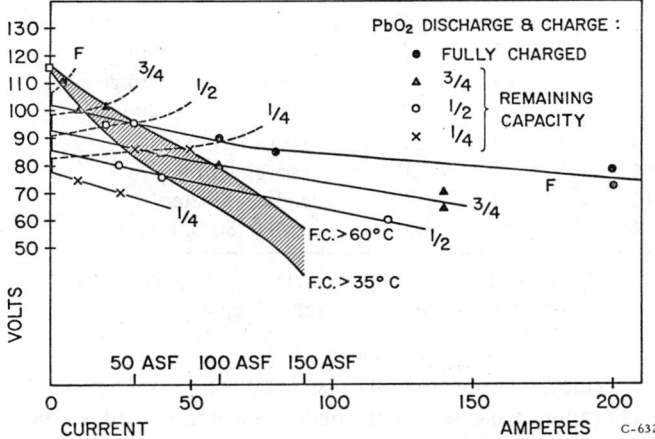

Figure 7-10. Characteristics of a Fuel Cell and Lead-Acid Battery System Operated in the Parallel Hybrid Mode.

The load take-over during driving is determined by the slope of the battery polarization curves. The temperature of the fuel cells and the state of charge of the lead-acid batteries determines the performance. 80 °C was a normal, easily-reached operating temperature, which also ensured the water removal balance up to 100 mA/cm^2. The loss of power at lower states of charge of the lead-acid battery was a very serious drawback.

Fortunately, the deficiency could be remedied by occasional separate recharging of the secondary battery from a power line. A 115 V, 10 A charger was installed in the car. The need for added power is also obvious from the following Table 7-3. Deep discharging the lead-acid battery reduces the cycle life-expectancy greatly (to about 50 cycles). The 6 kW output of the fuel cell is marginal, if continuous, unlimited driving is the objective. The performance of the car for city driving was quite acceptable.

Table 7-3. Examples of Recorded Hybrid Car Performance Data. -A: Discharge of the Lead-Acid Battery, +A: Fuel Cell Current, Amp

Hybrid	Output	Lead-Acid Battery			H_2-Air	Driving Conditions	
kW	V	A	St, of Ch	−A	+A	km/h, gear, grades	
10.6	85	125	4/4	− 80	45	80	Overdrive
9.2	80	115	3/4	− 60	55	75	(4th gear on)
7.9	75	105	1/2	− 40	65	70	(flat street)
6.7	70	95	1/4	− 25	70	65	(asphalt road)
19.5	75	260	4/4	−200	60	Moving from stand	
13.7	65	210	3/4	−140	70	(Driving up a 10%	
11.7	60	195	1/2	−120	75	grade in 2nd gear)	
8.1	90	90	4/4	− 60	30	67	4th gear
6.8	85	80	3/4	− 40	40	65	4th gear
5.6	80	70	1/2	− 20	50	60	4th gear
4.3	72	60	1/4	00	60	50	4th gear
4.1	70	58	1/2	+ 7	65	45	3rd gear
3.0	84	35	1/2	+ 8	48	25	3rd gear
2.0	88	22	1/4	+ 13	35	15	2nd gear
0.6	110	5	4/4	+ 5	5	(Parked Vehicle)	
2.0	100	20	3/4	+ 20	20	(charging of the	
2.9	95	30	1/2	+ 30	30	lead-acid battery	
4.3	85	50	1/4	+ 50	50	at 70 to 80 °C)	

Note: The consumption of the fuel cell accessories is ca. 200 W including a 110 V dc / 12 V dc converter charging another 12 V lead-acid battery serving the car lights & controls.

The conclusions gained from Table 7-3 were, that at least 6 kW of additional (spare) power should be available for charging of the secondary battery (which should not be a lead-acid battery). This requires that the fuel cell system should be able to produce 12 kW continuously.

Comparisons with gasoline or diesel battery hybrid systems gave essentially the same performance results. Later K. Kordesch experimented with a 350 cm^3 motor generator installed in place of a Air-H_2 fuel cell system. It delivered 120 V ac and, after rectification, could charge the lead-acid batteries on demand (depending on the state of charge) with up to 60 A. The output could also be controlled by changing the speed of the motor generator group. There was no transformer used in the ac circuit, and overloading was prevented by a thermal switch [29].

The need for a dc/dc power converter between fuel cell and secondary battery is frequently discussed. It definitely has an advantage of using a fuel cell system with a lower terminal voltage to charge a higher voltage secondary battery (e.g. for increased motor efficiency). The economy of the converter weight and its costs must be considered.

However, if used in the overload situation, it may fail even more rapidly in an avalange mode, if the fuel cell has no more voltage reserves. This can happen with air electrodes already working close to the gas diffusion limit (at 0.6 to 0.7 V per cell at 200 mA/cm^2, and far earlier with older electrodes). The author is of the opinion that a better solution is achieved by series-connecting a few (normally unused) reserve cells, already included in the stack, to boost the system voltage when required. This measure is also a solution for the aging problem of cells.

Control of the fuel cell system's sloping voltage/current characteristic can also be obtained by changing the air flow behind the cathodes. In the Kordesch vehicle, the air-fan speed had three, automatically-controlled steps. In this manner, the CO_2-Scrubber is also optimally used. Of course, this control mehtod is also inadequate, if the system is performing at the upper limit of its power capabilities for most of the time. A good self-regulation of the exchange currents between fuel cell and secondary battery existed at a cell voltage between 0.85 and 0.75 V. Current densities may vary between 20 and 100 mA/cm^2, adjusting to the state of charge of the secondary battery.

7.4.4. Improved Fuel Cell / Secondary Battery Hybrid Systems

Modeling of car performance with a zinc-bromine secondary battery, which has the same power output irrespective of state of charge and is not damaged by deep discharge, predicted a satisfactory vehicle behaviour. However, the weak point of the Zn/Br_2-system is its medium power capability and restrictive energy density per unit volume [30–32].

Improvements in the lead-acid battery field should be considered as possible achievements for the near future. The bipolar battery of the type developed by Dunlop in the 1980s would increase the power and energy density of present lead-acid batteries by about 30% [33, 34]. Recent work at Johnson Controls, Inc. (in a contract with NASA and the Jet Propulsion Laboratory, JPL) has produced experimental, bipolar lead-acid batteries with a power density of above 1 kW/kg at somewhat reduced energy densities (10 Wh/kg) [35].

Another future solution to the problem of finding a low-cost, hermetically-sealed secondary system as booster battery for a fuel cell hybrid may be the use of the rechargeable alkaline MnO_2-Zn battery [36]. This battery was marketed in 1993 in the USA by the Rayovac Co. as a cylindrical cell replacing the primary alkaline cells in consumer and electronic applications [37]. The cylindrical version has only a limited rechargeability if its initial energy density of about 90 Wh/kg is utilized. Complete discharge allows only 25 cycles, but hundreds of cycles are possible at a partial discharge to about 1/3 of its full capacity. This feature becomes especially pronounced if the cylindrical cells are used in parallel-series arrangements, which provide a very high power density, of up to several hundred W/kg. Such "bundle cell" arrays have been investigated for photovoltaic applications and for small vehicles like wheelchairs, shopping carts, etc. The study, named

"ENERSEARCH", was performed by Battery Technologies Inc., in a 1992–1994 contract with the Ontario Ministry of Energy[1].

The construction of this new, rechargeable battery (which is produced without mercury addition) in a bipolar version is expected to improve the energy density, cycle life and power output by a large factor. One reason is that the relatively thick, cylindrical cathodes are then replaced by thin-layer electrodes [38].

Table 7-4 gives the projected performance of a fuel cell hybrid vehicle using an alkaline air-hydrogen fuel cell and a rechargeable battery with 50 Wh/kg energy density and 300 W/ kg power density. Likely candidates; The Bipolar Lead-Acid or Bipolar Alkaline MnO_2-Systems.

Table 7-4. Projected Performance of a Hybrid Car System using an Alkaline Air-H_2 Fuel Cell and Improved Secondary Battery with a Projected Capability of 50 Wh/kg, 300 W/kg and 500 Cycles. The Candidates: The Bipolar Lead-Acid and the MnO_2–Zn Systems

12-15 kW Air-Hydrogen Fuel Cell Battery, weight:	220 kg
5 kWh rechargeable Battery, weight:	100 kg
Vehicle curb weight + 4 persons (or 2 + freight) *)	1500 kg
150 kWh lightweight H_2-Cylinder, weight: **)	150 kg
Hydrogen Consumption, H_2-weight per 100 km:	1.5 kg
Driving Range of the vehicle at 75 km/h: ***)	300 km
Efficiency of the system:	60%
Acceleration of the vehicle, 0 to 70 km/h:	10 sec
Maximum Speed:	100 km/h
High Power Level (20 min limit)	30 kW
Peak Power Output (1 min limit)	60 kW
Fuel cell working temperature range:	60 to 90 $^{\circ}$C
Readiness of the Secondary Battery	instantly
Cycle Life expectancy of the Secondary Battery	500 cycles
Full operation of the Air-Fuel Cell System in	1–3 min
Operating life the Fuel Cell electrodes:	5000 hrs
Max. Current density of the air electrodes:	150 mA/cm^2
Pt-metal catalyst on each electrode:	0.1 mg/cm^2

*)   An average sized car, to be used in the city and over land
**)   Recent reinforced hydrogen tanks are considerably lighter.
***) H_2-Storage as metal hydride may increase range to 500 km.

7.4.5. Other Fuels for Fuel Cell Systems

7.4.5.1. Ammonia as Fuel

Liquid ammonia remains an interesting carrier of hydrogen and its transportation in low pressure cylinders is common practice. Liquid hydrogen, with an energy density of 33 kWh/kg, is lowered to a useful energy density of 2.8 kWh/kg due to the use of a De-

---

1 Battery Technologies, Inc., (BTI) Richmond Hill, Ontario, L4B 1C3, Canada

war vessel. Liquid ammonia (5.4 kWh/kg) being transported in a low pressure steel container does not lose as much as highly compressed hydrogen, the energy density is only reduced to 3.3 kWh/kg. Another 15% are lost in the catalytical reactor which produces hydrogen.

Starting from hydrogen gas, only 10% energy content is lost for synthesizing ammonia, while 30% are needed for liquefying hydrogen. The boil-off is probably the biggest problem for the use of liquid hydrogen in relatively small containers. One percent per day is a practical figure.

An ammonia storage system was originally planned for the Kordesch vehicle, but cracking catalysts at that time needed 800 °C for the dissociation. Later (1981) the temperature was lowered to 450 °C and 10 bar pressure was used. Calculations predicted driving ranges for one-ton electric vehicles (fitted with an alkaline fuel cell system using ammonia as fuel) to be similar to those of cars with combustion engines operating on methanol. For comparison: methanol has an energy density of 5.6 kWh/kg, whereas isooctane has over twice as much at 12.7 kWh/kg [39].

The predicted advantages were therefore not exciting for a world using gasoline vehicles. For the military the use of fuel cells and a possibility of hydrogen storage in the form of ammonia was interesting because of the avoidance of heat detection, connected with any heat engine (motor generator) [40].

Today, the possible avoidance of CO_2 emission from vehicles may give environmentalists new arguments for the use of ammonia as fuel. The worldwide production is in the range of 30–40 million tons as a base for fertilizers, and its moderate cost is attractive.

Note: the strong smell of ammonia is a warning sign that some leakage exists. The exhaust of an ammonia fuel cell system should only contain nitrogen and water. Cu-salt absorbers can retain any traces of ammonia.

7.4.5.2. Hydrazine as Fuel

Starting with military objectives as stated above, hydrazine was also considered as a fuel for alkaline fuel cells in the early 70s. The Allis Chalmers Mfg. Co built a small, 3 kW golf cart operating on hydrazine and oxygen in 1963. N_2H_4 added to the KOH electrolyte forms hydrogen at the catalytic surface of a porous metal or carbon electrode. Nitrogen does not react and escapes as gas. A separator must be used to keep the hydrazine from the oxygen or air electrode surface, where it would also decompose and produce a "chemical short circuit". A system diagram is shown in Figure 7.11.

In cooperation between the US-Army and Monsanto Res. Corp., a 28 V, 20 kW hydrazine-air alkaline fuel cell system was developed in 1967 for an Army 3/4-ton truck (M-37). Two fuel cell battery modules were planned originally, but only one was used later in a hybrid combination with a lead-acid battery.

In 1966, Kordesch built an electric motorbike powered by two 16 V, 400 W hydrazine-air fuel cell modules and a Ni-Cd battery which could deliver peak currents up to 2.5 kW. The batteries could be switched in parallel or in series for speed control. The driving range of the vehicle was 100 km on 2 liters of a 64% aqueous hydrazine solution (monohydrate). A speed of 40 km/h was easily obtained with the 1 kW (continuous), 2 kW (peak power rating) dc motor. The motorbike was driven in public with the Ohio

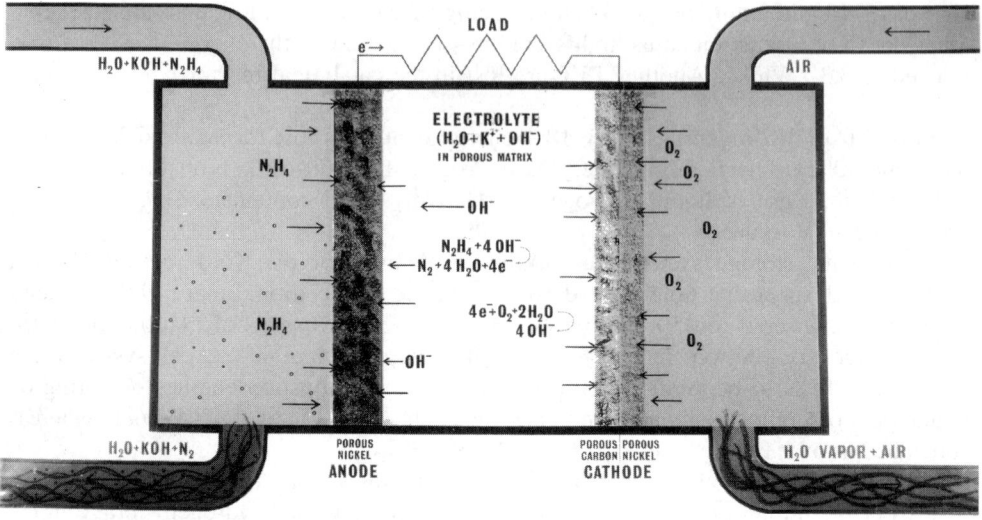

Figure 7-11. Principle of a Hydrazine-Air Alkaline Fuel Cell

Figure 7-12. The Kordesch Hydrazine Motorbike

Figure 7-13. The 300 W Hydrazine-Air System with KOH Circulation Loop and Cooling System. (The electrolyte contained only 1–2% of hydrazine, delivered on demand by an injector pump.)

license no. K-1815. It is now in the Technical Museum in Vienna, Austria [41], see Figure 7-12.

In 1972, a group of designers at Shell Research Ltd., Thornton, England, equipped a DAF-44 automobile with a 10 kW hydrazine-air fuel cell hybrid having a 72 V lead-acid battery in parallel. Vehicle weight: was 1380 kg and it had a top speed of 80 km/h [42].

Hydrazine was considered to be a practical fuel for fuel cells for electric vehicles, and at 50% efficiency the energy yield of the hydrazine monohydrate was about 1 kWh/liter. However, efforts to develop this fuel cell system were abruptly stopped after medical tests in the space industry discovered that even very small amounts of hydrazine were causing cancer. Trances of nitrose compounds are blamed for it.

Shin-Kobe Electric Machinery Co., Ltd., Japan, built a 4.24 kW hydrazine-air fuel cell stack for a 3 kW ac military power plant in 1982. As in previous designs for hydrazine batteries, the anodes were porous sintered Ni-plates, Pd-catalyzed, and the cathodes were PTFE-bonded carbon electrodes with Pt/Pd catalysts [43].

7.4.6. Overview – The Changes in Fuel Cell Technology

Two other major efforts in the development history of alkaline fuel cells for electric vehicles were started in the year 1972. One was the study of a French group (Institute Francais du Pétrol) interested in fitting a Renault 4-L with an 11 kW air-hydrogen fuel cell system (peak power 16 kW). Capable of a speed of 70 km/h, it should have had a range of 225 km. The vehicle was designed to weigh 814 kg, including a fuel cell weighing 165 kg. The motor weight was only 33 kg and the high-pressure H_2-cylinders had a weight of 40 kg, containing 2 kg H_2. This vehicle was supposed to operate without any secondary battery. Unfortunately, the project and market study was not continued [44].

A very significant European program started in 1972 was the alkaline hydrogen-air fuel cell hybrid bus project of Elenco NV in Belgium. Elenco also participated in the ESA/HERMES space H_2–O_2 fuel cell program which was cancelled in 1993. The bus program survived for two more years as an European effort ("EUREKA") and is discussed in the next section.

Alkaline fuel cell developments not connected with space suddenly declined in the early 1980s and phosphoric acid fuel cell systems were pushed to a high technological level.

They promised operation with reformers and CO_2-containing fuels, and alkaline cells were therefore largely discounted for electric automobiles and limited, with a few exceptions, to space vehicles. Companies building alkaline systems either went out of this business (ASEA in Sweden, Union Carbide in USA) or continued manufacturing alkaline space systems (United Technologies). Some produced special batteries (e.g. for submarines) like Siemens in Germany. "OXY" in France continued with H_2-air systems. The "OXY"-Alsthom project used a new light-weight, bipolar electrode assembly design and was supported by the Occidental Chemical Corp. in South Niagara Falls, NY., where it should have been used with hydrogen available from the electrolytic process for the manufacture of alkali and chlorine. Work was discontinued just when it had reached the multi-kW stage [45].

Phosphoric-acid fuel cell systems as propulsion batteries for electric vehicles had already been proposed in the mid-70s by the Energy Research Corporation. Steam-reformed methanol and propane were proposed as fuels. Demonstration vehicles were also constructed at the Los Alamos National Laboratories. A fuel cell powered vehicle "Workshop" was already held in 1977 [46–48].

Finally, in the early-1990s the phosphoric-acid system was used for several experimental vehicles (mainly buses). However by then expectations were reduced and progress has now been much slowed by the advances in improved membrane fuel cell technology. DuPont's "NAFION" and Dow Chem.'s membranes replaced some decades of General Electric's efforts with other "SPE" (Solid Polymer Electrolyte) membranes. Also, the new "PEM" (Polymer Electrolyte Membrane) fuel cells seem to outperform the 200°C phosphoric-acid systems in efficiency and can provide good performance even at 80°C. An electric city-bus demonstration program by Ballard in Canada turned out to be technically successful [49, 50].

However, the drawbacks were also revealed. One is the very high cost per installed kW (presently \$ 20000/kW), with no realistic signs of becoming economically feasible for electric road vehicles. For this reason, Siemens has decided to build only naval submarine batteries with such cells.

In Europe, alkaline H_2-air fuel cell batteries were receiving renewed attention for common vehicle propulsion purposes. The interest is partly a spin-off from Russian space technology, which is now demonstrating alkaline hydrogen fuel cells with excellent life and performance. The system ("PHOTON") was not yet ready for Russian space vehicles, but was tested in electric vehicle development projects. Tests performed by the European Space Agency (ESA) in 1993 show a high voltage level, high efficiency and very low degradation, outperforming many other systems in simplicity.

Dornier in Friedrichshafen, who recently started membrane fuel cell work after the earlier HERMES alkaline fuel cell project was cancelled, was successfully testing the Russian alkaline system originally started during the 1960s. At that time, Allis Chalmer's and Union Carbide's electrodes, the asbestos layers for KOH immobilization and General Electric's membranes were also tested in Russia as early as 1960. Prof. A. Frumkin was responsible for starting cooperation between the Russian Academy of Sciences and these US-companies [51, 52].

The US space program is still using the alkaline system and can demonstrate a very high state of reliability and performance in space shuttles. Some smaller, lightweigt NASA systems are performing even better and use other separators than asbestos (which is space approved and cannot be replaced unless an extensive testing procedure is conducted). The fuel cell stack is built from carbon-based electrodes with a very small amount of noble metal catalysts. Operation with air is also excellent [53, 54].

The lack of feasibility to operate on reformer fuels is therefore the only objection which can be mentioned against the use of alkaline fuel cells for electric vehicles. The complete removal of CO_2 from the air is necessary with immobilized electrolytes, but not with circulating liquid electrolyte. Gaseous hydrogen of technical purity is supposed to become a common fuel in the hydrogen technology of the 21st century. It can be stored as pressurized gas, as metal hydride and as liquefied gas. If combustion engines with hydrogen are envisioned for environmental reasons, why should alkaline fuel cells with at least twice the efficiency not do the job? In order to optimize the system for high acceleration and lower cost, a hybrid system would certainly be appropriate.

Following that sequence of events, the technology of fuel cells for electric vehicles has made a full cycle in system selection principles in the 30 years between 1964 and 1994.

The only competition for the alkaline vehicle propulsion system is therefore the membrane electrolyte fuel cell. This also existed in the 1960s and was considered the second choice in the US-space program, but confidence was lost due to its previous difficulties with reliability of the membrane and water removal problems. Cost was not a factor for space projects [55].

The exchangeability of components after a certain operating time is a new facet of distinction: the electrode stack is the part with the lowest cost in alkaline systems, but is the highest in membrane systems (still needing high Pt-metal catalyst levels). Alkaline stacks operate with technically pure gases, membrane cells need extremely pure gases. The water and heat removal accessories, pressurizing pumps (for air) and the control systems may finally be developed for mass production.

Fuel cell systems for power plants need the capability to operate on reformed fuels and to co-generate electricity and heat. The phosphoric acid, molten carbonate and solid oxide systems remain the preferred choices.

A summary of the characteristics of fuel cell systems operating on air and which may possibly be used in future electric road vehicles is given in Table 7-5. Worldwide efforts are being made to produce prototypes of electric vehicles operating on fuel cell power systems. Most of the projects are government-supported. Especially in the USA, the

Table 7-5. Characteristics of Air-Fuel Cell Systems for Electric Road Vehicles (Status of Technology: 1993).

System/Electrolyte	Temp.[°C]	Notes about the System:
Alkaline FC (AFC)	80–100	Circulating or immobilized KOH H_2-fuel only, needs CO_2 cleaner, carbon electrodes, low level Pt
Prot. Exch.Membr.(PEM)*	80–100	PE membrane is CO-sensitive but operates with CO_2 in air, needs very clean H_2, high level Pt.
Phosph.A.FC (PAFC)**	180–220	Gelled electrolyte, operates on reformed gas and air, low level Pt-catalyst on carbon.

* Polymer Electrolyte Fuel Cells (PEFC)
** Phosphoric Acid Fuel Cells (PAFC)

legislation concerning very low emission vehicles to be introduced at the end of the century has supported large programs, which also involved the automobile companies. Table 7-6 gives a survey about the mobile fuel cell programs which have been started in the nineties and are projected to be implemented before the year 2000.

Table 7-6. Fuel Cell Projects for Mobile Applications (year of effective start)

Country	Developer	Application	F.C.	kW	Fuel	Year
Belgium	Elenco, Eureka	Bus	ALK	78	Hydrogen	1994
USA	H-Power, Fuji	Bus	PAFC	50	Methanol	1994
Canada	Ballard	Bus	PEM	120	Hydrogen	1993
USA	EnergyPartner	Car	PEM	30	Hydrogen	1994
USA	Internatl. F.C.	Military	PEM	15	Hydrogen	1994
USA	Hamilton Stand.	Space	PEM	0.8	Hydrogen	?
USA	Analyt.Power	Military	PEM	10	Hydrogen	?
Germany	Siemens	Submarine	PEM	34	Hydrogen	1994
Italy	Ansaldo,DeNora	Bus	PEM	10	Hydrogen	?
Canada	Ballard	Car	PEM	tbd*	Methanol	1994
UK	Vickers	Marine	PEM	20	Methanol	?
Japan	RBD	Car	PEM	tbd*	tbd*	?
USA	General Motors	Car	PEM	60	Methanol	1997

* tbd – to be decided

7.5. Fuel Cell Vehicle Technology of the 1980s and 90s

7.5.1. Alkaline Fuel Cells for Electric Vehicles

7.5.1.1. The Elenco Fuel Cell System[2]

Elenco has developed multilayer gas diffusion electrodes based on PTFE-bonded graphite and carbon layers backed by a nickel-plated screen and a porous PTFE-layer on the gas side. The platinum content is about 0.3 mg per cm^2 and electrode thickness is less than 0.5 mm. In its automated manufacturing plant, the company can produce up 200 000 electrodes annually. This is equivalent to 2000 kW of fuel cell power output. The electrodes are set into frames by injection moulding, using ABS-type thermoplastic materials. The framed electrodes are friction-welded into 24 cell modules, each having an output of about 0.5 kW. This is the smallest standard unit which can be built as an operational module. The modules can be connected in series and/or parallel configurations to form the stacks needed for specific applications. The largest fuel cell installation built by

---

2 H. van den Broeck announced by the end of April 1995 that ELENCO N.V. discontinued their Fuel Cell Business.

Elenco contained 120 modules (52 kW) and it was a mobile power supply for the Belgian Geological Service. It was laid out for 220 V and was aimed at driving ground drilling machines.

The electrodes have a normal operational lifetime of 5000 hours, although 20000 hours of service have been obtained in some installations. The system uses pump-circulated KOH-electrolyte and a CO_2 -scrubber in the air inlet stream. The water is condensed out in the hydrogen loop, which contains a jet-pump[3].

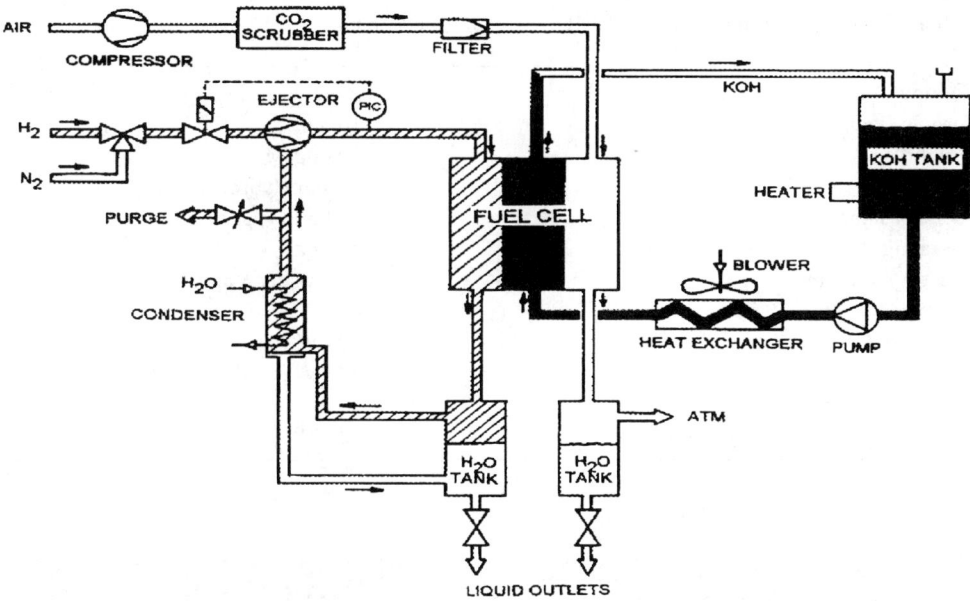

Figure 7-14. Flow Digram of the Elenco H_2/Air Fuel Cell System.

Elenco started in the electric vehicle field with a VW van powered by 32 modules (14 kW) of H_2-air fuel cells. The vehicle served as a "laboratory on wheels" and went through several modifications. The driving range with compressed hydrogen in steel cylinders was about 200 km. The vehicle used a dc-motor.

7.5.1.2. The EUREKA Bus Project

The project was initiated in 1989 with a study phase, and a cooperation agreement between the EUREKA partners was reached in May 1991. Bus construction started in October 1991. A demonstration of the completed fuel cell-operated vehicle was planned for late 1993, but has been postponed for 1994 [56].

---

3 H. van den Broeck, Alkaline Fuel Cells, ELENCO Brochure, 1993

The alkaline fuel cell system was the responsibility of Elenco N.V., Belgium, and the additional Ni-Cd booster battery was to be supplied by the SAFT Industrial Battery Group, France. (a subsidiary of Alcatel-Alsthom) The liquid-hydrogen supply was provided by Air Products, Industrial Gas Division, Netherlands. The dc-power converter system was constructed by Ansaldo Ricerche S.r.l., Italy. The estimated cost of the EUREKA project EU 201 was 4.25 million ECU. The design parameters of the demonstration bus are listed in the following Table 7-7.

Table 7-7. Design Parameters of the EUREKA demonstration bus

Configuration:	Articulated Design
Turning Radius:	10.9 m
Weight, empty:	20.600 kg
Fuel:	Liquid Hydrogen (LH_2)
Fuel Consumption:	15 kg LH_2 per 100 km
Operating Range:	300 km
Exhaust (emission):	Air, Water Vapour
Passengers:	80
Total Fuel Cell Power:	80 kW
Ni-Cd Buffer Battery:	775 V, 80 Ah
Traction System (ac):	800 V
Nominal Power of the Hybrid System:	180 kW

Figure 7-15. Picture of the EUREKA Demonstration Bus (1994) (Courtesy of ELENCO N.V.)

7.5.1.3. Fuel Cells for Underwater Propulsion

This application for fuel cells was sponsored by the Naval Sea System Command and produced a 700 kW System, which powered the Lockheed submersible Deep Quest during a series of 19 dives completed in 1980. The 30 kW hydrogen/oxygen PC-15 alkaline space power plant built by United Technologies was used in this endeavour. A phosphoric acid system and finally a SPE (Solid Polymer Electrolyte) from General Electric were subsequently considered by the US Navy. The reactants considered for the necessary reformers were JP-5 and methanol. Candidate oxidants for submarine purposes were either liquid oxygen or hydrogen peroxide [57].

7.5.2. Phosphoric Acid Fuel Cells for Electric Vehicles

7.5.2.1. The DOE-Program for Fuel Cells in Transportation

The program plan proposes a comprehensive, cost-shared partnership between government and industry to accelerate the R&D of fuel cell technologies and enable their commercialization by the auto industry soon after the year 2000. This mission focused program expands upon the present DOE fuel cell program. The goal is to establish Fuel Cell Vehicles (FCV) as a cost effective replacement for Internal Combustion Engine (ICE) vehicles as rapidly as possible [58].

The objectives are:

- Complete the demonstration of fuel cell Urban Transit Buses by the year 1995
- Establish the feasibility of multiple vehicle concepts to fabricate prototypes by 1998
- Establish commercial viability of FCV's in small fleets (50–200 vehicles) by 2000
- The first generation of commercial FCV's is provided by 2005 and
- begin sales of FCV's by 2011.

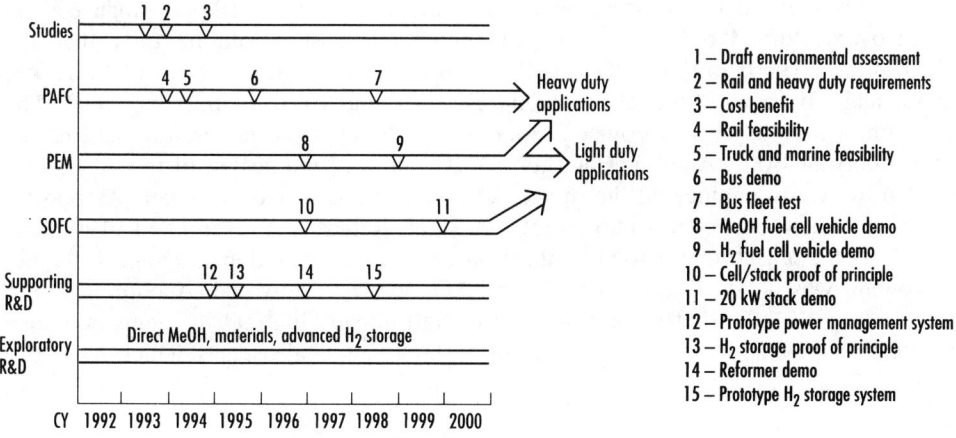

Figure 7-16. Schedule of the DOE-Activities for Fuel Cells in Transportation

The Technical Plan for Fuel Cells in Transportation includes:

- Phosphoric Acid Fuel Cell (PAFC) system designs for buses, locomotives and trucks
- Proton Exchange Membrane Fuel Cell (PEM) development for automobiles

7.5.2.2. The PAFC City Bus Program

This program is a multi-phase effort to develop and demonstrate a methanol-fueled PAFC propulsion system for urban transit buses. The reasons for selecting such an application for the development phases are many:

- A first generation FC-system needs a large vehicle to accommodate test instrumentation.
- The regularity of a city transit route permits testing under better controlled condition
- The environmental benefits are the largest in a city environment.
- The cost comparison is made with methanol driven (converted Diesel) urban buses.

The PAFC-System with a reformer (and in a battery hybrid configuration) was chosen because this FC-System was considered (in 1990) to be at a high prototype development stage. The PAFC should provide the average power, the secondary batteries the peak power.

At the start of PHASE I (1987) one of the industrial FC-contractors was Energy Research Corp. (ERC) and Bus Manufacturing, USA, Inc. (BMI), with technical advice provided by Los Alamos National Laboratory (LANL) and management by Georgetown University personnel. ERC had experience in building PAFC-systems for small vehicles (lift trucks). As secondary battery the plastic bonded Ni-Cd system was considered, used in a load leveling (shunt) mode (without electric power converter) to accommodate the hybrid set-up. Half-sized Brassboard Systems were built and evaluated.

The 240 cell-stack performance was shown to be 150 V, 240 A (36 kW). The methanol reformer used a commercial Cu-Zn catalyst for reforming and shift reactions, its weight was ca. 100 kg and could provide a hydrogen flow corresponding to a full 60 kW Fuel Cell output. The surge power (non-sintered) Ni-Cd Battery had 110 cells of 100 Ah capacity (5 h rate). Fuel cost estimates based on improved brassboard performance (with heat recovery) turned out that the diesel engine bus costs could be fully matched ($ 1.57/mile in 1990) [59]. In PHASE II (April 1991) the PAFC-system of FUJI was selected and a brassboard model of half the Bus-size underwent performance tests. The prime contractor was now H-Power Corporation[4]. The program integration became the responsibility of Booz, Allen & Hamilton. BMI continued the design of the 40 ft. City bus, SOLEC Corp. engineered the motor controller and power conditioner. As booster battery a lead acid battery was chosen [60]. A Status Report was given late in 1992 [61] In Phase II , the objective was to build three buses., to be delivered in 1993 and 1994, but the program was delayed, the full bus-size PAFC-was not delivered. According to the original plan, PHASE III (fleet testing) was to start already in 1994. It seems now also possible that the PEM-Fuel Cell Project may show technical aspects which lead to a review of the PAFC vehicle project.

---

4 H-Power Corp., 60 Montgommery Str., Belleville, NJ, 07109

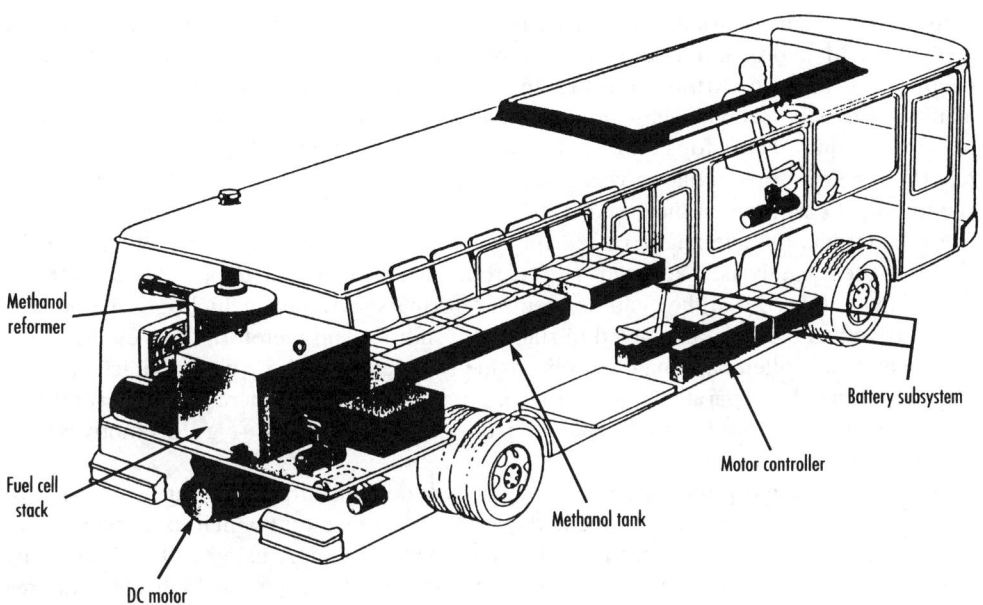

Figure 7-17. Lay-out of the Fuel Cell-powered Urban Bus for 25 passengers, sponsored jointly by the DOE and the California South Coast Air Quality Management District.

Figure 7-18. DOE-bus, showing the PAFC with reformer (San Diego, 1995)

In May 1992 a program was started by DOE to investigate the best way to design methanol and hydrocarbon reformers for vehicle fuel cells. A 30-month, cost shared contract was given to Arthur D. Little, Inc. of Cambridge, MA. The first phase is a feasibility study, the second phase is a fabrication and prototype testing effort. The best reformate requirements for PAFC and PEM fuel cells will be studied in the first part. A report on the reformer situation was given at the 1992 Fuel Cell Seminar [62]. Phase II is supposed to be finished at the end of 1994.

Methanol has the advantage of reacting with steam already at about 200 °C on suitable, low cost catalysts. Saturated hydrocarbons need far higher temperatures (700 °C) and it is not assured that they can be used in reformers suited for vehicles. Some processing difficulties can only be solved in stationary stations and therefore the view has been expressed that molten carbonate or solid oxide fuel cells may be a better choice because they do not need a separate reforming step. On the other hand, there is no question that these systems can only be used in large size mobile applications (e.g. locomotives) but not in street vehicles.

Other fuel storage possibilities like metal hydrides and liquid hydrogen are also to be compared with the reforming methods. Hydrogen generators which used magnesium and sea water as reactants were considered for Marine buoys, using SPE™-Fuel Cells [63]. Iron Sponge which reacts with water to produce hydrogen at high temperatures (over 700 °C) was suggested by H-POWER CORP. as a low cost hydrogen source. The basic idea is that iron sponge can be produced in large scale reduction furnaces from iron ores and can be stored and transported in small containers or in Truck-loads alike [64]. The thermal balances have not been clearly stated and therefore studies have been made at the Technical University Graz to establish the feasibility of this concept [65].

7.5.2.3. The Engelhard Industries "Liquid Cooled System"

Engelhard Industries has designed and built a 5 kW PAFC fuel cell stack and auxiliary component already in the early 80's. Different to the ERC System, which was air-cooled, it was liquid-cooled with plates built between the cells of the stack. Finally the system was installed in lift-trucks at Ft. Belvoir [66].

Prototypes were initially also considered to be used in the DOE (Georgetown). A special feature of the Engelhard PAFC system was the Bus Program. It was based on the 50 kW dielectric liquid cooled Engelhard PAFC system with methanol as reformer fuel. Each 25 kW stack was planned to contain 176 cells, operating at 180 to 204 °C. The original objective was an on-site integrated system which would only cost $ 200 /kW with a life time of 5 years. 0.5 mg of Pt per cm^2 of carbon electrode surface was used. A special feature of the Engelhard PAFC system was the development of low-cost bipolar electrodes which allowed exchange of the electrolyte by a wicking system, thereby reducing the degradation. The system was not pressurized and (in contrast to the gas cooled system of Westinghouse) had only a low power requirement for pumping of the "dielectric" liquid through the cooling plates interspersed in the stack between each 4th or 5th cell [67]. Engelhard discontinued production in the late 80's.

7.5.3. The Proton Exchange Membrane FC for Electric Vehicles

7.5.3.1. The DOE-PEM Fuel Cell Program

The Department of Energy's R&D Program sponsors basic and applied research on proton-exchange membrane (PEM) fuel cells which is believed to have significant advantages over the PAFC-system, once it is fully developed (see Figure 7-19):

- Reduced weight and size, due to high current density output.
- Fast start up from room temperatures and operation at 80–100 °C
- Lower cost over a longer operational life span.
- Operation on air (tolerating the CO_2) and simple reformates.
- The immobilized electrolyte should make the stack design (sealing) easy.

The principal difference to the alkaline system is that the proton (H^+) is the migrating species and the water is produced at the cathode (oxygen electrode). With the PAFC fuel cell it shares the possibility to use CO_2-containing hydrogen as fuel gas. However, there are also serious disadvantages which may not be easily eliminated:

- A high sensitivity to a few ppm CO in the fuel , requiring extreme cleaning efforts.
- The high cost of the platinum, which is presently used at the 4 mg/cm² level.
- The high cost of the membrane. in spite of great efforts to lower the price.
- The system must be pressurized, which leads to inefficiencies on the air side.
- The water content of the fuel gas must be kept high to prevent dehydration of the
- membrane and an increase in resistance.

The only point which has been solved recently at the Los Alamos Natl. Laboratory: in experimental cells the Pt-level has been lowered to 0.1 mg/cm² with a tolerable loss in cell performance [68]. The other points have been the subjects of development efforts: an oxidizer must be added to the fuel to eliminates the CO. Presently just air is added to the fuel to reduce he CO from about 0.5% to a few ppm. A turbocompressor pressurizes the air to at least 3 bar. An additional water removal system must be designed, because extra water is added to the gas entering the stacks for keeping the humidity above 400 mm H_2O [69].

The very high current densities, which are possibly drawn from the PEM electrodes led to suggestions to operate without a booster battery (Ballard). However, there is a trade-off between power output , need for expensive accessories, size and above all, **cost**. Heat management is also an important point. Therefore, the DOE proposal of a hybrid system is really justified.

The PEM-Program of DOE has been divided in four phases:

Phase I   Testing of the state of the art: Study of several 5 kW PEM Systems of Ballard Power Systems; New stack fabrication also by Ballard as Prime Contractor [70]
Reformer Development and fuel cell testing: Los Alamos National Laboratory
Membranes: Dow Chemicals Corp.
Catalysts: General Motors Corp.
Vehicle System: General Motors Advanced Engineering
Overall Management: Argonne National Laboratory

Phase II   Starting in 1993: Feasibility up to 25 kW systems
Phase III   Starting in 1995: Scale up to 50 kW
Phase IV   Starting in 1996: Vehicle testing.

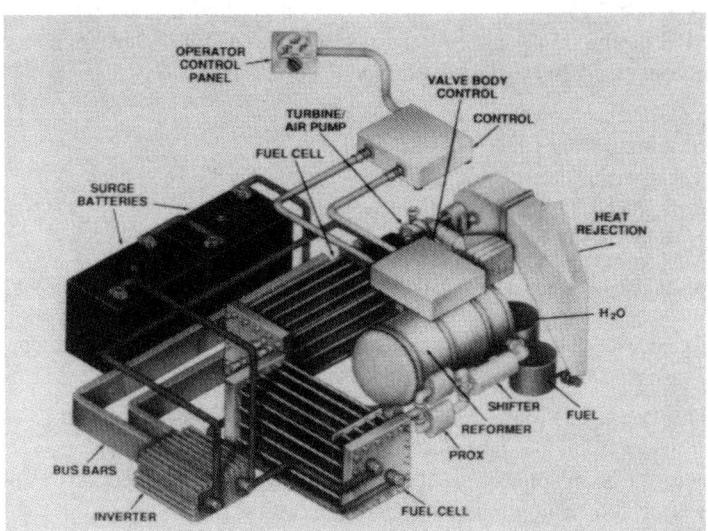

Figure 7-19. Conceptual Design of 10 kW PEM Fuel Cell System (Courtesy of the Department of Energy)

It should be recognized that the idea to use an organic ion exchange membrane as solid electrolyte and separator in an electrochemical cell was first described in the early 1960's by W. T. Grubb of General Electric Co. and that it took nearly 40 years until the first practical applications were found. The use in vehicles was always contemplated, research into membrane cells for space applications was a continuous project for General Electric Co and NASA Space efforts, but the key question was the reliability and the absence of "pinholes" in the membranes. The situation changed when DuPont Co. introduced the NAFION membranes. Further progress after the introduction of new membranes by DOW led to the general decision that electric vehicles should use PEM-Fuel cells [71]. The Company which was most active in the field and practically assured a covering of the field was BALLARD in Vancouver, BC, Canada.

7.5.3.2. The Ballard Power System PEFC Activities[5]

Ballard in conjunction with the Province of British Columbia and the Government of Canada has begun a program to develop a commercial transit bus powered by Ballard's PEM Fuel cells. PHASE I began in November 1990. The bus was designed to operate

---

5 Ballard Power Systems, 107–980 West 1st Street, North Vancouver BC, Canada V7P 3N4

solely on the Fuel Cell System (later with a secondary battery only to be used for the storage of energy from regenerative braking). The goal is to demonstrate the same performance capability as the corresponding commercial Diesel bus. The hydrogen is supplied from 6 steel cylinders stored under the bus floor [72]. Optimization is planned in PHASE II of the program.

Figure 7-20. Picture of the Ballard PEM-Fuel Cell operated Bus for 32 passengers. (Courtesy of Ballard Power Systems)

The bus power plant consists of 24 of the 5 kW stacks arranged in a series-parallel string providing 120 kW of gross power. Air is supplied by a combination of a motor-driven automotive supercharger and a turbocharger driven by the air exhaust stream. The fuel is compressed hydrogen stored in transportation-approved natural gas cylinders under the bus frame. This Bus has already driven over 2000 km in Vancouver, in Los Angeles, and in Sacramento, California[6].

The next tables 7-8 to 7-17 provide a summary about the technical features of the fuel cell powered ZEV bus from Ballard[7]. (ZEV = Zero Emission Vehicle)

---

6 until April 1994
7 Courtesy of Ballard Power Systems

Table 7-8. Full Load Performance

Gradability	Start on 20% grade
	Maintain 30 km/h (20 mph) on 8% grade
Acceleration	0–50 km/h (0–30 mph) in 20 sec
Top Speed	70 km/h (45 mph)
Range	160 km (100 mi)

Table 7-9. Fuel Cell Power Plant

Fuel Cell Type	24 Ballard PEM MK5D stacks
Arrangement	3 parallel strings of 8 in series
Voltage Range	160–280 Vdc
Gross Power	120 kW (160 HP)
Operating Fuel Pressure	207 kPa (30 psig)
Operating Air Pressure	207 kPa (30 psig)
Operating Water Temperature	70–80 °C (160–175 °F)
Emissions	Zero

Table 7-10. Bus Chassis

Manufacturer	National Coach Corporation Model RE-32
Body	Welded tubular steel floor, sidewalls and roof
Exterior Sidewall	Pre-baked galvannealed steel sheet
Roof	Full integral design with tubular steel roll cage
Bumpers	Steel reinforced rovel
Insulation	Sidewalls and roof: 38 mm (1.5 in) fiberglass
Wheels	Polished aluminium 571 × 171 mm (22.5 × 6.75 in)
Tires	229 × 571 m (9.0 × 22.5 in) – 12 ply
Brakes	Power hydraulic disc/dynamic braking from traction motor
Transmission	Allison AT-545 4-speed automatic
Seating Style	Transportation Seating Corp Model 1111
Seating Capacity	21
Wheelchair Accessibility	RICON S-2000
Gross Vehicle Weight	10 000 kg (22 000 lb)
Suspension Type	Leaf spring
Front Suspension Rating	3410 kg (7500 lb)
Rear Suspension Rating	6810 kg (15 000 lb)
Overall Length	9754 mm (32 ft)
Overall Width	2438 mm (96 in)
Overall Height	2870 mm (113 in)
Floor Height	896 mm (35.3 in)
Interior Height	1943 mm (75.6 in)
Wheelbase	4521 mm (178 in)
Steering	Hydraulic power steering
Turning Radius	9754 mm (32 ft)
Drive Axle Ratio	7.17:1

Table 7-11. Fuel System

Fuel Type	Gaseous hydrogen
Storage Pressure	20 680 kPa (3000 psig)
Delivery Pressure	Regulated to 207 kPa (30 psig)
Recirculation Method	Ejector
Storage Method	Fibreglass /Aluminium cylinders
Storage Capacity	150 scm (5300 scf)

Table 7-12. Air Compression System

Compression	Up to 207 kPa (30 psig)
Delivery Capacity	7362 lpm (260 cfm)
Drive Mator	Uniq Brushless DC motor / cntroller 24 kW @ 5000 rpm, 140–200 Vdc

Table 7-13. Traction Drive System

Motor Type	Nelon DC shunt field
Controller Type	Saftronics 2 quadrant DC / DC IGBT chopper
Controller Frequency	400 Hz
Controller Input Voltage	160–280 Vdc
Motzor / Controller Efficiency	80%
Power Output (continuous)	80 kW (107 HP)
Motor:	
Base Speed	1900 rpm
Field Weakened Speed	up to 3000 rpm

Table 7-14. Electrical System

Main DC Link Voltage	160–280 Vdc
Starting Battery and Charger	
Voltage	144 Vdc
Type	12 Lead /Acid
Capacity	20 Ah
Auxiliary Battery and Charger	
Voltage	12 Vdc
Type	Lead /Acid

Table 7-15. Cooling System

Fuel Cell Power Plant	Water cooled
Water Pump Delivery	265 lpm (70 gpm)
Motor Type	Nelco DC series field
Fan-cooled Radiator	1.39 m^2 (15 ft^2)

Table 7-16. Safety Features

Monitoring Power Plant	Low voltage shutdown
Gas Detection	Catalytic bead hydrogen sensors
Electrical	Single / double fault

Table 7-17. Environmental

Operating Temperature	3–40 °C (35–105 °F)
Storage Temperature	3–65 °C (35–125 °F)
Vibration	As per MIL-STD-180E Category 1

A program is now under way at Ballard to construct a full size transit bus also entirely powered by PEM fuel cells and operating on hydrogen fuel. The first bus was operational in 1995. The power plant for this is based upon an improved stack, which delivers twice the power of the MK-V stack. The 250 kW bus power plant will fit into the same space as is presently allocated to the diesel engine on the equivalent diesel bus. Hydrogen, stored on the roof, will provide a range of over 400 km.

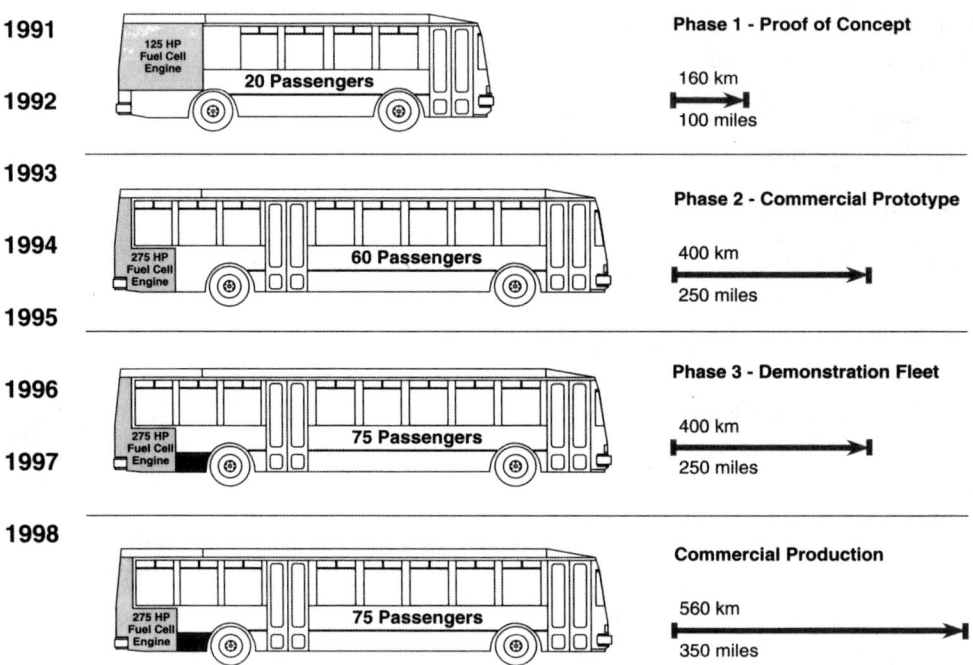

Figure 7-21. Ballard's Bus Commercialization Plan (Courtesy of Ballard Power Systems)

The second phase of Ballard's heavy duty engine development program commenced in 1994, as Ballard began work on a fuel cell engine for a full scale commercial prototype bus. California's South Coast Air Quality Management District joined the original sponsors to fund the Can-$ 6 million program. This second bus will be unveiled in mid-1995.

Figure 7-22. Ballard's prototype 275 horsepower (205 kW) fuel cell engine (phase 2)

To commercialize the Ballard Fuel Cell transit bus engine, Ballard entered into an alliance with New Flyer industries, one of the leading manufacturers of urban transit buses in North America. Ballard's current 275 horsepower (205 kW) fuel cell engine provides the same performance and fits in the space originally designed for the diesel engine.

In Phase 3, demonstration fleets are planned to be built and tested by transit operators in 1996–1997. Passenger capacity will be further increased to 75 with the smaller and lither third generation 25 kW fuel cell stacks. Range will be increased to 350 miles (560 km) by taking advantage of regenerative braking. Commercialization of ZEV electric buses is scheduled to begin in 1998.

Ballard has also collaboration agreements with different auto manufacturers with specific manufacturing expertise and an intimate knowledge of end-users. Ballard's cooperation with Daimler-Benz started in 1993 under a four-year agreement to develop a high power density PEM fuel cell stack using methanol and related manufacturing processes. The Daimler/Ballard collaboration successfully met all planned milestones for 1994. The agreement provides that Ballard receives US-$ 15 million from Daimler Benz for conti-

Figure 7-23. Ballard's 275 horsepower (205 kW) PEFC engine fits into the same space as the diesel engine and will meet the performance requirements of transit authorities without emitting any pollution. (Courtesy of Ballard Power Systems)

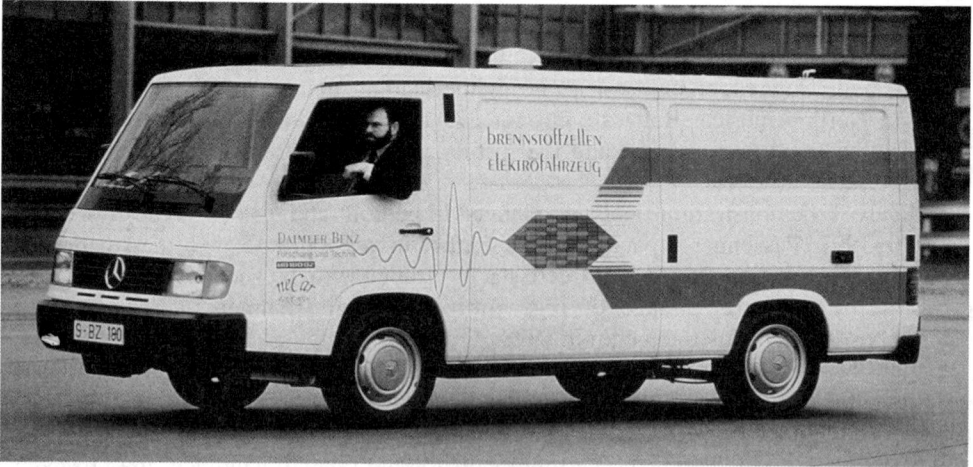

Figure 7-24. Daimler-Benz fuel cell powered Mercedes MB 180 van "NECAR" ("New Electric Car") (Courtesy of Daimler Benz)

nued development of the fuel cell stack, and both parties contribute an additional US-$ 10 million for the development of components [73–75].

General Motors is using Ballard Fuel Cells in a US DOE project to develop a methanol fueled hybrid electric vehicle. Ballard started the US-$ 4.3 million second phase of the project in 1994. This second phase has a goal of producing a 30 kW fuel cell engine for a 60 kW hybrid electric power system by 1996.

7.5.4. International Fuel Cells Corporation (IFC)

IFC, as a subsidiary of United Technologies Corp (UTC) acquired the rights to the solid polymer electrolyte (SPE) fuel cell technology as practiced by the General Electric Corp. in 1985. At that time the limitations of the PEM-technology were the life times of the membranes and the current densities achieved with cells, using membranes of Dow Chemicals. IFC could show that great improvements were possible over the General Electric stack performance (300 mA/cm^2 at 0.7 V) to nearly 800 mA/cm^2 at 0.7 V and 1500 mA/cm^2 at 0.6 V [76]. For comparison: Ballard reported already 2000 mA/cm^2 at 0.7 V at 3 bar pressure.

7.5.5. Other PEM-Fuel Cell Programs

7.5.5.1. Energy Partners

Energy Partners is a Company designing and developing proton exchange membrane fuel cells for applications in automobiles, buses, locomotives, submarines and military applications. One successful and widely publicized application was a two-man submersible, powered by a 1.5 kW PEM system. It has completed 16 dives by 1993. A 20 kW low pressure air breathing PEM Fuel Cell system is designed to power a special electric vehicle, the "Green Car" [77–79].

The Tables 7-18 to 7-20 summarize the technical features of this project [80].

Table 7-18. Fuel Cell Specifications

Type	Proton Exchange Membrane
Manufacturer	Energy Partners, Inc.
Membrane	Du Pont Nafion™
Collector Material	Graphite
Reactants	H$_2$/Air
Cell Active Area	0.84 sq. ft. (780.4 cm^2)
Reactant Operating Pressures	20 psig (1.4 bar)
Number of Stacks	3
Number of Active Cells	180

Figure 7-25. The Green Car™ (Courtesy of Energy Partners)

Table 7-19. Vehicle Specifications

Power Systems	Hybrid Fuel Cell / Battery
Motor Type	DC Brushless with Regenerative Braking
Motor Power	35 hp (26 kW)
Motor Maxiumum Torque	650 in-lbs at 3,300 rpm
Motor Maximum Speed	4000 rpm
Transmission	5 speed manual
Chassis/Body Type: built by Consulier Automative	Foam Core Advanced Composite Monocoque
Vehicle Weight	3000 lbs (1361 kg)
Hydrogen Storage	413 scf (11.7 m^3 STP)
Hydrogen Storage Pressure	3000 psig (207 bar)

Table 7-20. Vehicle Performance

Fuel Cell Power System Output	15 kW
Nominal Stack Voltage	125 V
Amperage at Nominal Stack Voltage	120 A
Maxiumum Vehicle Speed	60 mph (96.6 km/h)
Acceleration 0 to 30 mph (48.3 km/h)	10 s seconds
Range (city driving cycle)	60 miles (96.6 km)

7.5.5.2. United Technologies – Hamilton Standard Div.

In Navy operations the emphasis is on long life, high power and nearly perfect operation as far as sealing of all fluid compartment is concerned. Therefore, the same technology as used in nuclear submarines is applied. The difference to the carbon/graphite constructions of previous PEM-Fuel cells is the sheet metal approach, providing flexibility in situations where graphite may break. Solid Polymer Electrolyte (SPE) is the trademark of Hamilton Standard describing their membrane systems with NAFION membranes. SPE applies also to the hydrogen-oxygen electrolysers of that company. NAFION 120 and NAFION 117 were used at that time (1990). A special water removal system with a minimum of noise was used in the US- and ROYAL NAVY performance tests at Windsor Locks, Connecticut [81].

7.5.5.3. Ergenics Power Systems, Inc. (EPSI)

EPSI was formed in 1985 as a subsidiary of ERGENICS, Inc. and is developing fuel cells based on its ion exchange membrane type cells. The technology is intended to cover transportable devices in the 2 kW range. Industrial grade compressed hydrogen and H_2 from metalhydrides are the preferred fuels. NASA has contracted a 28 V, 7 A, 34 Ah power source for use in the space station. The energy density is comparable with AgO-Zn batteries. as a special feature, gas humidification is not needed due to a water removal system using wicks. The current density with hydrogen and oxygen at 0.8 V is 200 mA/cm^2 under ca. 1 bar pressure. at 60 °C. Platinum loading of the electrodes is 6 mg Pt/cm^2 With air as oxidants, at a 2 × stoechiometric flow, 0.67 V are obtained at 200 mA/cm^2. Peak currents up to 500 mA/cm^2 are possible [82]. The PEM-development efforts of IFC were geared at using the stack technology which was highly developed at UTC for PAFC systems in the (then) new field of improved proton exchange membranes. In 1988 the developments were not specifically directed at electric vehicle propulsion systems. Bipolar plates at IFC has already reached the size of square meters, to be used for power stations.

7.5.5.4. Siemens AG

Siemens has been developing alkaline fuel cell systems for submarine applications in the 1970's and 80, multiple stacks of hydrogen-oxygen systems with unit sizes of 20 kW had been built. Since 1990 the Siemens Technology changed to the PEM-Technology which it licensed from the UTC-Hamilton Division. Therefore, also membranes of DOW Chemicals are used [83]. Electric street vehicles are not considered in this program, cost is not the first point of limitations, reliability and long life of the oxygen-hydrogen system is of greatest importance. The electrode technology is similar to that of BALLARD, the Pt metal catalyst (several mg/cm^2) is deposited onto the membrane directly [84, 85].

7.6. Technological Outlook and Conclusion

The feasibility of electric vehicles to operate on fuel cell power plants is a fact. The choice of the type of fuel cells is still open but has been narrowed to the low and medium temperature fuel cell systems. The idea of oxidizing hydrocarbon fuels directly in high temperature systems, preferentially in solid oxide fuel cells is at least at the present time speculative. It should not be forgotten that the electrochemical conversion is most efficient at low temperature and that it is not very practical to use the waste-heat of such "dispersed power generation systems" as automobile propulsion systems. This is, of course, different with stationary power plants.

It appears that the type of fuel cell to be used in street vehicles will ultimately be decided by economic questions and answers. Such an investigation has recently been made by Olle Lindström [86].

Applying the teachings of the historical developments to predictions about electric vehicle fuel cell developments, the AFC may even have a better chance than the PEM-systems. This statement may appear prejudiced and colored by personal experiences, but it can stand to be challenged. Even simple "hindsight" (which is usually right) can come to similar conclusions. However, there are also hard facts:

The best performing fuel cells are the alkaline "Space Vehicle Systems". Since 1970 they are performing reliably on oxygen and hydrogen at high current densities (1 A/cm^2, at 0.75 V) The Russian PHOTON performance (Figure 7-26) needed 40 mg Pt/cm^2, NASA Shuttle units can do it at low Pt-loadings. Catalyzed carbon substrates have been successfully used in Air electrodes, which since 1960 deliver current densities between 100 and 150 mA/cm^2 at 0.75 V. Only PEM electrodes can do better now, but at a tremendous increase in cost and complexity of the system.

Practically AFC and all systems have arrived at the thin electrode and / or bipolar construction of the battery stacks. The alkaline "OXY" was an excellent prototype system, already geared for mass production before the company went out of business. The ELENCO alkaline system was the only publicly demonstrated fuel cell which looked economically affordable. In this connection it should also be mentioned that no large funding has gone into the development of alkaline systems for street vehicles during the last 15 or 20 years.

The life-time question is always brought up with the goal set at 40.000 hours. This is only a requirement for power stations. It is questionable if fuel cells which should operate in the field and not under strictly controlled conditions will be able to approach such time periods without the possibilities of replacement of components (electrode stacks, electrolytes or accessories).

Figure 7-26. "Photon" fuel cell generator (Courtesy of the European Space Agency "ESA")

The automobiles of the 1960's may provide a good comparison: oil changes were frequent, lubrication needs were apparent and the exchange of components (pumps, fans) was regularly anticipated. Still, today in the 1990's we do not expect a gasoline engine to operate for more than 2000 hours without an overhaul. Why should a street vehicle fuel cell stack be required to do much better, especially if it is built mainly from essentially low-cost materials (like carbon) and a small amount of catalysts, which can be easily recovered.

The key wording is therefore: **hours of "operation"**. In this matter the City Car of Kordesch has already shown proof of concept and its reliability details. The most significant difference, if compared with all previously and later built and driven fuel cell vehicles was the possibility to shut down completely between operational periods, like a combustion engine usually does. The "unavoidable" consequences were:

- The electrolyte could be exchanged when visibly "dirty" or partially carbonated (pH-change!).
- The gas manifolds could be washed when clogged.
- The well known process of "regeneration of catalysts" by exposing the electrodes occasionally to the opposite gas, (which meant polarization in the other direction) was performed at each shut-down. This method was patented by other researchers later.

The lay-out as a Hybrid Fuel Cell System was of great importance:

- The parallel connected (shunted) lead-acid battery prevented damaging parasitic currents and high-load electrode polarization into the voltage range where Pt or Pd-solubility increased.
- The parallel secondary battery eliminated all operational waiting periods and driving delays.

Last, but not least, the circulating KOH-system was self-regulating in its concentration range and cooling problems or gas cross leakage possibilities did not exist. The fact that the water was produced at the anode made the elimination easy by simple H_2-recirculation.

The design of both electrodes as hydrophobic porous carbon sheets with a gas permeable repellent backing was standard with Union Carbide electrode made in the 1970's. The electrodes were wetproofed in steps (by multiple spraying with increasingly repellent catalyzed active carbons). This procedure was probably responsible for the re-establishing of properly functioning interfaces, each time the stacks were re-filled with KOH after a shut-down. It should be noted that the carbonate formation was also minimal because the electrodes did not wet deeply, as is e.g. the case with Pt-catalized graphite/carbon PTFE layers. Surface wetting was also partly "reversible" and not "permanently established", like in cells with microporous asbestos layers or other separator-type immobilized KOH-electrolytes.

Experience and post-mortem studies of the stacks showed that the failure mode of the cells was mainly leakage at the frames, especially where the polysulfone sheets and metal terminals met. A non-uniformity of the electrode-interfaces related to gas- and KOH-manifolding was also noticeable. This was not surprising because the whole system was hand built. In 3 years only two 6 V H_2-air modules (out of 15) had to be exchanged for sudden loss of performance.

Unfortunately, the shut-down of electrode production and fuel cell testing at Union Carbide Corp. in the mid-1970's prevented a more detailed analysis and more experimentation with batteries and stacks in the intermittent operational mode of an electric automobile. At other companies the efforts were directed to immobilized electrolytes and finally to PAFC-systems.

CONCLUSION: It may pay to re-investigate alkaline fuel cells with circulating electrolyte from the view points of actual combustion engine duty cycles.

References for Chapter 7

[1] K. Kordesch, "The Electric Automobile", Chapter 2, in: "Lead Acid Batteries and Electric Vehicles", p. 201-413, K. V. Kordesch, ed., Marcel Dekker, New York 1977
[2] K. Kordesch, J. Oliveira, "Fuel Cells", Ullmann's Encyclopedia of Industrial Chemistry Vol. A12, pp. 55–83., VCH Verlagsgesellschaft, Weinheim, 1989
[3] K. Kordesch, "Brennstoffbatterien", Springer Verlag, Wien, New York, 1984
[4] P. G. Patil, Journal of Power Sources 37 (1992) 171
[5] E. J. Cairns, E. H. Hietbrink, "Electrochemical Power for Transportation", in Comprehensive Treatise of Electrochemistry, No 3, Electrochemical Energy Conversion and Storage, ed. by.: J. O'M Bockris, B. E. Conway, E. Yeager, R. E. White, Plenum Press, New York and London, 1981, pp. 421–504
[6] E. J. Cairns, "Power Sources for Electric Vehicles", 44th ISE Meeting, Berlin (Germany, Sept. 5–10, 1993, p. 412
[7] K. Kordesch, "Power Sources for Electric Vehicles". in Modern Aspects of Electrochemistry, Vol. 10, ed. by.: J'OM Bockris, B. E. Conway, Plenum Press, New York, 1975, pp. 339–443

[8] K. V. Kordesch, J. Electrochem. Soc. **125** (1978) 77C-91C

[9] K. Kordesch, "Brennstoffbatterien" Chapter 8, Brennstoffbatterien für den Fahrzeugantrieb, Springer Verlag 1984

[10] F. T. Bacon, BEAMA Journal **6** (1954) 61

[11] H. H. Chambers, A. D. S. Tantram, In: "Fuel Cells", Vol.1, (J. T. Young, ed.), Reinhold Publ. Co., New York, 1960

[12] J. L. Platner, D. Ghere, P. Heiss, 19th Annual Power Sources Conference, Atlantic City, 1965, pp. 32–35

[13] E. Justi, A. Winsel, "Kalte Verbrennung", Franz Steiner Verlag, Wiesbaden 1962. English Translation: "Fuel Cells, Pergamon Press, New York 1965

[14] E. Wendtland, A. Winsel, Chem.-Ing.-Tech. **39** (1967) 756

[15] O. Lindström, CHEMTECH (1988) 490–497

[16] O. Lindström, CHEMTECH (1988) 553–559

[17] O. Lindström, CHEMTECH (1988) 686–693

[18] O. Lindström, CHEMTECH (1989) 44–50

[19] O. Lindström, CHEMTECH (1989) 122–127

[20] K. Mund, F. v. Sturm, Electrochimica Acta **20** (1975) 463

[21] F. v. Sturm, "Elektrochemische Energieversorgung", Verlag Chemie, Weinheim, 1969

[22] W. Vielstich. "Brennstoffelemente – Fuel Cells", Verlag Chemie Weinheim/Bergstraße 1965, Engl.: J. Wiley & Sons, London 1970

[23] The General Motors Corp. "Electrovan", Papers No. 670176 and 670181, Soc. of Automotive Engineers Congress, Detroit 1967

[24] C. E. Winters, W. L. Morgan, SAE-Paper 670182. Detroit 1967

[25] M. B. Clark, W. C. Darland and K. V. Kordesch, Electrochem. Technology **3** (1965) 1966

[26] K. V. Kordesch, "Hydrogen/Lead Battery Hybrid System for Vehicle Propulsion", Union Carbide Corp., Consumer Products Div., Parma Res. Lab. Report CRM 244, (60 p., 44 Figs.), Aug. 11, 1970

[27] K. V. Kordesch, "City Car with H_2-Air Fuel Cell and Lead-Acid Battery", SAE-Congress 1971, Paper No. 719015

[28] K. V. Kordesch, "Fuel Cells for Electric Vehicles", International Society of Electrochemistry, 28th Meeting, Sept. 18–23, 1977, Varna, Bulgaria

[29] State of the Art Assessment of Electric and Hybrid Vehicles prepared by NASA Cleveland for the US. Deptm. of Energy, Div. of Transport, Energy Conv., HCP/M 1011-01, UC 96, Jan. 1978

[30] First International Zinc-Bromine-Battery-Symposium, January 22–24, 1985, SEA Mürzzuschlag, Austria; G. Tomazic, "Entwicklungsarbeiten an der Zink-Brom-Batterie", Sonderdruck aus Bulletin SEV/VSE **16** (1987), G. Tomazic ÖZE **40**, 1, (1987) 13

[31] G. Tomazic, "Die Zink-Brom Batterie auf dem Weg zur Serienreife", ÖChemZ, 1986/4

[32] G. Tomazic, "Erprobung der Zink-Brom-Batterie", Vortrag am 5. März 1991 in Essen, Haus der Technik, Fachveranstaltung "Batterien und Ladetechnik", March 5, 1991

[33] C. Robert Kline, Batteries International, January 1994, 40 (Horizon Battery Technologies Inc, 2809 Interstate 35 South, San Marcos, TX 78666, USA)

[34] K. Kordesch, "Neue Möglichkeiten zur Speicherung Elektrischer Energie", Berichte des Symposiums "Energieinnovation in der Elektrizitätswirtschaft und Elektroindustrie", TU-Graz, Kaprun (Austria), 15th January 1993, pp. 143–160

[35] R. A. Vidas, R. C. Miles, P. Korinek, "The Bipolar Lead-Acid Battery at Johnson Controls, Inc.", 34th International Power Sources Symposium, Cherry Hill, June 1990, IEEE Service center, 1990

[36] K. Kordesch, W. Harer, Y. Sharma, R. Ujj, K. Tomantschger and D. Freeman, "Rechargeable Alkaline MnO_2–Zn Battery", 33rd International Power Sources Symposium, 1988, The Electrochem. Soc., Inc., Pennington, 1988

[37] "The RENEWAL" Reusable alkaline Battery, RAYOVAC Corp., Madison, Wisconsin, Brochures 1993

[38] K. Kordesch, L. Binder, W. Taucher, Chr. Faistauer, J. Daniel-Ivad, "The Rechargeable Alkaline Zinc-Manganese Dioxide-System (New Theoretical and Technological Apsects), Power Sources 14, ed. by. A. Attewell, T. Keily, Stratford (UK), April 1993

[39] P. N. Ross, Proc. of the 16th Intersoc. Energy Conversion, Engineering Conf., Vol.1, pp. 726–733, IECEC-Atlanta, 1981

[40] "Ammonia-Air Fuel Cell System for Vehicle Propulsion" Union Carbide Corp., Techn. Proposal to the US-Army ERDL, Ft.Belvoir, Virginia RFP 64-3730-C, May 15, 1964. Engineering Conf., Vol.1, pp. 726–733, IECEC-Atlanta, 1981

[41] K. V. Kordesch, Hydrazin-Luft Batterien, Elektrotechnik und Maschinenbau 86 (1968) 451–456

[42] M. R. Andrews, W. J. Gressler, J. K. Johnson, R. T. Short and K. R. Williams, Automotive Engr. Cong. 1972, SAE Paper 720191

[43] Tatsushi Sugimoto, Kaoru Sohkura, "A–C 3 kW Hydrazine-Air Fuel cell Power Unit", 30th Power Sources Symposium, 7–12 June 1982, The Electrochem.Soc., Inc., Pennington, NJ, 1982

[44] Y. Breele, "Estimates for a Hydrogen-Air Fuel Cell Vehicle", Institute Francaise du Pétrole, 1972

[45] A. T. Emery, "The "OXY" Hydrogen-Air Fuel Cell", demonstration and paper at the 1983 Natl. Fuel Cell Seminar, Orlando, FL., 1983, 98

[46] Proceedings of the Fuel Cell in Transportation Applications Workshop, Aug. 15–17, 1977, Los Alamos Scientific Laboratory, Los Alamos, NM 87545

[47] H. C. Maru, L. G. Christner, S. G. Abens and B. S. Baker, ERC., "Phosphoric-Acid Fuel Cell Stack and Technology Program", National Fuel Cell Seminar, Bethesda, Maryland, 1979

[48] Assessment of Phosphoric Acid Fuel cells for Vehicular Power Systems, United Technologies Corp., Final Report, Los Alamos National Laboratory, PO 4-L61-3862 V-1 (1982)

[49] General Electric Co., Aircraft Equipment Div. Feasibility Study of SPE Fuel cell Power Plants for Automotive Applications, Final Study Report PO 9-L61-3863 V-1, prepared for Los Alamos National Laboratories, Nov. 17, 1981

[50] K. B. Prater, Journal Power Sources 29 (1990) 239

[51] M. Schautz et al., "Results of tests on the PHOTON Fuel Cell generator", Proc. of the European Space Power Conf., Graz, Austria, Aug. 23–27, 1993

[52] F. Baron, "Fuel Cells for European Space Vehicle", Status of Activities, Proc. of the European Space Power Conference, Graz, Austria, Aug. 23–27, 1993

[53] R. B. Ferguson, "Apollo Fuel Cell Power System", Proceedings, 23rd Annual Power Sources Conference, 1969, pp. 11–13

[54] C. C. Morrill, "Apollo Fuel Cell System", 19th Annual Power Sources Conference, 19–20th May 1965, pp. 38–41

[55] R. Cohen, "Gemini Fuel Cell System", Proceedings 20th Power Sources Conference, 24–26th May 1966, pp. 21–24

[56] H. Van den Broeck, et al., "Hybrid Fuel Cell Battery City Bus Eureka Project", 1994 – Fuel Cell Seminar, San Diego (CA), 212

[57] J. A. Woerner, David Taylor, Lloyd Nielsen, "Fuel Cells for Underwater Propulsion Applications", 30th Power Sources Symposium, 7–12 June 1982, Atlantic City, NJ., The Electrochem. Soc., Inc., Pennington, 1982

[58] P. G. Patil, Journal of Power Sources 37 (1992) 171

[59] C. V. Chi, D. R. Glenn, S. G. Abens (ERC), "Methanol Fuel Cell Power Source for City Bus", 34th Internatl. Power Sources Symposium, June 1990, Proc., pp. 399-402 IEEE Service Center, Piscataway, NJ 08854 (1990). Phase I Final Technical Rep DOE/CH/10714-01, Feb 1990

[60] Booz, Allen and Hamilton, Inc., Research and Development of a Fuel Cell/Battery Powered Bus Systems, Phase I, Final Technical Report, DOE/CH/10650-01, Feb.1990

[61] R. J. Kevala, "Status of Fuel cell Bus Development Program", Proc. of the SAE 1992 Future Transportation Technology Conference, Costa Mesa, CA, August 10–13, 1992

[62] J. Bentley, P. Teagan, "Comparison of on Board Reforming and Hydrogen Storage for Fuel Cell Powered Vehicles". 1992 Fuel Cell Seminar, Tucson, Arizona, Nov. 29–Dec. 2, 1992

[63] E. Michelena et al., "Use of Prototype Fuel cells on NOAA Data Buoys", 30th Power Sources Symp., Atlantic City, 1980, Proceedings pp. 20–22. The Electrochem. Soc., Inc., Pennington, NJ., 1980

[64] The H-Power Literature, Information brochures. 1993

[65] G. Simader, "Investigations about using iron sponge as hydrogen supply material". Dissertation, Technical University Graz, 1994

[66] A. Kaufmann and P. L. Terry, "Fuel Cell Power System for Forklift Truck", 30th Power Sources Symp., Atlantic City, 1980, pp. 3–5, .The Electrochem. Soc., Pennington, NJ, 1980

[67] W.Buchanan, A. Kaufman, et al., 1983 National Fuel Cell Seminar Abstracts, p. 15, EPRI, Palo Alto, 1983

[68] S. Mukerjee, S. Srinivasan, A. J. Appleby, Electrochimica Acta **38** (12) (1993) 1661

[69] H. F. Creveling, "Research and Development of a Proton-Exchange-Membrane (PEM) Fuel cell System for Transportation Applications": Abstracts, 1992 Fuel Cell Seminar, Tucson, AZ, Nov. 29–Dec. 2, 1992

[70] Fuel Cell News, **X**, 283, Summer, Fall, 1993

[71] R. A. Lemons, Journal Power Sources **29** (1990) 239

[72] K. B. Prater, Journal of Power Sources **37** (1992) 181

[73] Daimler Benz, Media Information, "Daimler-Benz report breakthrough in alternate power technology – Presentation electric automobile", Stuttgart/Ulm, Germany, April 13, 1994

[74] Daimler Benz, High Tech Report, 3/1994

[75] Ballard Power Systems Inc, Annual report 1994

[76] A. P. Meyer, J. V. Clausi, J. C. Trocciola, "The IFC Proton Exchange Membrane Fuel Cells", 33rd International Power Sources Symposium, Proc., June 1988, pp. 799–802, The Electrochem. Soc., Inc., Pennington 1988

[77] Energy Partners, "The Green Car™ Program", Overview, 1993

[78] Energy Partners, "Energy Partners Corporate Capabilities", Information brochure 1993

[79] Energy Partners, "Brief on the Closed Loop Fuel Cell Power System", 1993

[80] Fuel Cell News, Vol. X, Nos. 4, Winter 1994, page 9

[81] J. F. McElroy et al., "SPE Hydrogen-Oxygen Fuel Cells for Rrigorous Naval Applications", 34th International Power Sources Symp., Cherry Hill, NJ, 1990. Proceedings pp. 403–407, Publ. by IEEE Service Center, 1990

[82] M. J. Rosso, P. M. Golben, J. J. Orlando, O. J. Adlhart, 33rd Power Sources Symposium, 13–16 June 1988, The Electrochem. Soc., Inc., Pennington, NJ

[83] K. Strasser, Journal of Power Sources **37** (1992) 209

[84] W. Drenckhahn, K. Hassmann, "Brennstoffzellen als Energiewandler", Sonderdruck aus Energiewirtschaftliche Tagesfragen **43** (6) (1993) pp. 382–390

[85] Siemens, PEM-Brennstoffzellen, Wirtschaftlichkeit an Bord – mit Elektrotechnik von Siemens, Brochure, 1994

[86] O. Lindström, "A Critical Assessment of Fuel Cell Technology, Dept. of Chem. Engineering and Technology, S 10044, Stockholm, 1993

8. Fuels for the Fuel Cell Technology

8.1. Hydrogen

Hydrogen is the only practical fuel for use in the present generation of fuel cells. The main reason for this is its high electrochemical reactivity compared with that of the more common fuels from which it is derived, such as hydrocarbons, alcohols, or coal. Also, its reaction mechanisms are well understood, and are characterized by the relative simplicity of its reaction steps, which do not lead to side products. In both respects, hydrogen is more desirable than hydrocarbons or alcohols (e.g. methanol), whose catalytic reaction rates are not only very slow at temperatures under 300 °C, but also lead to the formation of poisonous side-products. At higher temperatures, particularly at 1000 °C in the SOFC, the oxidation rates of these compounds are sufficiently rapid[1]. In the MCFC, they may irreversibly decompose, yielding carbon. This reaction can be prevented by the addition of steam to the fuel gas, whereby the carbonaceous material is decomposed by the reforming reaction to H_2, CO, and CO_2. Hence, hydrogen is still the effective fuel even under these conditions.

The characteristics of hydrogen as an energy carrier in a hydrogen energy economy have been extensively discussed in the past. Reviews of this subject are given in the following papers [1–5].

8.1.1. Hydrogen Production

Hydrocarbons are the main source for the production of hydrogen on an industrial scale. An even larger source, however, is water. When hydrogen is used as fuel, water is again produced. Thus, a cycle is involved, and water can be regarded as an unlimited source of hydrogen. With the help of an external energy source, hydrogen can be produced from either of these sources alone or from mixtures.

However, the many production routes for hydrogen are by no means of equal economic importance. Most of the hydrogen for industrial use is produced from natural gas and oil, either as a main product or as a by-product, in processes involving a chemical conversion. A smaller percentage is produced electrolytically or occurs as a by-product of electrochemical processes. The energy here is supplied by the electricity, which itself may be produced using hydroelectric, nuclear power or, still predominant, the energy stored in fossil fuels.

---

1 No carbon at 1000 °C

Thermal, thermochemical, biochemical and photochemical processes have not found industrial applications as yet. The preference for producing hydrogen from hydrocarbons is due to the fact that the energy demand using hydrocarbons is much lower than that for electrolysis of water.

Table 8-1. Theoretical Energy Consumption for the Production of Hydrogen from Various Hydrocarbons, Coal and Water

	Natural Gas (CH_4)	LPG $(CH_{2.6})$	Naphta $(CH_{2.2})$	Heavy Oil $(CH_{1.4})$	Coal $(CH_{0.7})$	Water H_2O
Higher Heating Value[2]						
(kJ/kg)	50050	46130	44300	41300	38100	
(kJ/mol)	890	2480				
By-products	CO_2	CO_2	CO_2	CO_2/S	CO_2/S	O_2
Specific production of by-products kmol/kmol H_2	0.25	0.30	0.32	0.37 / 0.003–0.0015	0.43 / 0.002–0.001	0.5
Theoretical energy consumption kJ/kmol H_2	41280	37500	38350	59300	57150	242000

Considering the strong correlation between energy prices and the prices for hydrocarbons, it is unlikely that the high percentage of the hydrocarbon processes for producing industrial hydrogen will change substantially in the near future. Considering environmental pollution by emission of carbon dioxide, sulfur compounds, nitrogen oxides and heat, this aspect may reverse the situation some day [6].

8.1.2. Safety Aspects of Hydrogen

Giving sufficient thermal activation (ignition energy), hydrogen reacts more or less violently with oxidizing agents such as oxygen (air), fluorine or chlorine, and N_2O. Depending on conditions, combustion, deflagration or detonation may occur.

The ignition and detonation properties of hydrogen – air mixtures are particularly important from the safety aspect. The flammability limits (i. e. the difference between the rich and lean limit concentration) are exceptionally wide for hydrogen. Table 8-2 shows a comparison of safety-relevant thermo-physical and combustion properties of hydrogen with those of methane, propane and gasoline.

---

2   Calculated from the heat of formation, oxidizing carbon completely to carbon dioxide (at 298 K and 0,1 MPa)

Table 8-2. Combustion and explosion properties of hydrogen, methane, propane and gasoline

	Hydrogen	Methane	Propane	Gasoline
Density of gas at standard conditions [kg / m^3 (STP)]	0.084	0.65	0.42	4.4[a]
Heat of vaporization [J/g]	445.6	509.9		250–400
Lower heating value [kJ/g]	119.93	50.02	46.35	44.5
Higher Heating value [kJ/g]	141.8	55.3	50.41	48
Thermal conductivity of gas at standard conditions [mW/cmK]	1.897	0.33	0.18	0.112
Diffusion coefficient in air at standard conditions [cm^2/s]	0.61	0.16	0.12	0.05
Flammability limits in air [vol.-%]	4.0–75	5.3–15	2.1–9.5	1–7.6
Detonation limits in air [vol.-%]	18.3–59	6.3–13.5		1.1–3.3
Limiting oxygen index [vol.-%]	5	12.1		11.6[b]
Stoichiometric composition in air [vol.-%]	29.53	9.48	4.03	1.76
Minimum energy for ignition in air [mJ]	0.02	0.29	0.26	0.24
Autoignition temperature [K]	858	813	760	500–744
Flame temperature in air [K]	2318	2148	2385	2470
Maximum burning velocity in air at standard conditions [m/s]	3.46	0.45	0.47	1.76
Detonation velocity in air at standard conditions [km/s]	1.48–2.15	1.40–1.64	1.85	1.4–1.7[c]
Energy[d] of explosion, mass-related [gTNT/g]	24	11	10	10
Energy[d] of explosion, volume-related [gTNT/m^3]	2.02	7.03	20.5	44.2

a 100 kPa and 15.5 °C: b Average value for a mixture of C1–C4 and higher hydrocarbons including benzene; c Based on the properties of n-pentane and benzene; d Theoretical explosive yields

The flammability limits are affected by temperature, so that a preheated mixture has considerably wider limits for coherent flames. An increase in pressure up to 10 kPa has only a small effect. Water vapour has a strongly inhibiting influence on the oxyhydrogen reaction (see ternary diagram, Figure 8-1). The flammability limits of hydrogen in pure oxygen are 4,65–93,9 vol.-% hydrogen.

Volumetric leakage of hydrogen gas will be 1.3–2.8 times as large as gaseous methane leakage and approximately four times that of air under the same conditions ("air-proof is not hydrogen-proof")

Released hydrogen disperses rapidly by turbulent convection, drift and buoyancy, thus shortening the duration of hazardous conditions. The prompt dispersion, however, favours the formation of gas mixtures within the wide flammability and detonation limits; the lower limit is the critical one in most applications and is comparable to that of other fuels.

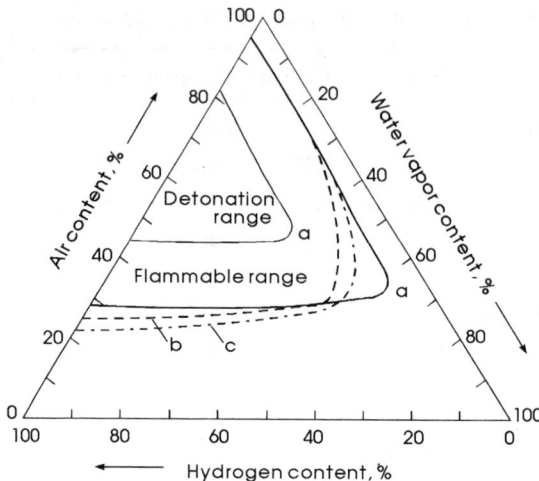

Figure 8-1. Flammability and Detonation Limits of the Three Component System Hydrogen – Air – Water Vapour
a) 42 °C, 100 kPa, b) 167 °C, 100 kPa, c) 167 °C, 800 kPa

The minimum energy for ignition of hydrogen – air mixtures is extremely low. However, like the inherent ignition energy of every other fuel-air mixture it is theoretically sufficient for ignition. Ignition starting or retardation by catalytic action is also possible.

Toxicology of Hydrogen

Hydrogen does not show any harmful physiological effects. It is non-poisonous. Inhalation of the gas leads to sleepiness and a high-pitched voice. A danger of asphyxiation exists, as with any other gas mixture, if the oxygen content sinks below 18 vol.-% due to hydrogen accumulation in the air. Direct skin contact with cold, gaseous or liquid hydrogen leads to numbness, a whitish colouring of the skin and to frost bite. The risk is higher than, for example, with liquid nitrogen, because of the greater temperature differences and the higher thermal conductivity of hydrogen.

8.2. The Industrial Production of Hydrogen

Hydrogen can be produced in large amounts from primary energy sources, such as fossil fuels (coal, oil, or natural gas), from a variety of chemical intermediates (refinery products, ammonia, methanol) and from alternative resources such as biomass, biogas and waste materials. Hydrogen can also be produced by water electrolysis, which uses electricity to split the hydrogen and oxygen, and can be regarded as a secondary energy source. Other possibilities, such as coal electrolysis or reforming with sponge iron are also mentioned in this chapter.

The next figure illustrates the most obvious pathways for the production of hydrogen from various energy sources. From the figure, the close link between energy resources (nuclear, fossil, alternative or renewable) and hydrogen is evident: hydrogen can, to some extent, be considered as an intermediate energy vector, whose potential applications span from substituting many fuel and electric tasks to upgrading raw materials. Fuel cells constitute just one of many hydrogen utilization possibilities in future energy applications.

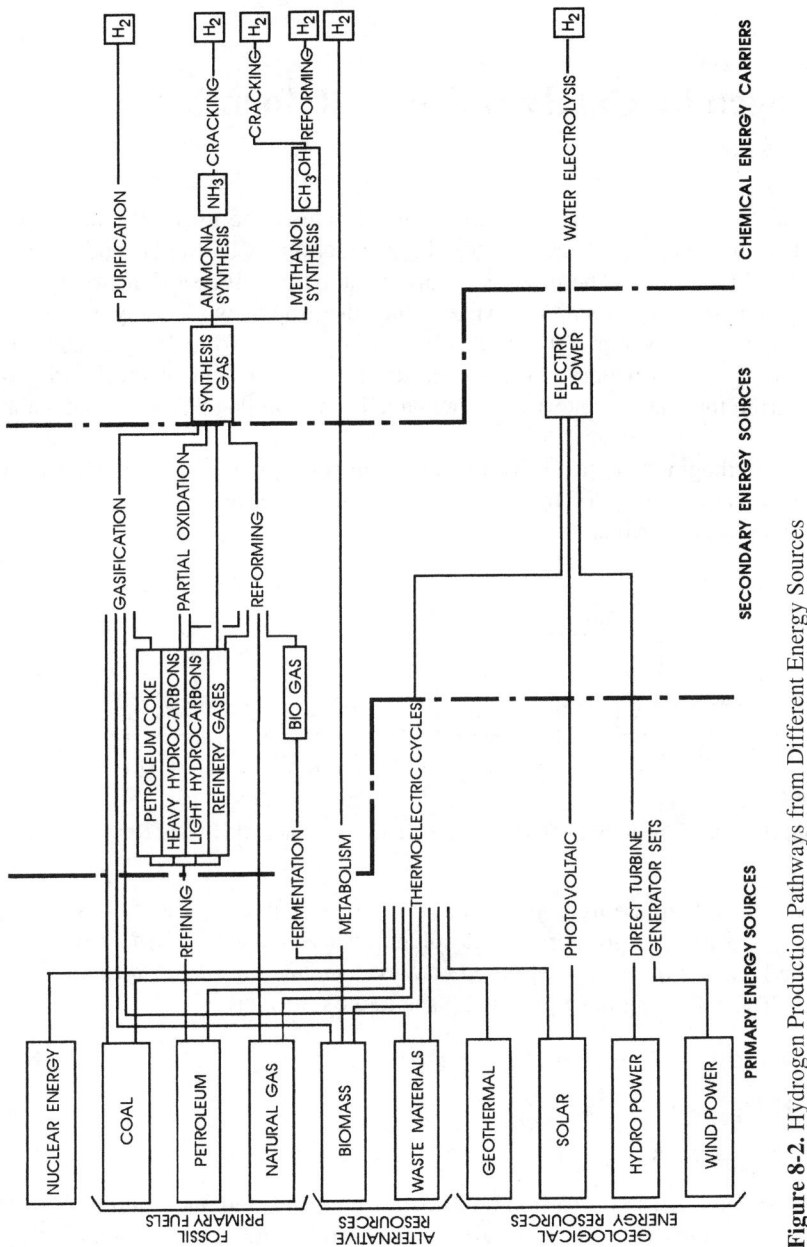

Figure 8-2. Hydrogen Production Pathways from Different Energy Sources

Each box in Figure 8-2 represents either a primary or intermediate source of energy, while the lines represent processing options. Not all the pathways shown are technically and economically feasible; for example, bacteria can produce hydrogen, but it would be unrealistic at present to assume that appreciable amounts of industrial-quality hydrogen can be produced from biomass by the utilization of micro-organisms.

8.3. Hydrogen by Catalytic Steam Reforming of Natural Gas

For several decades, steam reforming of hydrocarbons has been the most efficient, most economical and most widely-used process for the production of hydrogen and hydrogen/carbon monoxide mixtures. The process involves catalytic conversion of the hydrocarbon and steam to hydrogen and carbon oxides. Since the process works only with light hydrocarbons which can be vaporized completely without carbon formation, a range of feedstocks from methane, to naphta [7] has been used. Fifty per cent of the hydrogen produced comes from the water (steam) when methane is used and 64.5% when naphta is used [8–10].

A simplified flow diagram of the conventional steam reforming process is shown in Figure 8-3. The process consists of three main steps: (1) synthesis-gas generation, (2) water-gas shift and (3) gas purification.

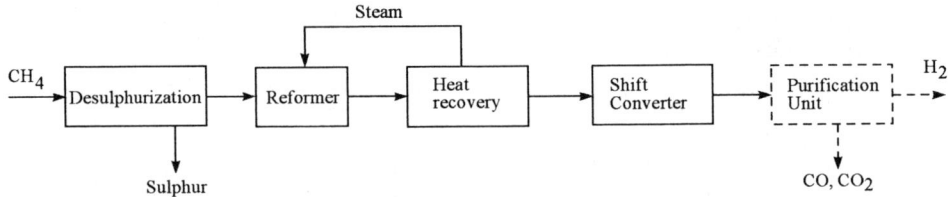

Figure 8-3. Simplified Flow Diagram of Hydrogen Production by Catalytic Steam Reforming

To protect the catalysts in the hydrogen plant, the hydrocarbons have to be desulphurized before being fed to the reformer. The desulphurized feedstock is then mixed with process steam and reacted over a nickel based catalyst contained inside a system of high alloy steel tubes. The following reactions take place in the reformer:

$$C_nH_m + n\ H_2O \longrightarrow n\ CO + (n + m/2)\ H_2 \tag{8-1}$$

$$CO + H_2O \longrightarrow CO_2 + H_2 \tag{8-2}$$

$$CO + 3H_2 \longrightarrow CH_4 + H_2O \tag{8-3}$$

where n = 1 and m = 4 if the hydrocarbon feedstock is methane, and n = 1 and m = 2.2 if it is naphta. The reforming reaction is strongly endothermic, and energy is supplied by the combustion process of fuel gas or oil. At the same time, the CO shift and methanation reactions quickly reach equilibrium at all points in the catalyst bed. The equilibrium composition of the reformed gas is favoured by a high steam to carbon ratio, low pressure and high temperature. To ensure a minimum concentration of CH_4 in the product gas, the process generally employs a steam to carbon ration of 3 to 5 at a temperature of around 815 °C and pressures up to 3.5 MPa. These conditions also allow a conversion of more than 80% of hydrocarbon to oxides of carbon at the outlet of the reformer. The typical composition of a synthesis gas leaving a steam-methane reformer is as follows [11]:

Table 8-3. Composition of Reformed Gas [12] (S/C = 2.5; Reaction Temperature = 684 °C)

Gas composition	Reformed Gas [%]	Calculated value [%]
H_2	72.04	71.95
CO	15.17	14.19
CO_2	10.82	10.92
CH_4	1.96	1.88

There are three types of reforming furnaces used in industry. They differ with respect to the position of the burners and the types of flow pattern. The first type, the top-fired furnace design, is characterized by cocurrent flows of process and synthesis gases. This ensures that maximum cooling from the endothermic reaction is achieved in the area where the tube is exposed to maximum radiation from the flame. At the exit of the tube, where most of the methane has already reacted and the process gas is hotter, flue gas is cooler and the tube-skin temperature is still acceptable. The skin temperature of the catalyst tube is almost constant, while the heat flux varies significantly.

The second type of furnace, the bottom-fired furnace, offers countercurrent flue gas and process gas flows. Maximum heat fluxes are obtained close to the bottom of the furnace, corresponding to maximum process gas temperatures, when most of the methane conversion has already occurred. This process carries the risk of tube-skin overheating if high methane conversions are to be achieved. It is caused by insufficient heat absorption by the endothermic reaction which has already taken place. Such reformers are therefore only utilized in those processes where the gas leaving the reactor tubes still contains appreciable levels of unreacted methane – for example, the primary reformers of ammonia plants.

The third type of furnace, the side-fired design, offers a very homogeneous heat flux along the entire length of the tube. However, the tube-skin temperature reaches a maximum at the lower part of the reaction tubes, forcing the designer to select the tube material on the basis of this specific hot portion. Side-fired reformers require a larger number of burners compared to top or bottom-fired types. This is considered as a disadvantage for all applications where forced air is used (as in preheated air or gas turbine exhaust), and it also requires more careful adjustment of fuel firing. As a result, the trend has been

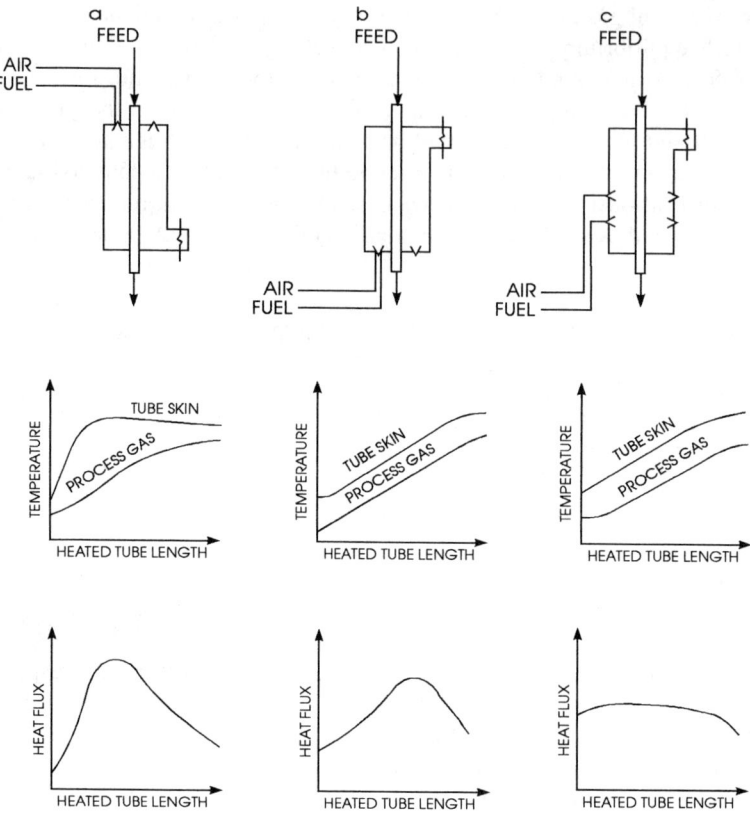

Figure 8-4. Three Typical Configurations for Steam-Reforming Furnace: (a) Top-Fired, Co-current; (b) Bottom-Fired, Countercurrent; (c) Side Fired

toward top-fired furnaces, which combine the cheaper design with multi-purpose application and smaller furnace dimensions.

In contrast to large chemical and petrochemical plants, fuel cells produce their own saturated steam, which may be utilized as process steam in the fuel-processing section. This is the case for most currently-available phosphoric acid stacks. To reduce energy rejection from the steam reforming process, regenerative tube designs and pressurized combustion reformers have been used in such cases.

A typical reformer for a PAFC is shown in Figure 8-5, where the catalyst-filled annulus is exposed on one side to the radiant heat transfer from the flue gas, and on the other side to a mostly convective heat transfer from the process gas. With this geometry, the reformer effluent is cooled to 500 to 600 °C before leaving the reactor. This reduces complexity and size of downstream gas cooling equipment, while at the same time supplying process heat to the reforming reaction. The net effect is a heat-saving (less firing time), because part of the reaction heat is reconverted by the regenerative tube exchanger, resulting in a smaller flue-gas rate and thereby reduced convection section duty-time. This design has one other major advantage: the tube is free to expand linearly, and all connections are positioned in the cold-end section. This is quite important for those operating conditions requiring frequent thermal cycles, and it reduces the cost of conventional reformer "pigtails". In practice, the design "pigtails" (a tube location in the reformer where the effluent is at temperatures in excess of 700 °C) constitutes a complex technology, for which the know-how is specific to each major equipment supplier. This problem is significantly reduced with regenerative-type tubes, thereby improving the overall on-stream availability of hydrogen production equipment. The equipment shown in Figure 8-5 shows a regenerative-tube reformer similiar to that developed by United Technology Corporation (UTC, IFC) for fuel cell applications.

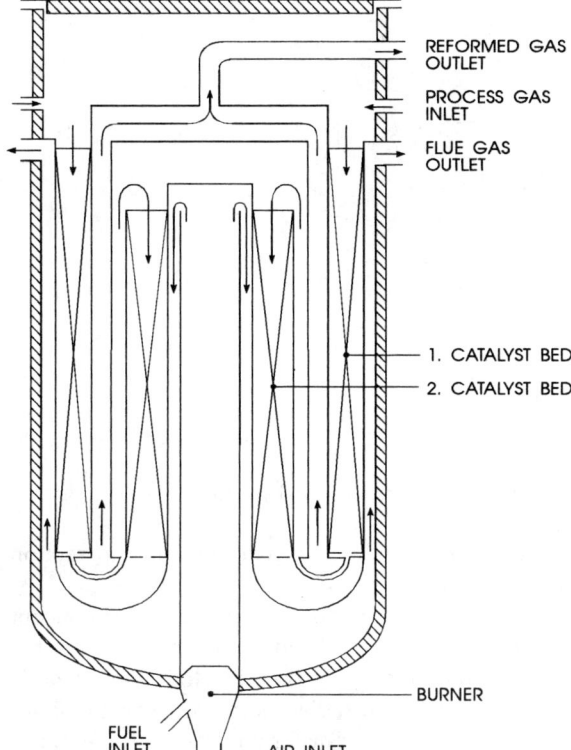

REFORMED GAS OUTLET

PROCESS GAS INLET

FLUE GAS OUTLET

1. CATALYST BED
2. CATALYST BED

BURNER

FUEL INLET

AIR INLET

Figure 8-5. Schematic of the Haldor Topsoe Reformer, Featuring a Combination of Counter-Current and Co-current Arrangement [13] (Courtesy Haldor Topsoe A/S)

Similar reformers have been utilized in commercial hydrogen plants with some success. The schematic of Figure 8-6 shows a countercurrent flow between flue gas and process gas. A critical point with such reactors having conversion efficiencies close to 100%, is heat transfer in the hottest section of the tube, as discussed above. To reduce the risk of unacceptably high skin temperatures, this type of reformer tube is often protected at its hottest section by a ceramic cup. One way to avoid excessive tube skin temperatures is to use a different geometry, like that utilized by Haldor Topsoe for the pilot plant built in Houston (see Figure 8-5). In this configuration, flue gas and process gas are ex-

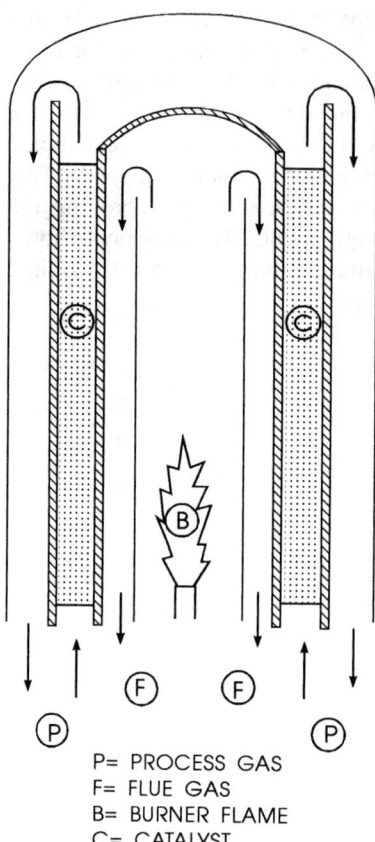

P= PROCESS GAS
F= FLUE GAS
B= BURNER FLAME
C= CATALYST

Figure 8-6. Regenerative-Type Reformer

changing heat concurrently in the hottest section, and countercurrently in the section where tube skin temperatures are no longer critical.

The use of innovative combustion technology is also shown in the reformer concept schematic of Figure 8-7, where a radiant fiber burner is utilized. This reformer is of the type developed by KTI, and its heat profile represents a side-fired reactor, where heat is homogeneously supplied along the entire length of the radiant section. Tube impingement is prevented here as part of the radiant heat being transferred by gray-body radiation between the burner body and the tube, which is not directly exposed to the flame. The flue-gas temperatures on leaving the burner surface are therefore much lower and cannot damage the tube. As a consequence of the very constant temperature profile, the burner and reformer tube can be extremely close to each other, thereby reducing overall external dimensions.

Such solutions make it possible to recover heat at high temperatures for process duties, reducing fuel firing-rates and flue gas flow-rates. The systems shown are normally coupled with a section where flue gas convective heat transfer is enhanced by special geometries or promoters. This results in further heat recovery, reduced firing rates and lower flue-gas exit temperatures.

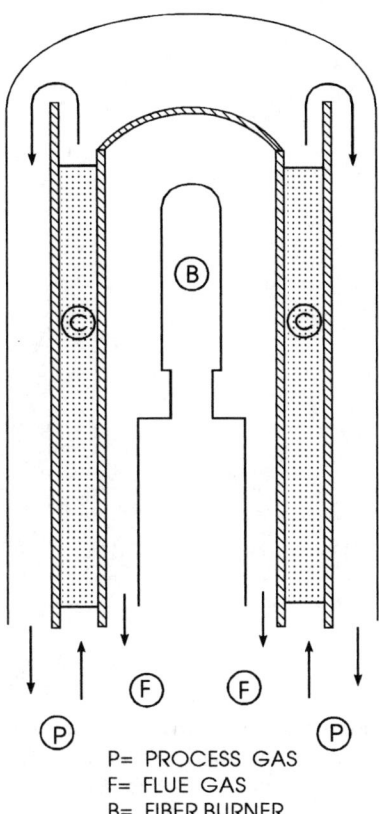

P= PROCESS GAS
F= FLUE GAS
B= FIBER BURNER
C= CATALYST

Figure 8-7. KTI Compact Regenerative Reformer (Courtesy of Kinetics Technology International, "KTI")

To further improve system net efficiency when no useable steam can be produced, many reformers designed for fuel cell applications utilise pressurized combustion. This is favoured by the need for pressurized airflow for the fuel cell stacks and the requirement of utilising the exhaust gas from the stacks as fuel. The integration of compressor and power-recovery turbine is such that pressurization of the combustion chamber results in cost reduction (more compact design) and energy recovery (the enthalpy of the flue gas leaving the reformer is utilized in the expander to produce mechanical work instead of heat). Pressurized combustion is a possible development in technology for high temperature, heat-transfer equipment, and there are already examples of commercial application in hydrogen plants, such as the one in Figure 8-8. This reformer has been in operation since 1991 and uses a regenerative-type tube arrangement.

After the reformer, the process gas mixture containing CO and H_2 passes through a heat recovery step and is fed into a water-gas shift reactor to produce additional H_2. Normally, two stages of conversion (high temperature and low temperature) are used to reduce the CO content to a level of approximately 0.2–0.4% by volume. This exothermic reaction occurs at temperatures ranging from 200 to 400 °C.

Figure 8-8. Pressurised Combustion Reformer, Toshiba Design, Goi (Japan) (Courtesy of Toshiba Corp.)

Next, the cold, raw gases pass through gas purification units to remove CO_2, the remaining CO and other impurities to deliver the required purified H_2 product. Several commercial processes for removing CO_2 exist. One of these systems uses wet scrubbing with monoethanolamine (MEA), hot potassium carbonate and sulfinol™. The residual CO and CO_2 remaining in the H_2 stream after CO_2 removal are converted to CH_4 by a methanation reaction. The product, H_2, leaves with a purity of approximately 97 to 98%.

An alternating technology for wet scrubbing is the use of the pressure swing adsorption (PSA) technique. This process reduces the number of unit processes and complexity of the operation by replacing the low temperature shift, CO_2 removal and methanation with one PSA process unit. In this process, the raw gas is passed through a series of beds of molecular sieves or activated carbon, where all components except H_2 are preferentially absorbed. The beds are regenerated by adiabatic depressurization at ambient temperature. The purge gas, which contains water vapour, CO_2, CO and CH_4, is then fed to the furnace for supplying heat to the reformer. The purity of H_2 from the PSA system can be 99% or higher, and can be used advantagageuously for a wide variety of chemical and petrochemical processes.

8.3.1. The Reforming Process for Phosphoric Acid Fuel Cells

The last (methanation) step mentioned above is only important for the production of pure hydrogen in large scale production plants. There is no need for such an intensive purification step for small scale, on-site phosphoric acid fuel cells. Only a low-temperature CO conversion is necessary in order to decrease the carbon monoxide below 1.5%.

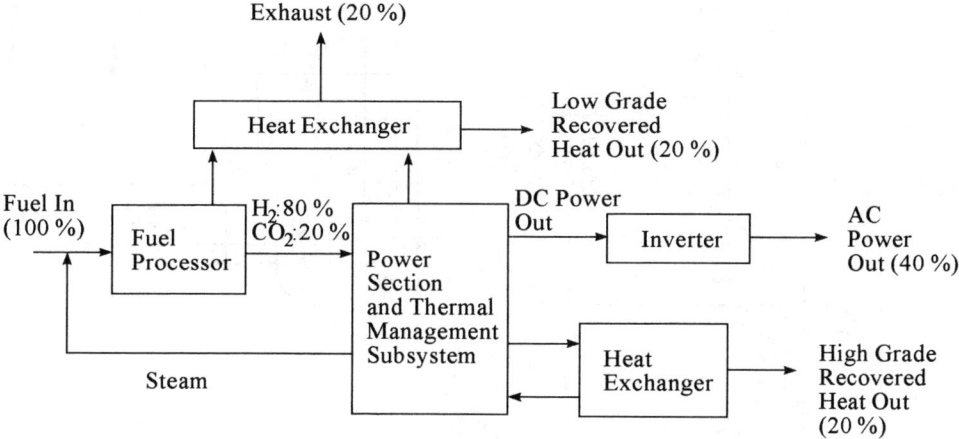

Figure 8-9. Flowsheet of a 200 kW On-Site PAFC Power Plant (Atmospheric Pressure)

As shown in Figures 8-9 and 8-10, the fuel cell power plant consists of the cell stack, the reforming system, the waste heat recovery system, the inverter and the control system. The waste heat recovery system includes two systems:

i) a system for 40 to 65 °C low-grade heat, which is recovered from the combustion exhaust gas leaving the reforming system and from the air passing through the cell stack;
ii) a system for 150 to 170 °C high-grade heat, which is mainly recovered through cooling water from the electrochemical reaction in the cell stack and can be recovered as steam.

8.3.2. Problem Areas in the Reforming Process

Desulfurization

For hydro-desulfuration, the temperature must be increased at the start of the process because this technique requires a reaction temperature of 350 °C. Osaka Gas stated that the steam reformer can operate with a ppm-level sulphur content [14]. However, R&D activities are continuing to find high-performance catalysts capable of achieving ppb-level desulfurization. A decrease to room temperature level is necessary.

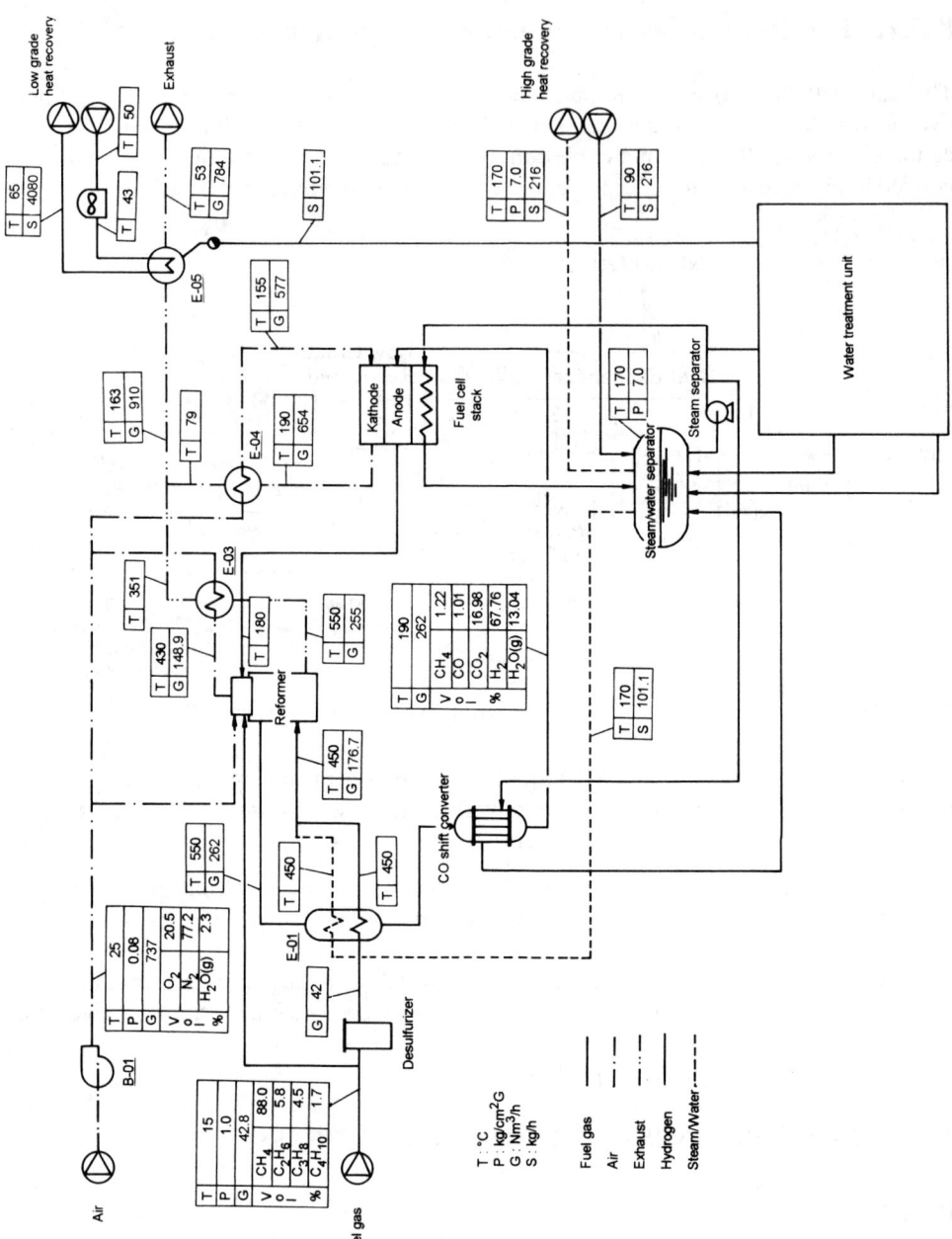

Figure 8-10. Mass Flows of a 200 kW Plant with above Mentioned Energy Efficiencies (Figure 8-9)

This problem is especially evident with the use of naphta or kerosene. If petroleum products are used for fuel cells, it is generally expected that the usage of fuel cells will be effectively promoted, as petroleum products can be easily supplied to any area where city-gas pipeline networks are not existing. Even petroleum products such as kerosene with sulphur contents up to 10 to 100 wt.-ppm can be used. A combination process of conventional hydro-desulfurization (HDS) and adsorption desulfurization (ADS) may be needed in order to guarantee the long-term, low sulphur content in the feed gas [15]. Another hydrogen production method which involves the reaction between natural gas and hydrogen sulphide ($CH_4 + 2H_2S \longrightarrow 4H_2 + CS_2$) has been also already discussed [16].

Reforming

A Ni catalyst is usually used for reforming. The traces of sulphur remaining after the desulfurization process can lead to poisoning of the catalyst, which results in a shorter life, and to avoid carbon deposition on the catalyst, the reforming process must be operated at an S/C ratio (ratio of reforming steam to carbon in the fuel gas) of 3 or more. This requires a large quantity of steam for reaction. The combination of a highly-active, reforming catalyst and a completely effective desulfurization catalyst eliminates potential carbon deposition on the reforming catalyst and allows operation at a low S/C ratio. As a consequence, the required quantity of reforming catalyst can be relatively small. An increase in high-grade waste heat, a reduction in start-up time, and a compact and light structure are the results of this optimal combination.

Naphta produces carbon more easily than natural gas under reforming conditions. A long-term operation is known to be very difficult. Therefore, the development of catalysts suppressing carbon formation is the most important problem to be solved for the use of naphta. A new candidate for the solution of this problem has been developed by the Fuel Research Laboratory [17]. Ru-based catalysts with ZrO_2 as support have a long life (about 9000 hours) compared with Ni-based catalysts, because neither a decrease in activity nor an increase of pressure in the catalyst bed occurs even at the low S/C ratio of 2.

CO shift conversion

In CO shift conversion, a combination of high-temperature and low-temperature shift conversion or, to simplify the system, only low-temperature shift conversion is applied. In either case, the CO shift converter requires considerable space in the system because of the large amount of catalyst required. Efforts are being made to ensure availability even at a low S/C ration, minimising the consumption of reforming steam and increasing the amount of high-grade waste heat that can be externally extracted. These features will help to establish the desirable waste-heat recovery system for fuel cells as co-generation systems.

8.3.3. Substitution of Gaseous Fuels in PAFC Systems

Nippon Telegraph and Telephone Corporation (NTT) [18, 19] is developing a PAFC system to provide power for telecommunication equipment and to make use of its waste heat in cooling telecommunication equipment. NTT suggests that methanol and liquefied

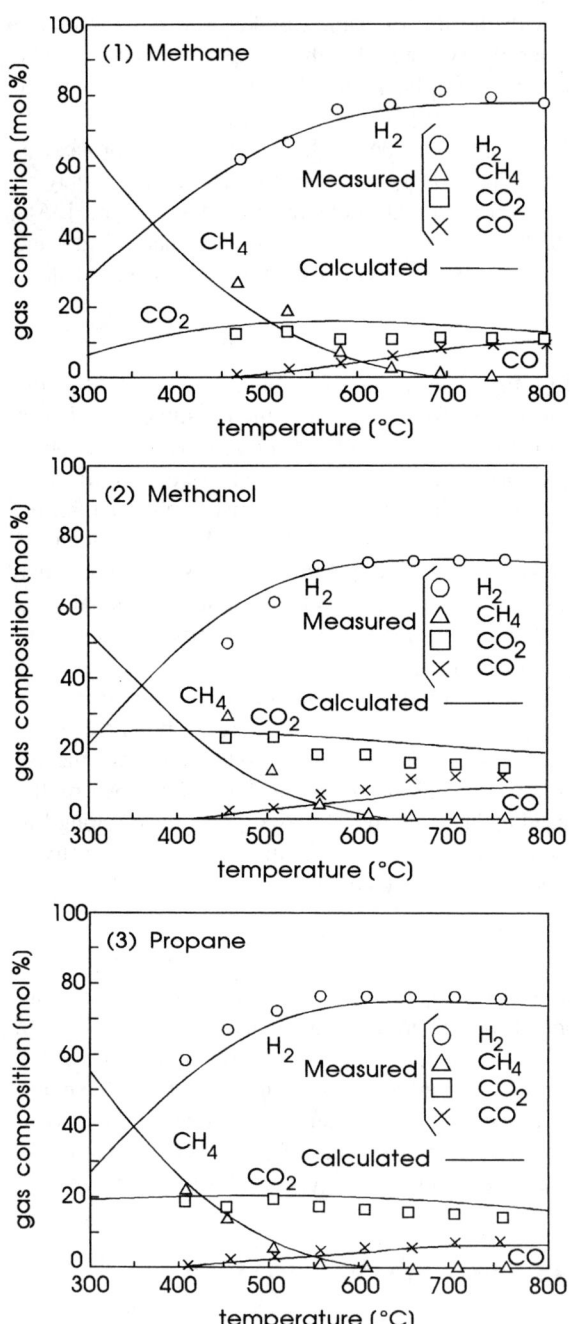

Figure 8-11. Dependence of the Steam-Reformed Gas on the Reforming Temperature

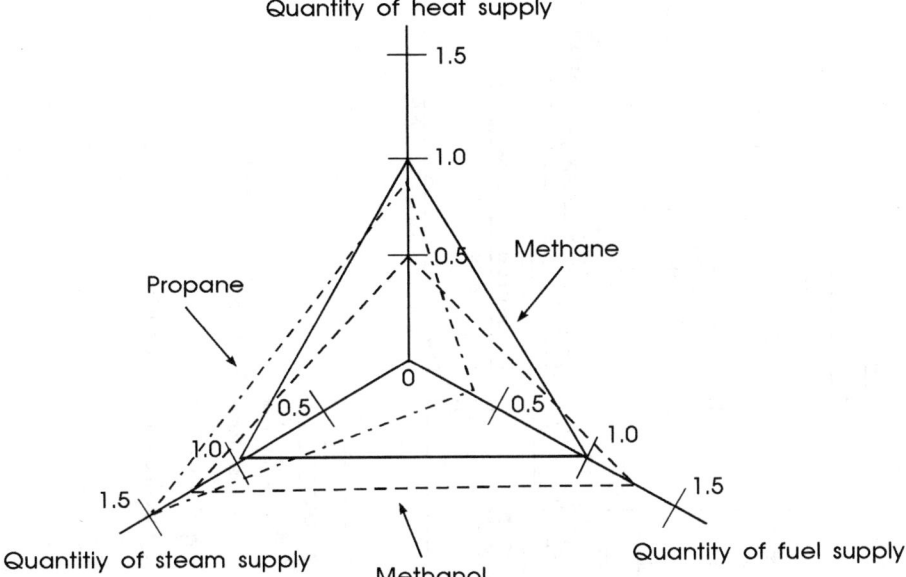

Figure 8-12. Comparison of Steam-Reforming Conditions to Form the Same Quantity of Hydrogen by Methane, Methanol, and Propane.

propane gas are suitable as emergency fuels for the fuel cells. Pipeline gas (which is the usual fuel), methanol and liquified propane gas can be steam-reformed in the same reformer, filled with a $Ni-Al_2O_3$ catalyst, by appropriate control the fuel supply, steam supply and reformer temperature. Figure 8-11 shows the effect of temperature on the gas compositions.

Figure 8-12 shows the steam-reforming conditions for methane, methanol and propane to form the same quantity of hydrogen. Pipeline gas is taken as reference baseline. When the fuel is changed from methane to methanol or propane, the heat supply must be reduced to avoid a rise in reformer temperature and to prolong the catalyst life. On the other hand, the steam supply must be increased when the fuel is changed from methane to methanol or liquefied propane gas, and the fuel cell system requires a larger steam separator and larger diameter plumbing to maintain the same pressure drop.

Figure 8-13 shows a diagram of a fuel cell system which can operate on alternative fuels. When a pressure drop is detected in the gas pipeline, the flow control valves for pipeline gas, methanol or propane gas, and steam are immediately adjusted and the fuel cells can keep their output constant.

8.4. Hydrogen by Methane Decomposition

Thermal decomposition of methane has been known for many years. It was developed originally for the production of high-quality carbon products (carbon blacks, graphite etc.) [20] and not for hydrogen production. This method of hydrogen production has

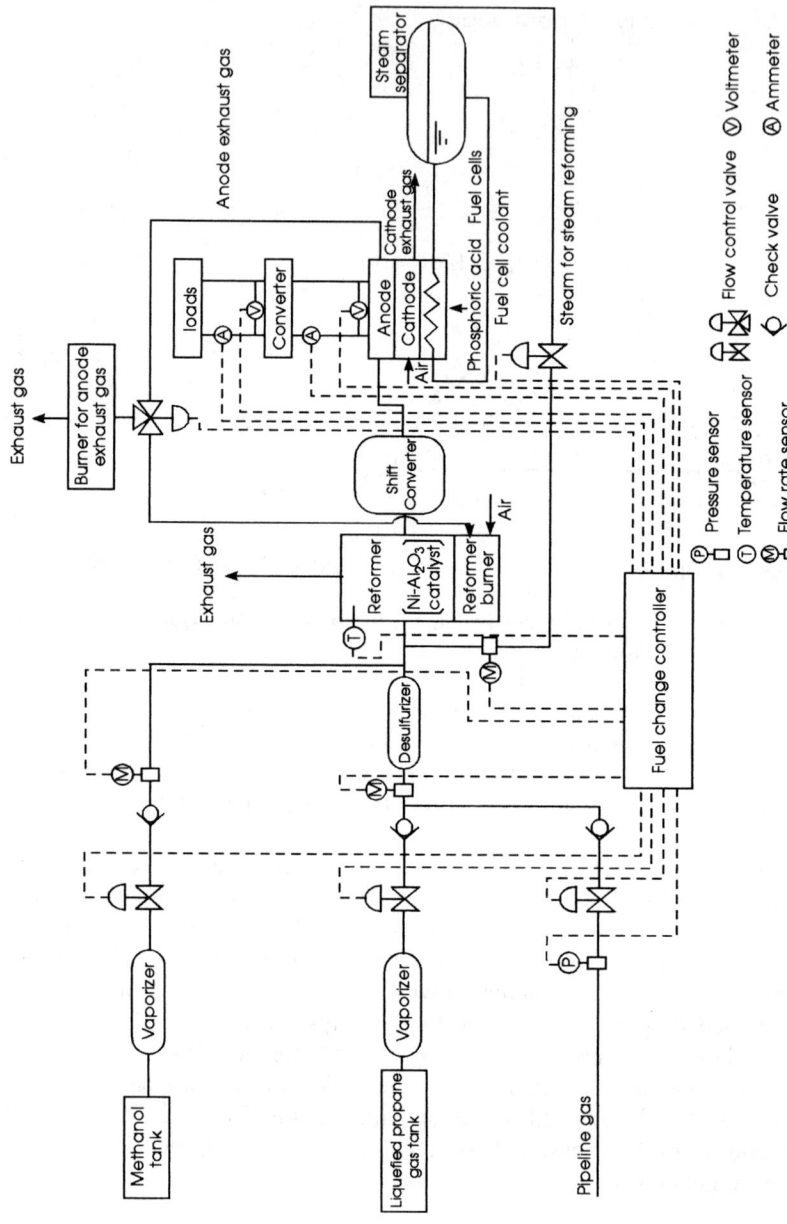

Figure 8-13. System Diagram of a PAFC with Fuel Supply from Pipeline Gas to Methanol or Liquefied Propane Gas

been suggested as a non-polluting use of natural gas by a study made by five Austrian utilities under the leadership of OMV AG [21]. Summarising the study's findings, it was established that such an objective can actually be achieved by employing fuel cell technology. The remaining problem is how to avoid or dispose of the carbon dioxide generated in the process. The study pointed to a method of methane decomposition which yields pure carbon instead of carbon dioxide. The carbon can then be put to other uses. A total predicted efficiency of 60% [22] seems to be very optimistic, however a closer thermodynamic inspection confirms this opinion[3].

$$CH_4 + 2O_2 \longrightarrow CO_2 + 2H_2O \qquad \Delta H = -803.0 \text{ kJ/mol} \qquad (8\text{-}4)$$

$$CH_4 \longrightarrow C + 2H_2 \qquad \Delta H = 74.9 \text{ kJ/mol} \qquad (8\text{-}5)$$

$$C + O_2 \longrightarrow CO_2 \qquad \Delta H = -393.8 \text{ kJ/mol} \qquad (8\text{-}6)$$

$$2H_2 + O_2 \longrightarrow 2H_2O \qquad \Delta H = -484.1 \text{ kJ/mol} \qquad (8\text{-}7)$$

The required high reaction-temperature decreases the process efficiency. Neglecting potential usage of carbon oxidation seems to be very wasteful at present. A better, non-polluting use of natural gas would be a cyclic process of carbon dioxide and methanol, or the reforming of methane with carbon dioxide.

8.5. Partial Oxidation of Heavy Hydrocarbons

The partial oxidation of hydrocarbon liquids is a process used to produce hydrogen where natural gas [23] is not economically available or excess heavy oil is available at low cost. The hydrocarbon liquid reacts with steam to form synthesis gas according to the reaction:

$$C_n H_m + n H_2O \longrightarrow n CO + (n + 1/2m) H_2 \qquad (8\text{-}8)$$

At high temperatures, this endothermic reaction produces appreciable quantities of hydrogen. The heat of reaction, rather than passing through the wall of the catalyst-tube (as in hydrocarbon steam reformers), is produced directly by the combustion of a portion of the fuel-feed with controlled quantities of oxygen (or air) in a high-temperature, re-fractory-lined reactor:

$$C_n H_m + \left(n + \frac{m}{4}\right) O_2 \longrightarrow n CO_2 + \frac{1}{2} m H_2O \qquad (8\text{-}9)$$

Structurally, the partial oxidation reactor is very similar to a steam/gas/O_2 reformer, but it does not contain any catalyst (Figure 8-14).

---

3   ΔH at 0 °C, 0.1 MPa

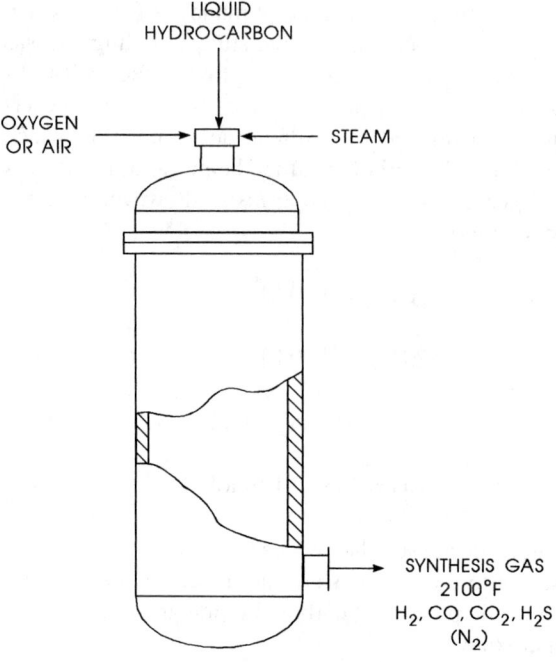

Figure 8-14. Partial Oxidation Reactor

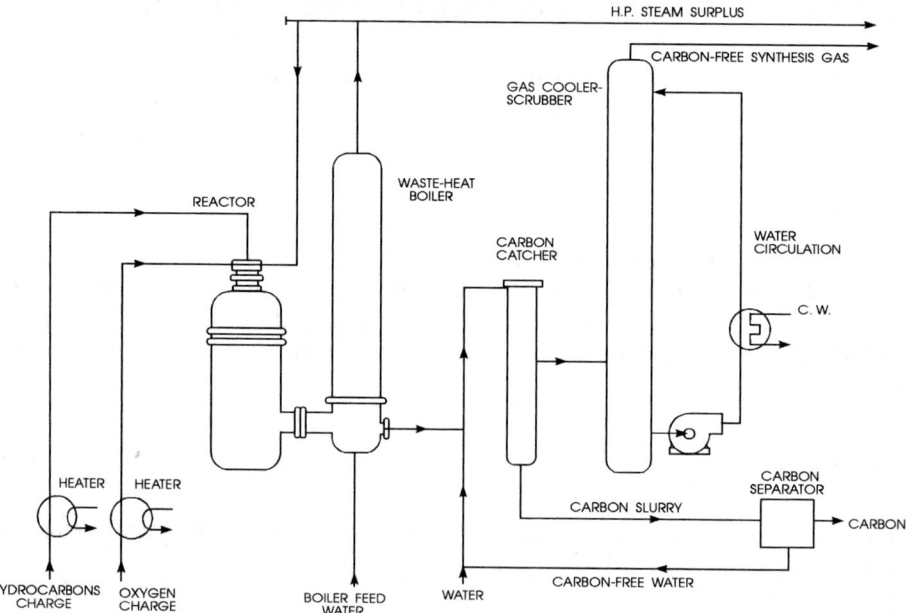

Figure 8-15. Basis Shell Partial Oxidation Process Configuration

The two processes are applied commercially throughout the world by Texaco and Shell. They are similar in some respects, but use different mechanical designs for the reactor and the cooling systems. Schematics of the Shell reactor design and of the overall Shell system are shown in Figure 8-14 and 8-15.

Liquid hydrocarbons, and specifically oil residues, can contain sulphur compounds, which react in the reducing (cracking) atmosphere of the reactor to produce COS and H_2S. The presence of sulphur compounds in the syngas requires additional downstream equipment to remove the sulphur before final use of the gas, and demands a careful selection of materials and catalysts.

The design of systems for heat recovery from partial oxidation reactors is critical due to the high temperature of the effluent and the presence of uncombusted carbon and other solid particles, which can foul the cold surfaces of the waste-heat boiler. The schematic design of the Shell partial oxidation syngas cooler is shown in Figure 8-16.

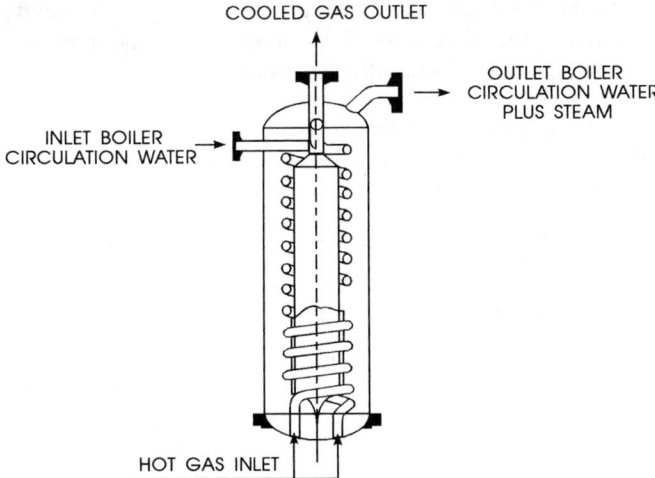

Figure 8-16. Helical Tube Waste Heat Boiler

The effluent from partial oxidation systems contains several contaminants as well as carbon and ash. The treatment of such effluent utilises well established techniques which require substantial capital investment.

8.6. Hydrogen from Coal

8.6.1. Hydrogen from Coal Gasification

World-wide, large-scale programs are developing MCFCs and SOFCs, and these fuel cells are intended to operate on hydrogen produced by the gasification of typical coals. In coal gasification, slurried coal is reacted with oxygen and steam to produce a mixture consisting primarily of carbon monoxide and hydrogen. The main types of coal gasifiers under study include those using high-temperature fluidized beds, low-temperature fixed or fluidized beds, together with high-temperature entrained flow systems and molten salt reactors [24–26].

Coal Gasification

Coal gasification is the endothermic reaction of carbon atoms contained in coal with water vapour, according to the equation

$$C + H_2O \longrightarrow CO + H_2 \qquad\qquad \Delta H = +131.38 \text{ kJ/mol} \qquad (8.10)$$

For the reaction to proceed from left to right, it is necessary that heat be transferred at high temperatures.

Most coal gasification processes are based on the concept of direct combustion of a portion of coal feedstock with oxygen or air introduced into the same vessel where the gasification takes place. As combustion and gasification occur under steady-state condition, the energy balance of the system is therefore zero. The combination of the reactions ·is listed in Table 8-4 and, theoretically, there is no net flow of energy.

Table 8-4. Basic Reactions Occurring Inside a Coal Gasifier

Reaction	Reaction Enthalpy [kJ/mol]
$C + \frac{1}{2}O_2 \longrightarrow CO$	−110.6
$C + O_2 \longrightarrow CO_2$	−393.8
$C + H_2O \longrightarrow CO + H_2$	+131.4
$C + 2H_2O \longrightarrow CO + 2H_2$	+ 96.7
$3C + 2H_2O \longrightarrow 2CO + CH_4$	+185.5
$2C + 2H_2O \longrightarrow CO_2 + CH_4$	+ 12.1

The characteristics of coal, such as its ash content and composition, tendency to agglomerate, sulphur content, etc., make coal gasification a complex process with many alternative routes. Figure 8-17 summarises the available coal gasification processes grouped according to the type of reactor. Moving-bed reactors have been utilized for almost a century to generate hydrogen-rich gas from the bottom of a vessel filled with coal or lignite by injecting steam and air. Continuous rather than batch operation has been achieved by the use of a rotating grid at the bottom removing the ash. Improvements in metallurgy, process control and flow distribution made it possible to develop modern, oxygen-blown pressurised gasifiers based on moving-bed technology.

The need to utilise caking coals (which tend to agglomerate in the upper portion of moving-bed gasifiers), and the complications presented by tar and oils condensing during gas cooling, have resulted in alternative processes, where gas is formed at much higher temperatures and coal is either suspended in a fluid bed or injected into burners. Fur-

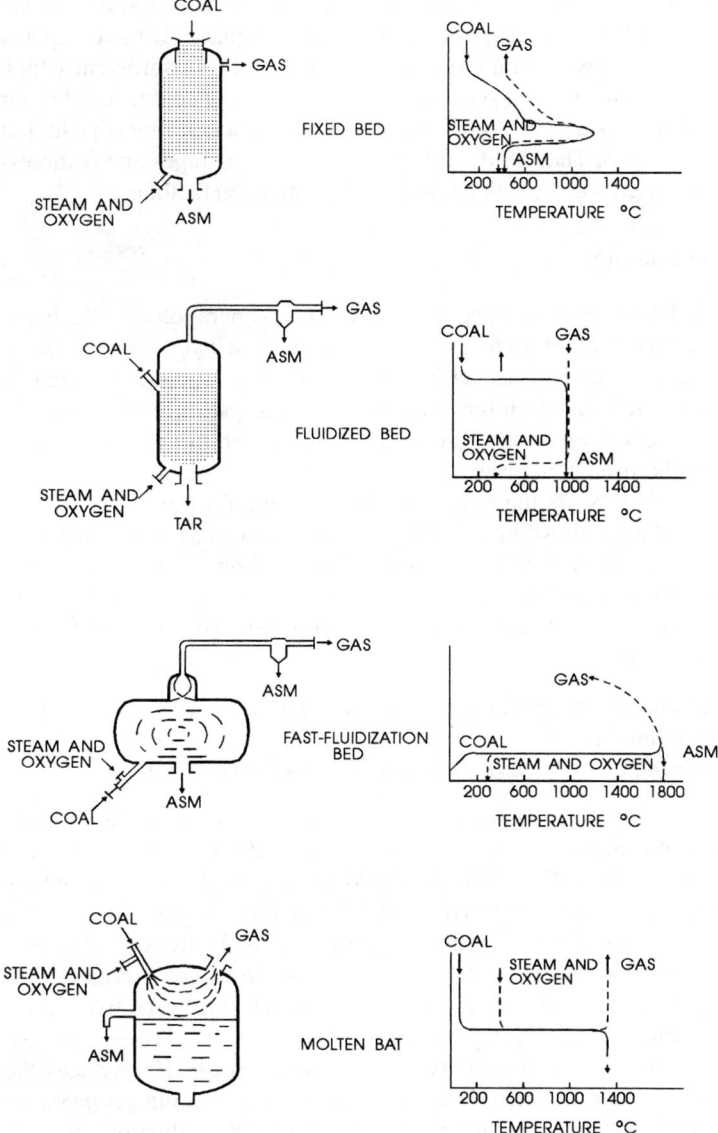

Figure 8-17. Coal Gasification Routes

thermore, steel production technology, which is a main user of coal, can be applied to develop even higher temperature processes aimed at producing clean gas from less reactive coals. Figure 8-17 shows schematically the four routes of coal gasification processes. The adjacent temperature profiles illustrate the main characteristics of each process.

It is important to note that not just the temperature, but also the average residence time of gas-vapour molecules have an effect on the gas composition. Coal is an agglomerate of carbon, heavy hydrocarbon compounds, ash and other elements. When exposed to heat, the portion termed "volatile matter" will evaporate and join the gas phase. If these vapours are not given enough time to react with steam, or if the temperature is not sufficiently high (above 900 °C), the vapours remain unreacted and will eventually condense, forming tar and oils. For most applications, tar and oils represent not only a reduction in gas yield, but also a major process complication. They tend to plug heat exchangers, pipes and compressor blades, and pollute condensed water obtained from coal gasification facilities.

Removal of Sulphur Compounds

Coal gasification has found many applications in electric power generation (IGCC, Internal Gas Combined Cycle Power Plants) and in the manufacture of hydrogen, synthetic gas, methanol, methane and gasoline (Fischer-Tropsch-synthesis). There are technical and economical advantages for the high-temperature and high-pressure processes in removing corrosive and abrasive compounds (such as sulphur and nitrogen compounds, particulates and alkali metals) from coal gases.

Sulphur compounds, with H_2S being the most abundant, account for most of the acidic gas species formed during the combustion of coal gas. For power generation purposes, the H_2S-content of coal gas, from several thousand parts per million (ppm) down to the order of 100 ppm for gas turbines, and down to 1 ppm for fuel cells. Coal gas H_2S can be removed at low temperatures [27, 28]. Two options exist for the removal of H_2S from coal gas at elevated temperatures:

- reduce the gas temperature to a range (500–800 °C) where H_2S can be removed using a suitable external absorbent bed, or
- use an absorbent which can be injected directly into the gasifier (internal removal).

External removal of H_2S from coal gas can be achieved by metal oxide sorbents (especially for fuel cell applications). Oxides of metals such as Zn, Cu, Mn, Ti, Fe and their derivatives (Zn-Fe-O, Zn-Ti-O, Cu-Al-O, Cu-Fe-Al-O etc.) are thermodynamically the most suitable sorbents [25]. They also provide sufficiently high reaction kinetics for capturing H_2S and for regeneration with air, air-steam mixture or hydrogen. The most commonly-used metal oxides are Zn, Cu or Fe based. Among these metal oxides, ZnO provides the best thermodynamic equilibrium for sulphidisation, with effective desulphurization down to a few ppm H_2S. Sintering and reduction of ZnO by H_2 to volatile Zn is limited in its operating range to approximately 600 °C. Addition of TiO_2 reduces the kinetics of ZnO reduction by hydrogen, thereby increasing the operating temperature limit. The presence of steam drastically reduces the kinetics of ZnO reduction. No significant difference in sulphidisation rate is observed by addition of TiO_2. The measured activation energy of 37.6 $kJmol^{-1}$ for the ZnO sulphidisation reaction suggests that chemisorption of H_2S is the rate-controlling step.

Commercialisation of the H_2S removal process using metal oxides still remains a challenge for the chemical industry. Internal removal of H_2S from coal gas, especially for IGCC applications, can be achieved by using relatively inexpensive sorbents, such as limestone ($CaCO_3$), dolomite ($CaCO_3$, $MgCO_3$) and ferric oxide. The performance of these sorbents depends on gasifier operating conditions as well as coal gas composition.

Overview of Commercial Gasifiers

Coal gasification has been the object of considerable commercial research and development efforts for electric power generation in IGCC plants, for chemical feedstock (CO, H_2 and methanol) and synthetic natural-gas production. In fact, commercial-scale coal gasification plants (1000 tons per day) for these applications are currently in operation. At least five commercial-process developers are offering various types of coal gasification technology for licensing and several other types are under development.

Coal gasification involves the high-temperature (>900 °C) reaction of coal/char with an air/steam, O_2/steam or pure O_2 feed stream, depending on the gasification technology used. As the process operates under reducing conditions, the product gases are a mixture of CO, H_2, CO_2, H_2O, H_2S, NH_3, CH_4 and N_2 (plus tar and light hydrocarbons in some cases). The relative concentrations in the product gas are dependent on the coal type, feed rates, gasification reactor type, operating temperature and pressure of the gasifier. Commercial coal gasifiers can be classified on the basis of reactor type and coal feed properties. The major commercial processes available or in demonstration are listed below:

British Gas Lurghi Process. This technology is based on the original Lurghi gasifier, equipped with a counter-flow, moving bed reactor with O_2/steam or air/steam-blow. This is one of the most efficient processes thermally, due to the counter-flow design. It produces the highest amount of CH_4, and also tar, and has limitations with regard to coal type (handling of fines). This process is being used commercially in South Africa for synthetic crude and in the United States for synthetic natural gas production.

Texaco Process. This technology has a slurry-fed, oxygen-blow, entrained flow reactor. The reactor operates at temperatures above the coal-ash melting temperature. The coal/water slurry feed contains 30–40% (by weight) water. The process has almost no limits with regard to feedstocks, as long as a pumpable slurry with sufficient fuel content can be prepared. Due to the extremely high reactor temperatures, essentially no tars or CH_4 are present in the product. The efficiency of the process relative to other technologies is reduced by its high-temperature operation and high water feed-rate. This process is used commercially in Japan, Europe and in the United States for chemical synthesis, and a demonstration plant is in operation in the United States for IGCC application.

Dow Process. Dow technology uses a slurry-fed, oxygen-blow, two-stage entrained flow reactor. Operating conditions are similar to the Texaco process, but the reactor is designed with a secondary coal/water/O_2 reaction zone. This modification increases overall process efficiency, as more of the thermal energy released in the primary zone is converted to chemical energy in the secondary zone. As with the Texaco process, due to the extremely high reactor temperatures, essentially no tars or CH_4 are present in the product. A commercial-scale plant is in operation in the United States for IGCC application.

Shell Process. This process has a dry-feed, oxygen-blow, entrained flow reactor, and operating conditions are similar to the Texaco and Dow processes. The efficiency of this process is greatly increased by the absence of water in the feedstream, although some of this gain is offset by the necessity of drying the feedstock. As with the Dow and Texaco processes, CH_4 and tar yields are minimal. It is currently being demonstrated in Europe and in the United States for IGCC application.

KRW process. This technology has a dry-feed, air/steam-blown, fluidized-bed reactor. Gasifier operating temperatures are below the coal-ash melting temperature. The efficiency of this process is increased by its relatively low-temperature operation and the ability to carry out some product gas cleaning (sulphur removal) using in-bed sorbents. CH_4 and tar yields are higher than in the entrained flow processes. Some feedstock limitations exist due to carbon burn-out and bed-clinkering situations. This technology is basically limited to IGCC applications due to its air-blow operation and has been demonstrated on a pilot scale in the United States.

Differences in the various gasification processes result from the differences in reactor design and feed properties, which, in turn, cause differences in reactor operating temperature, pressure and product gas composition. These are summarised in Table 8-5.

Table 8-5. Comparison of Commercial Gasifiers and Product Gas Compositions for Western Canadian Subbituminous Coals [30]

	Texaco	British Gas	Shell	Dow
Reactor type	Entrained	Moving	Entrained	Entrained
Pressure [MPa]	1–3	1–3	2–7	1–3
Temperature [°C]	1600	1800	1800	1800
Feed [kg]				
Coal (as received)	1.0	1.0	1.0	1.0
Oxygen	0.736	0.441	0.638	0.634
Steam/H_2O	0.776	0.273	0.026	0.667
Flux	0.009	0.025	0.0	0.0
Slurry concentration [wt.-%, dry solids]	56.3	–	–	60.0
Product gas				
Dry gas [m^3 at 15 °C]	1.617	1.540	1.651	1.850
Composition [Vol.-%]				
CO	38.79	63.09	66.31	43.39
H_2	34.60	25.07	27.03	37.33
CO_2	25.99	3.20	1.55	17.39
CH_4	0.0	5.27	0.02	0.17
Others	0.62	3.37	5.09	2.72
HHV dry gas [MJ/m^3]	8.81	12.54	11.19	9.63
Cold gas efficiency [% based on HHV]	68.3	91.4	82.9	68.5
Advantages	Advanced stage of development	High efficiency	No flux, Good efficiency	No flux
Disadvantages	Low gas efficiency, flux required	Low fines tolerance, high flux, tar handling	Requires finer feed, coal drying required	Low gas efficiency

8.6.2. Coal-Assisted Electrolysis

In the late 1970s, Coughlin and Faroque proposed a novel electrochemical process for producing hydrogen, involving the anodic oxidation of coal and the cathodic evolution of H_2 in an electrochemical cell. Their studies instigated numerous publications on the electrochemical oxidation of coal in aqueous electrolytes. An excellent review on the electrochemistry of coal and other carbon materials is available [31].

Electrochemical Oxidation [32]

The mechanism for the electrochemical oxidation of coal in an electrolysis cell is postulated to be

$$C + 2H_2O \longrightarrow CO_2 + 4H^+ + 4e^-. \tag{8.11}$$

It should be noted that this reaction is the same as that used to describe the electrochemical oxidation of carbon electrodes in low temperature Fuel Cells. The corresponding electrochemical reaction at the cathode is the evolution of hydrogen:

$$4H^+ + 4e^- \longrightarrow 2H_2. \tag{8.12}$$

The overall reaction is

$$C + 2H_2O \longrightarrow CO_2 + 2H_2. \tag{8.13}$$

In theory, less electrical energy is required for this electrochemical process than that required for conventional water electrolysis to produce hydrogen; in practice anode potentials of 0.7 to 0.9 V (vs. NHE[4]) are required to oxidize coal. At 25 °C, the reversible potential for Reaction 8.13 is only 0.21 V, and the corresponding reversible potential for water electrolysis to produce H_2 and O_2 is 1.23 V. About six times more electrical energy is theoretically required to produce one mole of hydrogen by conventional water electrolysis. In effect, the purpose of the coal is to serve as an anode depolarizer in the electrochemical cell to form hydrogen.

Many coals and carbon powders have been evaluated, and the majority of the experiments were conducted in H_2SO_4 electrolyte. In a typical experiment, the particles of carbon material (e.g. particle size < 200 μm) are suspended in an electrolyte (concentration of particles is 0.01 to 1.0 g/cm^3 of solution) containing an anode current collector, usually Pt or graphite, which is separated from the cathode by a porous separator. It has been observed that the oxidation current increases with an increase in the concentration of carbon particles in suspension, higher anode potential and temperature of the electrolyte, and also with a smaller carbon particle size.

In addition to the formation of gaseous carbon oxides (CO_2 and CO) and surface oxides, evidence for the formation of an organic oxidation product has been reported by numerous investigators. Humic acid, fulvic acid, mellitic acid, benzene carboxylic acids and C_8-C_{18} hydrocarbons are some of the compounds that are produced by the electro-

---

4   NHE = Normal Hydrogen Electrode

chemical oxidation of coal. A tar-like substance, which is a dark reddish brown extract in acetone, is obtained from the electrolysis of lignite in H_2SO_4. With prolonged electrolysis of bituminous coal at 1.5 V (vs. NHE) in H_2SO_4 containing Ce^{4+}, the residual electrolyte assumes a yellow-brown colour, which is attributed to aromatic carboxylic acids. The anodic product obtained from anthracite coal, which was electrolysed for 88 h in 1 N NaOH, has an IR spectrum similar to that of humic acid obtained from lignite. It has also been reported that various humic acids are formed during the electrolysis of lignite in 1 N NaOH. Analysis of one of the humic acids (e.g. 55.98% C, 3.74% H, 1.21% N, 32.57% O, 6.50% ash) obtained by electrolysis at 1.2 V (vs. SCE[5]) for 24 h indicates a composition similar to that of the parental coal (55.1% C, 5.2% H, 1.0% N, 22.7% O, 20.1% ash), except for the larger amount of oxygen in the acid. The composition of this humic acid changes with the oxidation potential at which it is formed; the oxygen content in the humic acid decreases with an increase in the potential.

8.7. Internal Hydrocarbon Processing

As mentioned in Section 4.5. "Molten Carbonate Fuel Cells", metallic components in the MCFC anode chamber catalyze the shift reaction at 650 °C, so that the cell consumes carbon monoxide in the presence of water vapour under the same kinetic conditions as hydrogen.

8.8. Hydrogen as a By-Product from Other Conventional Processes

There are a number of processes which produce hydrogen in varying amounts as a by-product. Some of these are discussed below.

1. Hydrogen is a by-product of the production of chlorine from the electrolysis ofaqueous NaCl:

$$2\,NaCl + 2\,H_2O \longrightarrow 2\,NaOH + Cl_2 + H_2 \tag{8.14}$$

2. Hydrogen is produced as a component of the light gases from crude oil refinery processes.
3. Hydrogen is produced during the production of coke from coal in coke oven gases:

$$CH_{0.8}O_{0.2} \longrightarrow 0.8\,C + 0.2\,CO + 0.4\,H_2 \tag{8.15}$$

4. Hydrogen is released in chemical dehydrogenation processes (e.g. ethylene plant flue-gas; ammonia dissociation, hydro-dealkylation):

$$C_2H_6 \longrightarrow C_2H_4 + H_2 \tag{8.16}$$

---

5   SCE = Standard Calomel Electrode

8.9. Methanol Steam Reforming

Methanol is a promising substitute for petroleum products if they become too expensive for use as motor fuels. As a result of the oil crisis in the early 1970s, a number of projects were started, based on the assumption that the use of methanol produced from coal would be more economical than the use of petroleum products. The estimates made at the beginning of the 1980s proved to be too optimistic with regard to costs and ways to overcome the technical or environmental problems involved in producing synthesis gas from coal. They were too pessimistic with regard to the price and availability of crude oil [23]. The importance of methanol for use in Direct Methanol Fuel Cells was recognized in optimistic predictions because a break-through in this kind of technology would mean a break-through in transportation applications [34–37].

Production

Hydrogen and carbon monoxide can be obtained from the decomposition of methanol, a reversal of the synthesis reaction. The endothermic decomposition occurs at temperatures >700°C without a catalyst, or at 300–450°C with catalysts based on copper – nickel or zinc – chromium alloys. Higher hydrogen yields are obtained by catalytic steam reforming of methanol. The process proceeds via an initial methanol decomposition, followed by a water-gas shift reaction as shown in the overall equation:

$$CH_3OH \longrightarrow CO + 2\,H_2 \qquad \Delta H_{298} = +91.7 \text{ kJ/mol} \qquad (8.17)$$

$$CO + H_2O \longrightarrow CO_2 + H_2 \qquad \Delta H_{298} = -41.0 \text{ kJ/mol} \qquad (8.18)$$

$$CH_3OH + H_2O \longrightarrow CO_2 + 3\,H_2 \qquad \Delta H_{298} = +50.7 \text{ kJ/mol} \qquad (8.19)$$

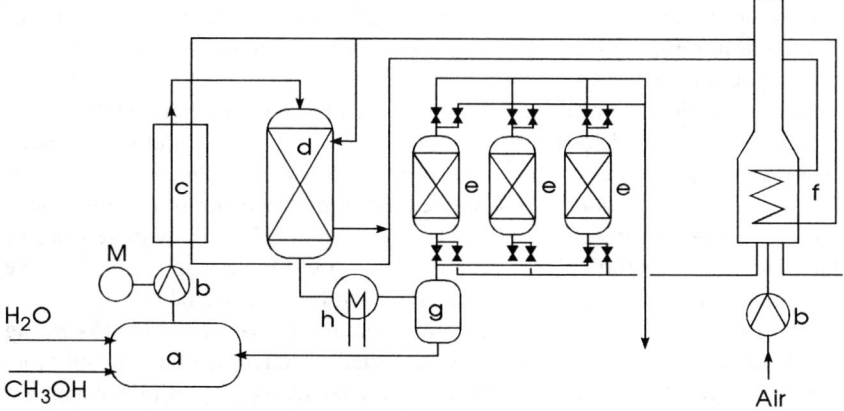

Figure 8-18. Methanol Reformer with Pressure Swing Adsorption Purification for the Production of 50 m^3 (STP) Hydrogen per Hour, Purity 99.9999 vol.-% (Linde, Germany)
a) Methanol/water vessel; b) Pump, c) Preheater, d) Reformer, e) Adsorber, f) Heater for heat transfer medium, g) Condensate feedback, h) Cooling water

The reaction is carried out technically at pressures of 0.7–3 MPa on CuO–ZnO or CuO–Cr_2O_3 catalysts. The molar ratio of water to methanol will vary from 0.67 to 1.5, and excess steam lowers the risk of carbon formation.

Hydrogen production from methanol is carried out in small to medium-sized plants (some of which are mobile) for the production of 1–2000 m^3 (STP)/h. Purification of the product gas is performed by carbon dioxide scrubbing, pressure-swing adsorption or (especially for small units and high purity), by diffusion through a palladium-silver membrane. Figure 8-18 shows the schematic of a methanol cracker for 50 m^3 (STP)/h of hydrogen at 1.4 MPa using a PSA purification unit to give 99.9999% purity. 1.5 m^3 (STP) of hydrogen are produced per kilogram of methanol.

The production costs for hydrogen by methanol steam-reforming are given in Section 8.11. "Hydrogen from Electrolysis of Water".

8.10. The Production of Hydrogen and Methanol from Solid Biomass

The focussing of public attention on environmental conservation and the greenhouse effect has resulted in a controversy regarding present technologies for energy supply. As it is impossible to switch quickly from fossil fuels to regenerative energy sources for current energy supplies and the present standard of living, the development of suitable transitional technologies has become an important objective. A renewable energy source, which might be used with today's technology, is the biomass. The available amount of biomass differs widely from country to country. The alpine sections and the Scandinavian countries in Europe have a low population density and large areas of forest. For example, approximately 10% of the total energy consumption could be covered by energy from biomass in Austria [38]. This might be increased to a certain extent, but biomass is not available in unlimited amounts. It is, therefore, extremely important to optimize the efficiency of conversion technologies.

Interest in the large-scale use of biomass for energy, and in methanol and hydrogen, is motivated largely by global and local environmental issues. If grown and used renewably, biomass would make no net contribution to atmospheric, greenhouse-gas concentrations. Biomass-derived fuels would reduce greenhouse gas concentrations to the same extent as they could replace present fossil fuel usage. Methanol and hydrogen are of interest for large-scale transportation use, because they would have significant, positive local environmental impacts over fossil fuels when used in fuel cell vehicles.

Methanol or hydrogen can be produced from biomass by comparable processing routes. The steps involved are similar to the established production of methanol from coal. All process steps could use commercially-available technology – with the exception of the first step: thermochemical gasification. A number of biomass gasifier designs, at various stages of development, can be considered for the initial step. Some biomass gasifiers have been developed based on coal gasifier designs. Others are being developed to take advantage of the higher reactivity of biomass [39, 40].

Basic Process Descriptions

In the production of methanol (CH_3OH) or hydrogen (H_2) from biomass, the biomass is first converted in a gasifier to a synthesis gas containing carbon monoxide, hydrogen, methane, carbon dioxide, water vapour and additional hydrocarbons in proportions that vary with the gasifier type used. The gas is cooled and then quenched to remove particulates and other contaminants. Sulphur (if still present) is then removed to prevent poisoning of catalysts downstream.

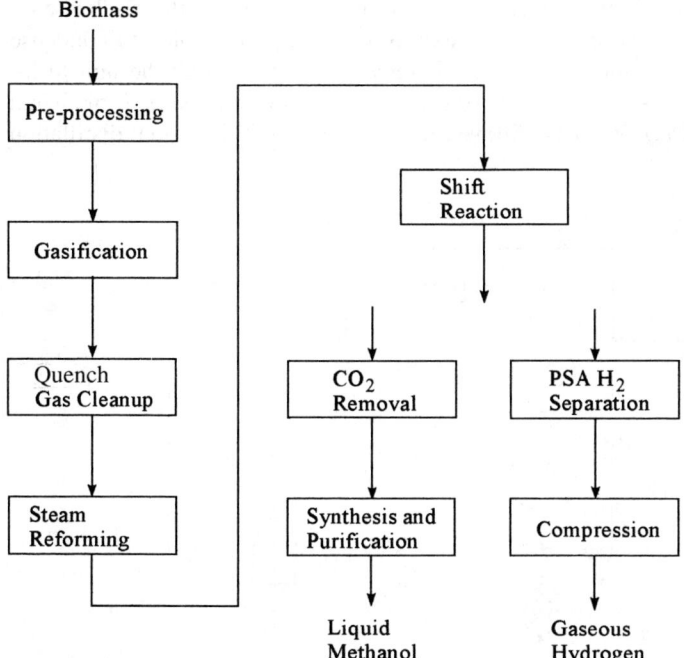

Figure 8-19. Block diagram of Process Configurations for the Production of Methanol or Hydrogen from Biomass [41]

The synthesis gas undergoes one or more catalyzed chemical reactions designed to adjust the carbon-to-hydrogen ratio to the level needed for effective conversion into the desired end product. The gas from most types of gasifiers considered for biomass contains some hydrocarbons – primarily methane (CH_4) – so the initial reaction step reforms these hydrocarbons into carbon monoxide (CO) and molecular hydrogen (H_2) by reacting with steam:

$$CH_4 + H_2O \longrightarrow CO + 3H_2 \qquad \Delta H = +206.28 \text{ kJ/mol} \qquad (8\text{-}20)$$

The CO:H_2 ratio is then adjusted to the desired level in a "shift" reactor, wherein one mole of H_2 is produced from each mole of CO via the water-gas shift reaction:

$$CO + H_2O \longrightarrow CO_2 + H_2 \qquad \Delta H = -41.16 \text{ kJ/mol} \qquad (8\text{-}21)$$

Steam added to the reformer and the shift reactor promotes the desired reactions and helps prevent the formation of coke (solid carbon). After the shift reactor, the processing steps to produce methanol differ from those for hydrogen.

For methanol, only part of the CO leaving the reformer undergoes the shift reaction to produce H_2, Carbon dioxide and water vapour are then separated out, and the remaining gas is compressed and fed to a methanol synthesis reactor, where CO and H_2 combine over a catalyst to form methanol. The principal reaction is:

$$CO + 2H_2 \longrightarrow CH_3OH \qquad (8\text{-}22)$$

With commercial methanol synthesis technologies, only a fraction of the feed is converted in each pass through this unit. The gas exiting the reactor is cooled to condense most of the methanol, and the unconverted gas is recycled back through the unit to increase the yield of the conversion. (Overall conversion of carbon monoxide to methanol is typically in excess of 98%.) Water is removed from the methanol in a final distillation step.

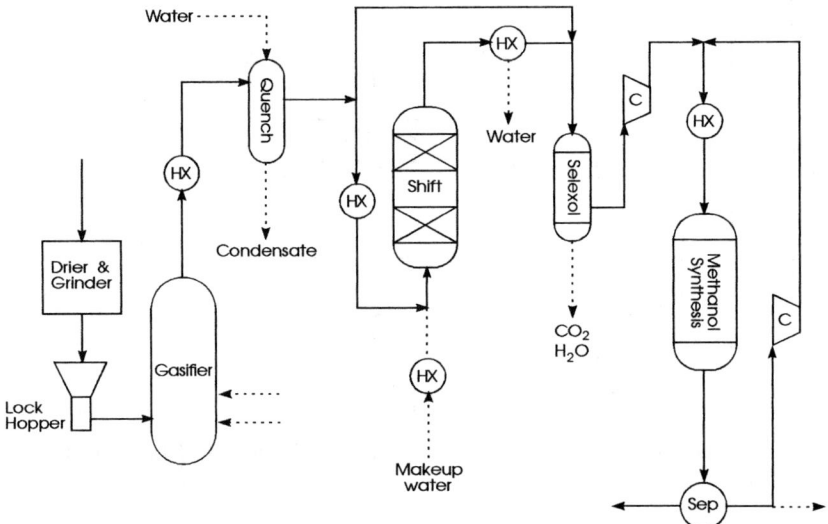

Figure 8-20. Process Configuration for Methanol Production from Biomass based on the Shell Gasifier. (The main process flows are shown as solid lines. "HX" indicates "Heat exchange" and "C" indicates compression.) [42]

For hydrogen production, most of the CO leaving the reformer is converted in the shift reactor. The gas then enters a pressure swing adsorption unit (PSA), in which first CO_2 and H_2O are removed by physical adsorption, followed by adsorption of all remaining gas components except hydrogen. Up to about 97% of the hydrogen fed to the PSA can be recovered as a final product with greater than 99.99% purity. The product may then be compressed for subsequent storage or transport.

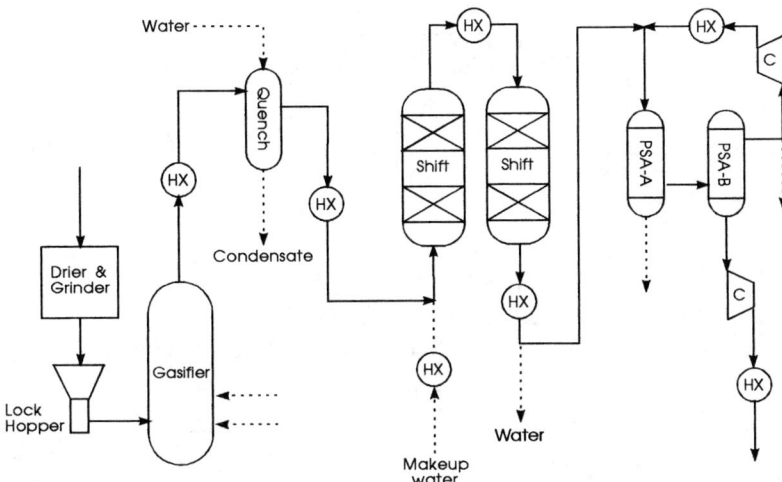

Figure 8-21. Process Configuration for a Hydrogen Production from Biomass Based on the Shell Gasifier. (The main process flows are shown as solid lines. "HX" indicates heat exchange and "C" indicates compression.) [43]

8.11. Hydrogen from Electrolysis of Water

Water electrolysis based on current technology is the only industrial, hydrogen production process which does not necessarily rely on fossil energy. The required electrical energy can be derived from fossil, solar or nuclear energy. Although presently not economically competitive with hydrogen produced from hydrocarbons, electrolytic hydrogen production has several advantages such as high product purity and flexibility of operation. An electrolysis plant can operate at a wide range of power outputs which is attractive for electricity load levelling – either at the generation site or anywhere on the grid. In addition, the technology is convenient for small operating capacities [44, 45].

Water electrolysis involves splitting water into hydrogen and oxygen by passing a direct current through water that has been made electrically conductive by the addition of hydrogen or hydroxyl ions (e.g. potassium hydroxide). The simplified block flow diagram for water electrolysis is shown in Figure 8-22. The process consists of a water purification plant, an electrolyser section, gas separation, power transformer, regulator and rectifiers as well as compressor systems. The current state-of-the-art offers very reliable electrolyzers which have an energy efficiency (based on the H_2 enthalpy of the combustion) of up to 80% with current densities of about 2000 A/m² [46]. Water electrolysis is a modular operation technology. The oxygen can be used technologically or for the enrichment of the cathodic air supply.

The rate of hydrogen generation per unit electrode area is related to the current density. In general, the higher the cell voltage, the higher is the current density and the power cost per unit of hydrogen produced. The size of the electrolysis plant, and hence the capital investment, decreases with increasing current density. The designer of a new

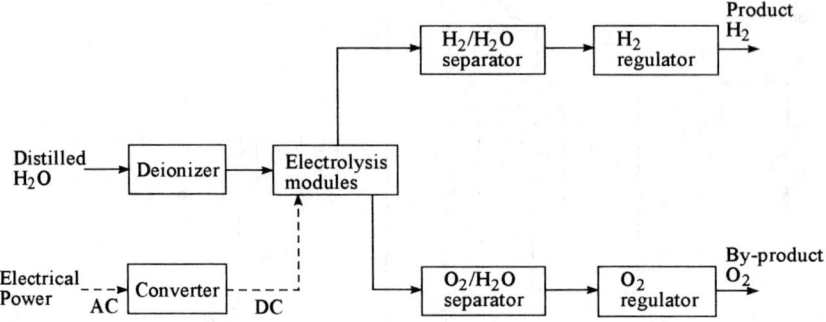

Figure 8-22. Simplified Flow Diagram of Hydrogen Production from Water Electrolysis

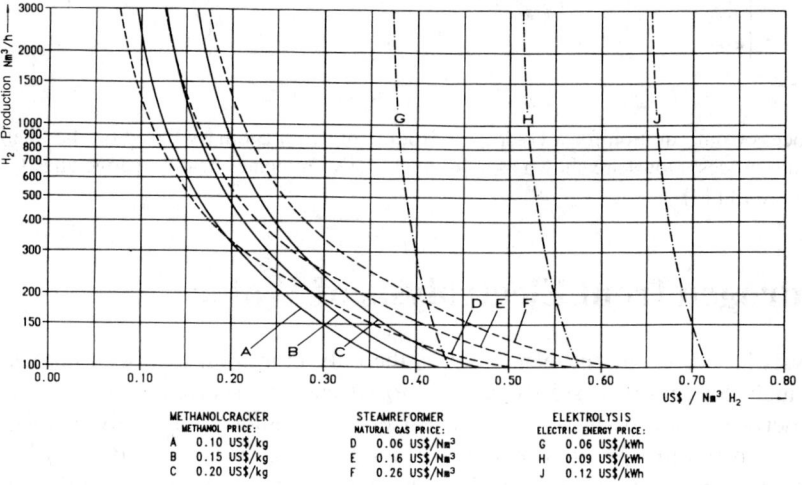

METHANOLCRACKER METHANOL PRICE:	STEAMREFORMER NATURAL GAS PRICE:	ELEKTROLYSIS ELECTRIC ENERGY PRICE:
A 0.10 US$/kg	D 0.06 US$/Nm³	G 0.06 US$/kWh
B 0.15 US$/kg	E 0.16 US$/Nm³	H 0.09 US$/kWh
C 0.20 US$/kg	F 0.26 US$/Nm³	J 0.12 US$/kWh

Figure 8-23. Production Costs per Nm³/h H_2 [47]

electrolysis plant must make a trade-off between capital cost and operating cost to minimise the total.

Similar to fuel cells, there are two possible arrangements for the electrodes: unipolar and bipolar. Hydrogen production takes place either at ambient or at increased pressure (commercial plants operate at up to 3 MPa).

Normal Pressure Electrolysis Units

The advantages of a process at near ambient pressure are:

1. Low mechanical strains on pipe fittings, seals and instruments.
2. Simple operation, because feed water is not required under pressure, quick start-up, and easy maintenance.
3. Safer operation, because of the avoidance of high-pressure hydrogen, small pressure differences between the evolved gases, and simple blanketing of the electrolysis units and the piping network with nitrogen under low pressure

Pressure Electrolysis

Pressure electrolysis units in the range of 0.6–20 MPa are available. Because hydrogen is usually required at high pressure, and because the power consumption for a pressure electrolysis does not increase significantly, these units save both equipment and energy for hydrogen compression. The construction and sealing of the electrolysis cells is complicated. To accurately compare the data, the energy consumption for hydrogen compression has to be considered.

Advanced Conventional Electrolysis

The conventional, potassium hydroxide electrolysis can be improved considerably. Some of the suggestions made have been incorporated in commercial electrolysis units. Other necessary components are still in the development stage. Areas of development are:

Zero-Gap Cell Geometry. By placing the electrodes directly on the surface of the diaphragm, the space between the electrodes can be reduced and the voltage drop in the electrolyte, and thereby the heat loss caused by the current, can be minimized. Figure 8-24 shows perforated foil electrodes in a "zero-gap" construction.

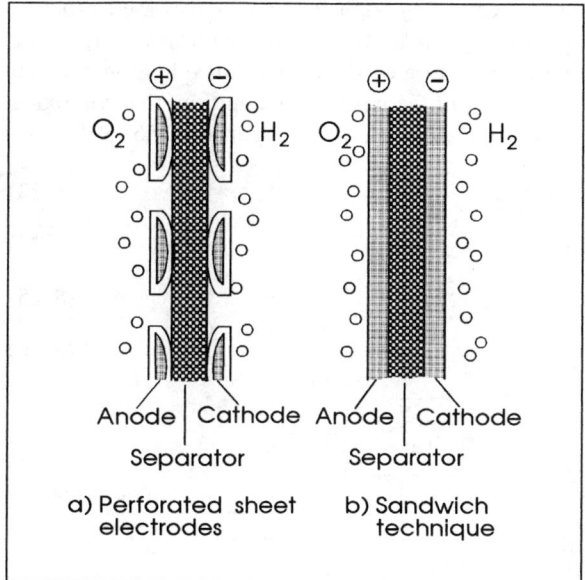

Figure 8-24. "Zero gap" Cell Geometry. Perforated Cathodes and Anodes, in Direct Contact with the Diaphragm and the Membrane. (The conical apertures, with different diameters for O_2 and H_2 facilitate the detachment of the developing gas bubbles.)

In particular, the construction with perforated electrodes has already been used in commercial plants. The disadvantage of this type is that in these electrolysis units, operating at high currents and at higher temperatures, corrosion is greatly increased. Products of corrosion, which may be formed at the anode, can grow through the diaphragm and lead to dangerous short-circuits.

New Materials for Diaphragms. Polysulfones, with their hydrophobic character decreased by means of additives (e.g., polyantimonic acid), offers an alternative to asbestos diaphragms. Another line of development is the use of oxide-ceramic diaphragms based on barium or calcium titanate, or nickel oxide. A hydrophilic, gas-tight oxide layer is deposited on a metallic net. The thickness of the layer is < 500 μm, which leads to a low resistance of ca. 2×10^{-5} Ω/m^2. The stability against electrolytes at high temperatures is also favourable.

Detailed articles pertaining to this kind of technology have been published in the literature [48–55].

8.12. Hydrogen from the Steam-Iron Process

The steam-iron process, which is one of the oldest ways of producing hydrogen, has been subject to renewed interest. Principally, this process involves reacting steam with hot iron to produce an H_2-rich gas and iron oxide. Although it is a syngas based process, hydrogen is derived from the decomposition of steam by reaction with iron oxide. The synthesis gas is mainly used for the reduction of iron oxide to iron. Since hydrogen is not derived from the synthesis gas, air can be used in the gasifier. In addition, the iron oxide "barrier" prevents nitrogen from significantly contaminating the hydrogen [56–59].

$$Fe_3O_4 + H_2 \longrightarrow 3\,FeO + H_2O \tag{8-23}$$

$$Fe_3O_4 + CO \longrightarrow 3\,FeO + CO_2 \tag{8-24}$$

$$FeO + H_2 \longrightarrow Fe + H_2O \tag{8-25}$$

$$FeO + CO \longrightarrow Fe + CO_2 \tag{8-26}$$

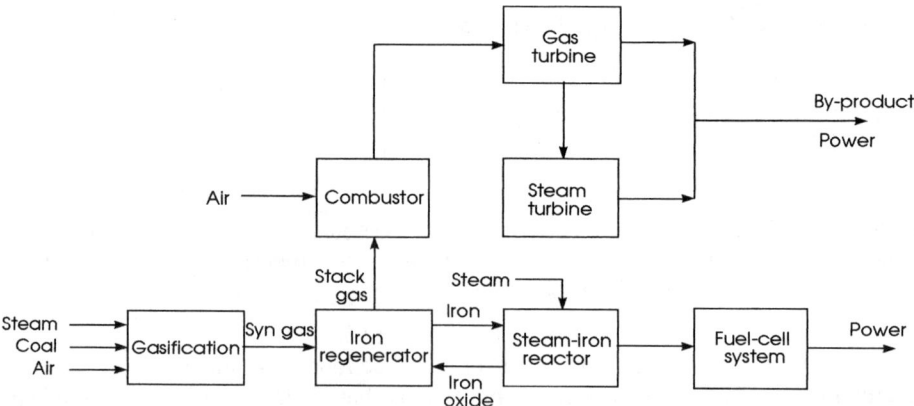

Figure 8-25. Simplified Flow Diagram of Hydrogen Production from Steam-Iron Process [60]

Figure 8-25 shows the process cycle with the usage of coal in combination with Gas- and Steam-Turbines. The gasified coal converts iron oxide (Fe_3O_4) in a sponge iron bed (Iron regenerator) to carbon dioxide, water vapour, sponge iron etc., while water vapour converts sponge iron to hydrogen and Fe_3O_4 in the other bed (Steam-iron reactor). The beds can be made large enough, if desired, so that the fuel cell can run on one bed for hours, days, and even weeks. The wet hydrogen produced from sponge iron and water vapour is free of, CO, CO_2, sulphur, and other contaminations and can be directly used by any type of fuel cell without further purification. Heat is needed and supplied to the sponge iron beds.

In summary, the sponge iron acts both as a purifier of the gasified wood and as a hydrogen storage medium enabling the fuel cell to run for long periods without a new supply of wood gas. The cost of sponge iron is probably much lower than that of a gasified-wood storage tank combined with a sophisticated gas purifier.

8.13. Carbon-Free Energy Carriers

8.13.1. Ammonia

World-wide use of fossil fuels for energy production is creating an increase in the carbon dioxide content of the atmosphere, which, if continued will cause significant changes in world climate and ocean levels. It is important to find ways to reduce atmospheric pollution through energy conservation and the use of energy sources which do not generate CO_2 and other absorbers of infra-red radiation.

Compared to hydrogen ammonia, offers significant advantages in cost and convenience as a vehicular fuel due to its higher density and ease of storage and distribution, both of which are of great importance in transportation applications [61–63].

Ammonia is an easily-liquefied gas, which is made by combining three molecules of hydrogen with one of nitrogen. The liquid density is 0.60 kg dm^{-3} vs. 0.07 kg dm^{-3} for liquid hydrogen, and the heat of combustion is 11.2 MJ dm^{-3} vs. 8.58 MJ dm^{-3} for liquid hydrogen. NH_3 is the primary source for nitrogen fertilisers and it is produced and distributed world-wide in quantities of millions of tons per year. The United States ammonia production was 18 million tons in 1987, and the world production was 125 million tons. In heating value, the latter amount is equivalent to 1.3 million barrels of gasoline per day. Procedures for safe handling of ammonia have been developed in every country. Facilities for storage and transport by barges, tank-cars, trucks and/or pipeline from producer to ultimate consumer are available throughout the world.

Ammonia is presently commercially manufactured in a series of processes which involve reaction of methane with steam and air. Water and carbon oxides are then removed to create a feed gas mix consisting of nitrogen and hydrogen in the proper ratio for ammonia synthesis. Special equipment is used to remove sulphur and other impurities from the gas stream to avoid damage to the catalyst used for the synthesis. The pure nitrogen and hydrogen mixture passes through a catalytic converter, where ammonia is formed. The emerging gases are chilled, allowing the ammonia to liquify and to be separated. The

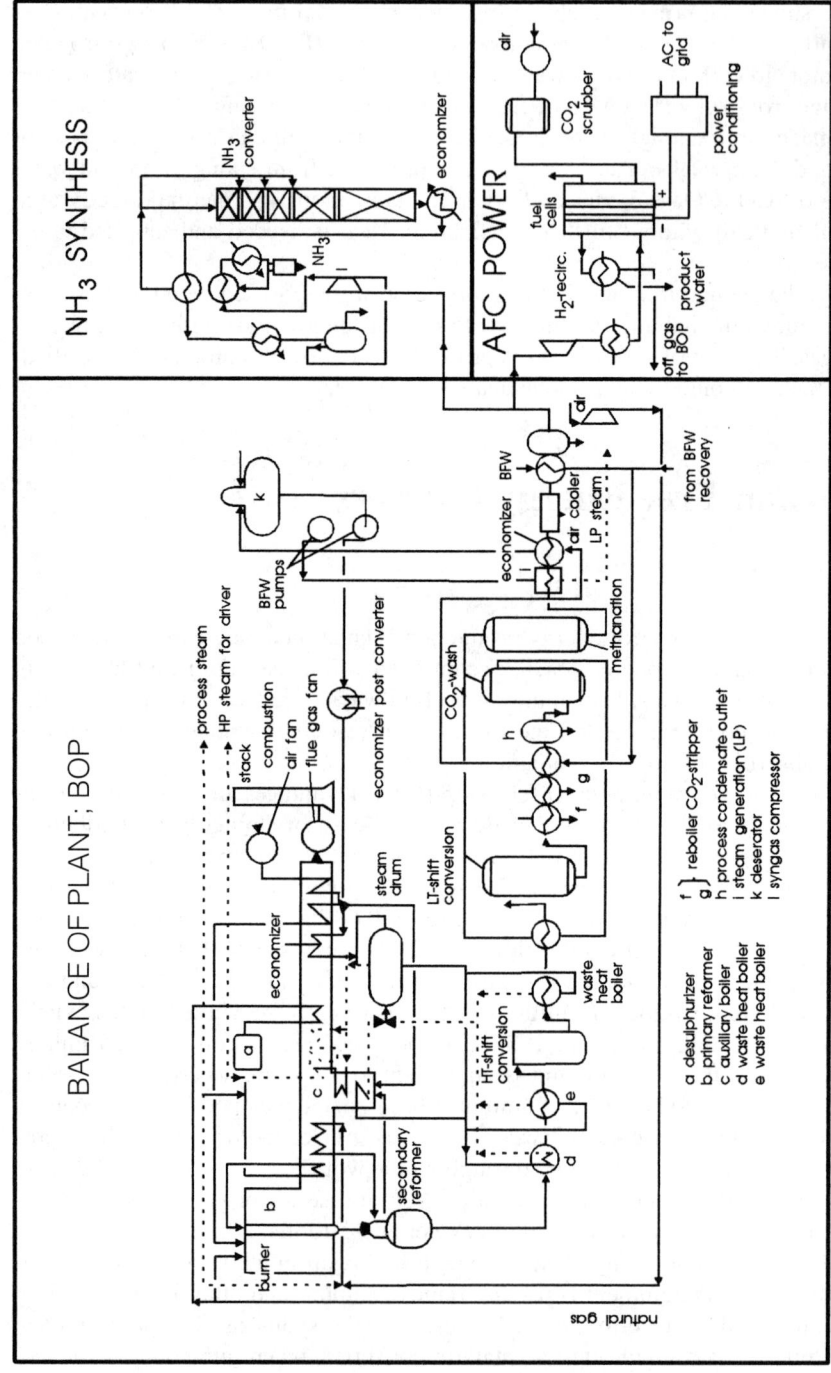

Figure 8-26. Simplified flow sheet for a large single-train steam reforming ammonia plant with an add-on fuel cell power plant [64]

unreacted gases, supplemented with additional hydrogen and nitrogen produced in the first phase, are then recycled. The catalytic reaction takes place under equilibrium conditions with low heat evolution, so that roughly 88% of the heating value of the hydrogen is conserved in the formation of ammonia. The thermal efficiency can be higher, if the heat evolved in ammonia formation is used for steam production. A schematic diagram of the process is shown in Figure 8-26.

Ammonia synthesis is greatly simplified, if electrolytic hydrogen is available as a feed stock. In this case, the nitrogen for the reaction is produced by air liquefaction and distillation. The pure reactants are mixed, and then enter the catalytic converter directly. A conceptual design of an on-land ammonia plant based on the use of electrolytic hydrogen has been proposed by Crawford and Benzima [65].

Ammonia in general use is stored and transported at ambient temperatures as a liquid ...) K). Trucks, tank-cars and pipelines are ... or tanker. In this mode, the ammonia is ... which allows the use of unpressurized con- ... monia as a fertilizer involves spraying the ... afe local storage and handling, including ... al farmer, have been solved and are stan- ... k with liquid ammonia for his farm use, ... e, if one were available, from the same ... required.

... re characteristics almost identical to those ... pour phase, except at temperatures below ... dily cracked into hydrogen and nitrogen, ... ents that react with alkaline electrolyte, as ... e other hand, cracked ammonia is difficult ... ount of ammonia in the feed stream reacts ... to CO_2 with alkaline electrolytes. Because ... an alkaline electrolyte, but its reactivity is ... el cells is normally not very effective.

... rogen and recovering the hydrogen has the

$$\Delta H_{298} = +91.86 \text{ kJ/mol} \qquad (8\text{-}27)$$

... received little attention and are usually not ... luction, except as specialised remote, low- ... ammonia makes them attractive. However, ... lytic converters, may be of future interest.

... ion (up to 64%). The only application for ... s a monopropellant for satellites and space- ... converted into organic derivatives. All other

uses are based on hydrazine as a reducing agent or as an energy-rich compound with a high hydrogen storage capacity. The most important uses of hydrazine and its derivatives are as blowing agents for foamed plastics and for the production of pesticides. Other uses include: synthetic building blocks, pharmaceuticals and propellants.

Fuel cells based on oxidation of hydrazine either with oxygen or hydrogen peroxide have been extensively studied. Their use is restricted to military applications and, as hydrazine is poisonous in small amounts, further use in vehicle development has been abandoned [66–68].

Hydrazine must also be catalytically decomposed to obtain hydrogen and an aqueous solution of hydrazine is contacted with a nickel or palladium coated electrode at room temperature for this purpose. An addition of 1–2% of hydrazine to the alkaline electrolyte is sufficient. The theoretical energy output of the hydrogen produced from hydrazine is 2 kWh from 1 kg fuel.

The actual anode reaction suggests a higher cell voltage, but in reality it is only that of a hydrogen-fuelled AFC.

$$N_2H_4 + 4OH^- \longrightarrow N_2 + 4H_2O + 4e^- \quad E_0 = 1.56 \text{ V} \tag{8-28}$$

This fact indicates that hydrazine is only acting as a source of hydrogen. A practical anode is a Pd-activated, porous Ni electrode, and the electrolyte is a 25% KOH solution with 2–3% of hydrazine. The fuel cell operates at between 20 and 80 °C. Hydrazine is produced as a very diluted solution, and the concentration to monohydrate is very expensive manufacturing processes avoiding poisonous nitrose compounds as side products have been suggested in patents.

Toxicology of hydrazine

Hydrazine is highly toxic, even in dilute solutions, and must be handled with care. No antidote to hydrazine is known; appropriate treatment must be based on the symptoms observed.

Contact of the liquid with the skin and the eyes should be avoided. Irritation, alkali burns, dermatitis or allergies may result. Inhalation may cause irritation or injury to the nose, throat and lungs. Ingestion may induce convulsions and depression of the central nervous system. Repeated exposure to hydrazine may result in damage to the lungs, liver or kidneys.

References to Chapter 8

[1] J. v. Linde, "Captive Generation or Outsourcing of Hydrogen", Sonderdruck aus Gas Wärme International, Band 9 (November 1989)
[2] G. Sandstede, Chem.-Ing.-Tech. **63** (6) (1991) 575
[3] H. J. Derdully, K. Wellner, Nachr. Chem. Tech. Lab. **39** (5) (1991) 503
[4] M. Fischer, Chem.-Ing.-Tech. **61** (2) (1989) 124
[5] D. Wöhrle, Nachr. Chem. Tech. Lab. **39** (11) (1991) 1256
[6] J. Rothenstein, Int. J. Hydrogen Energy **20** (4) (1995) 283

[7] R. G. Minet, K. Desai, Int. J. Hydrogen Energy **8** (4) (1983) 265

[8] M. Steinberg, H. C. Cheng, Int. J. Hydrogen Energy **14** (11) (1989) 797

[9] Gas Production, Vol. A 12, Ullmann's Encyclopaedia of Industry Chemistry, Vol. A 12, pp. 169–306, 1989

[10] M. A. Rosen, Int. J. Hydrogen Energy **16** (3) (1991) 207

[11] R. B. Moore, Int. J. Hydrogen Energy **8** (11/12) (1983) 905

[12] N. Iwasa et al., "Development of a new reforming process for fuel cells", International fuel Cell Conference (IFCC), Makuhari, 1992, p. 83

[13] Haldor Topsoe A/S, Heat Exchange Reformer, Concept-3, 1994; Haldor Topsoe A/S, Nymollevey 55, DK-2800 Lynby, Denmark

[14] N. Iwasa et al., Development of a New Reforming for FC, International Fuel Cell Conference (IFCC), Makuhari, Japan, pp. 83–90

[15] T. Shiori, "Development of PAFC Utilizing Petroleum Products", International Fuel Cell Conference (IFCC), Makuhari, Japan, 1992, p. 75

[16] S. K. Megalopfonos, N. G. Papayannakos, Int. J. Hydrogen Energy, **16** (5) (1991) 319

[17] T. Yanagino, et. al., "Development of PAFC using Naphta", International Fuel Cell Conference (IFCC), Makuhari, Japan, 1992, p. 79

[18] T. Take et al., "Fuel Substitution in PAFC for Telecommunication Use", International Fuel Cell Conference (IFCC), Makuhari, Japan, 1992, p. 63

[19] T. Take et al.; "Fuel Substitution Characteristics in Phosphoric Acid Fuel Cells", 1992 – Fuel Cell Seminar, Tucson (AZ), pp. 477–480

[20] K. Kinoshita, "Electrochemical and physicochemical properties of Carbon", Lawrence Berkeley Laboratory, Berkeley, California, 1988

[21] K. Ledjeff, J. Giesshoff, "Schadstofffreie Erdgasnutzung", Wien (Austria), January 1991

[22] Schadstofffreie Erdgasnutzung, Marketing Gas, no 16, summer 1994, pp. 13–14

[23] A. Jess, K. Hedden, Erdöl Ergas Kohle **110** (9) (1994) 365

[24] B. Özüm et al., Int. J. Hydrogen Energy **18** (10) (1993) 847

[25] Gas Production, Vol. A 12, Ullmann's Encyclopedia of Industrial Chemistry, (1989), p. 169

[26] M. A. Rosen, D. S. Scott, Int. J. Hydrogen Energy **12** (12) (1987) 837

[27] R. A. Douglas, Am. Chem. Soc., Div. of Fuel Chem. **35** (1990) 136

[28] B. M. Wilson and Newell, Chem. Engng. Prog., October (1984), 40

[29] P. R. Westmoreland, D. P. Harrison, Environ. Sci. Tech. **10** (1976) 659

[30] B. Özüm, et. al., Int. J. Hydrogen Energy **18** (10) (1993) 847

[31] Su-Moon Park, "Electrochemistry of Carbonaceous Materials and Coal", J. Electrochemical Society, 131, 363 C, (1984)

[32] O. J. Murphy, J. O'M Bockris, Int. J. Hydrogen Energy, **10** (7/8) 453

[33] E. Fischer et al., Methanol, Vol A16, Ullmann's Encyclopedia of Industrial Chemistry, Weinheim, 1990, pp. 465–486

[34] M. A. Rosen, D.S. Scott, Int. J. Hydrogen Energy **13** (10) (1988) 617

[35] C. B. van der Decken et al., Energie-Alkohole, Kernforschungsanlage Jülich GmbH, März 1987

[36] B. Ganser, Verfahrensanalyse: Wasserstoff aus Methanol und dessen Einsatz in Brennstoffzellen für Fahrzeugantriebe, Forschungszentrum Jülich GmbH, März 1993

[37] C. Sylwan, "50 W Methanol-Air Fuel Cell with Hydrophobic Air Electrodes", Energy Conversion and Management, Vol. 20, No 1., pp. 1–7

[38] Werner Schaller, "Dezentrales Biomasse-Heizkraftwerk in der Steiermark", in Berichte des Symposiums "Energieinnovation in der Elektrizitätswirtschaft und Energieversorgung", January 15, 1993, Kaprun, Austria

[39] E. Larson, R. E. Katofsky, "Production of Methanol and Hydrogen from Biomass", PU/CEES, Report No. 271, July 1992, center for energy and environment studies, Princeton university

[40] Y. Wang, C. M. Kinoshita, Solar Energy **49** (3) (1992) 153

[41] R. H. Williams, E. D. Larson, R. E. Katofsky, J. Chen, "Methanol and Hydrogen from Biomass for Transportation", Bioresources '94 Biomass Resources; a means to sustainable development, 3–7 October, Bangalore, India,

[42] E. D. Larson, R. E. Katofsky, "Production of Methanol and Hydrogen from Biomass", Center for Energy and Environmental Studies, Princeton University, Princeton (NJ) 1992.

[43] R. E. Katofsky, "Thermodynamics of Fluid Fuels Production from Biomass", MSE Thesis, Mech. and Aerospace Engng. Dept., Princeton University, Princeton (NJ) 1992.

[44] M. Steinberg, H. C. Cheng, Int. J. Hydrogen Energy **14** (11) (1989) 797

[45] G. Sandstede, Chem.-Ing.-Tech. **61** (5) (1989) 349

[46] M. Hammerli, Int. J. Hydrogen Energy **9** (1984) 25

[47] P. Daum, Inform, Vol. 4., No. 12, pp. 1394–1399

[48] M. Fischer, Chem.-Ing.-Tech. **61** (2) (1984) 124

[49] G. Sandstede, Chem.-Ing.-Tech. **61** (5) (1989) 349

[50] D. Wöhrle, Nachr. Chem. Tech. Lab. **39** (11) (1991) 1256

[51] H. J. Derdully, K. Wellner, Nachr. Chem. Tech. Lab **39** (5) (1991) 503

[52] "Wasserstoff-Energietechnik III", VDI-Berichte **912**, Teil "Fortgeschrittene Elektrolysen", 1992

[53] K.-F. Knoche, "Thermochemische Kreisprozesse zur Wasserspaltung", Wasserstoff-Energietechnik II, VDI-Berichte **725**, 235–248

[54] H. Wendt, "Technik der Wasserstofferzeugung durch alkalische Wasserelektrolyse", Wasserstoff-Energietechnik, VDI Berichte **602**, 21–34

[55] E. Erdle. W. Dönitz, "Hot Elly Technologie – Aktueller Entwicklungsstand und künftige Einsatzperspektiven", Wasserstoff-Energietechnik, VDI Berichte **602**, 51–62

[56] M. Steinberg, H. C. Cheng, Int. J. Hydrogen Energy **14** (11) (1989) 797

[57] H-Power, Hydrogen Energy Systems, Non-published Reports

[58] A. Steinfeld, P. Kuhn, "Iron and Syngas in direct solar irradiation of Fe_3O_4-Particles fluidized in methane", Paul Scherrer Institut, Annual Report 1992, pp. 28–32

[59] G. Simader, Dissertation, Section "Wasserstoff aus Eisen und Wasserdampf", Technical University of Graz, April 1994

[60] M. Steinberg, H. C. Cheng, Int. J. Hydrogen Energy **14** (11) (1989) 797

[61] W. Lavers, O. Lindström, "Life after ammonia", Nitrogen **196**, March–April (1992) 36–42

[62] O. Lindström, Ammonia and/or fuel cell electrons?, unpublished report, 1992

[63] W. H. Avery, Int. J. Hydrogen Energy, **13** (1988) 761

[64] O. Lindstroem, Ammonia and/or fuel cell electrons? report, 1992

[65] G. A. Crawford, S. Benzima, Int. J. Hydrogen Energy **11** (1986) 691

[66] K. Kordesch, Brennstoffbatterien, Wien. Springer Verlag, 1984, pp. 65

[67] J. Perry, "Electrodes for Hydrazine Fuel Cells", 21st Annual Power Sources Conference, May 16–18, 1967, pp. 16–18

[68] E. A. Gillis, "Hydrazine-Air Fuel Cell Power Sources", Proceedings of the 20th Annual Power Sources Conference, May 24–26, 1966, pp. 41–45

9. Fuel Cells:
Summary and Outlook beyond 2000

Introduction

In Section 1.6 we have attempted to write a short outlook for the fuel cell technology as we saw it in 1994 with the intention to prepare the reader for the subsequent chapters which were supposed to give the presently existing information. Now it should be possible to project this knowledge into the future and estimate the developments expected at the beginning of the 21st century. The task is to estimate and forecast the "Future Fuel Cell Potential" in the shifting energy scenario represented by the rising electricity demand due to the increasing world population and the development of the Third World countries (see Fig. 9-1).

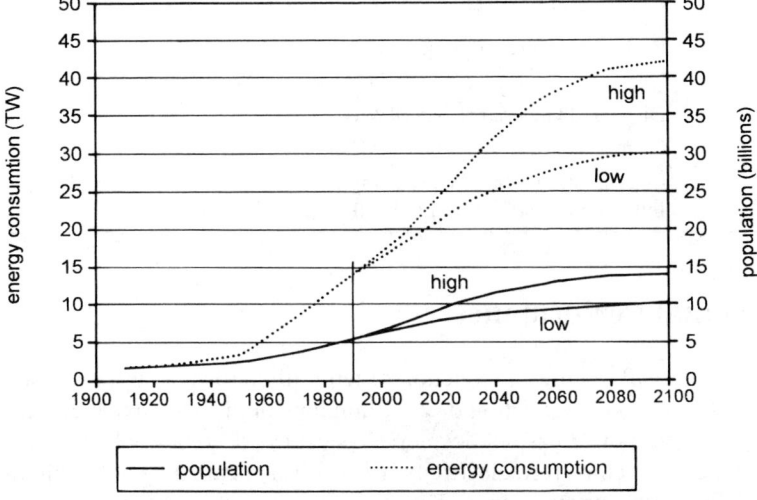

Figure 9-1. Forecast of Energy demand as a function of the increasing world population [1]

Environmental and economic forces must be recognized in a world which is in a transition from the fossil energy supply state to a mixed nuclear and renewable energy concept, made possible by a hydrogen-linked industrial complex. Economics may ultimately decide the fate of the fuel cell technology. Its role will either be as a widely applicable energy technology for utilities and a leading technology for mobility or only as a niche-technology restricted to special applications.

9.1 Fuel Cell Systems as Energy Converters, Applications

9.1.1 Dispersed Systems Partly Replacing Central Systems

The Electric Power Research Institute (EPRI) has intensively researched the possibilities of decentralized power generation and published the results in a paper, "The vision of decentralized Energy production" [2]. The feasibility of replacing gas turbines, gas motors and diesel engines by fuel cell power plants was critically investigated. Fuel cells, storage batteries, solar and wind power (as renewable energy sources) could be combined in a network like computers.

Together with the existing central power stations, a highly efficient and environmentally more advantageous energy production was envisioned. Fuel cells in the 50 kW to several MW classes would be used. Fuels like gas from biomasses could be added in specifically selected local situations. Table 9-1 shows the commercial availability of fuel cell types as expected by EPRI.

Table 9-1. Fuel Cell Power Plants Commercial Availability (EPRI 1994)[1]

Year	Fuel Cell Type	Power plant type , Output and Manufacturers
1992	PAFC	Packaged Co-generation: 50 kW (Fuji); 200 kW: IFC
1993	PAFC	Dispersed Generators: 5 MW (Fuji); 15 MW (Toshiba)
1996	MCFC	Dispersed Generator: 2–5 MW (ERC)
1997	MCFC	Cogeneration: 200–500 kW (MC-Power / IHI)
1997	SOFC	Cogeneration: 100 kW (Westinghouse)
2000	MCFC	Integrated Gasification Fuel cell Demonstration (ERC)

9.1.2 Environmental Performance

The U.S. Clean Air Act of 1990 requires a reduction of ten million tons of sulfur dioxide emissions from the 1980 level and a reduction of 2 million tons of nitrogen oxide emissions by the year 2000. Nuclear power has not developed as expected. Renewable energies are not as abundant and easy to convert to electricity. Fuel cells are called upon to help in the power generation and transportation area. The far lower emission of noxious pollutant makes them suitable for vehicles. Fuel cells have obvious capabilities to emit less CO_2 per kWh produced even from fossil fuels and biomass. Recently a "Renewables Fuel Cell System" was proposed by Sacramento Municipal Utility District (SMUD) [3]. It was suggested to use hydrogen containing biogas from a digester for electricity generation and off-heat production. A 2 MW fuel cell system was envisioned.

---

1   Electric Power Research Institute, Inc., Generation and Storage Div., Palo Alto, CA, 94303.

9.1.3 Economic Aspects

- **Cost Reduction**

Current (1995) fuel cell power plants, such as the 200 kW International Fuel Cell units sell for about $ 2500 per installed kW. For market competition this price must be reduced to $ 1000 to be ready for the utility and the on-site markets. Power plant systems for locomotives could tolerate the $ 1000 cost range, but this is too high for street vehicles. Fuel cell costs of $ 200 to 300 per kW could be accepted for automotive vehicle systems in the 20–50 kW sizes. In this application the use of an additional rechargeable battery or power capacitor system would be required in order to supply peak acceleration. The PAFC system is in the commercialization stage, prices are therefore more realistic than estimates for the systems in development: MCFC and SOFC Systems are presently estimated to cost about 2 to 4 times the price of PAFC systems and the predicted costs of PEM-systems are over a very wide range, from several hundred dollars to $ 10000 per kW, depending on who makes the estimates for whom.

- **Reliability and Durability**

These properties must be demonstrated over a minimum of five years (ca. 45000 hours) of plant operations. The longest lifetimes achieved up to now are in the 20000 hour range (ca. 2 years). The life expectancy of a present day utility plant is stated to be 40 years, but not for all components. Five years of life could be managed right now by twice exchanging the fuel cell stacks, which represent about 1/4th of the plant. Optimism is justified, because the reliability curve has improved greatly since 1992, when the larger trial units were started, (see Fig. 9-2). The outing rates dropped to 1/5th in the two years between 1992 and 1994.

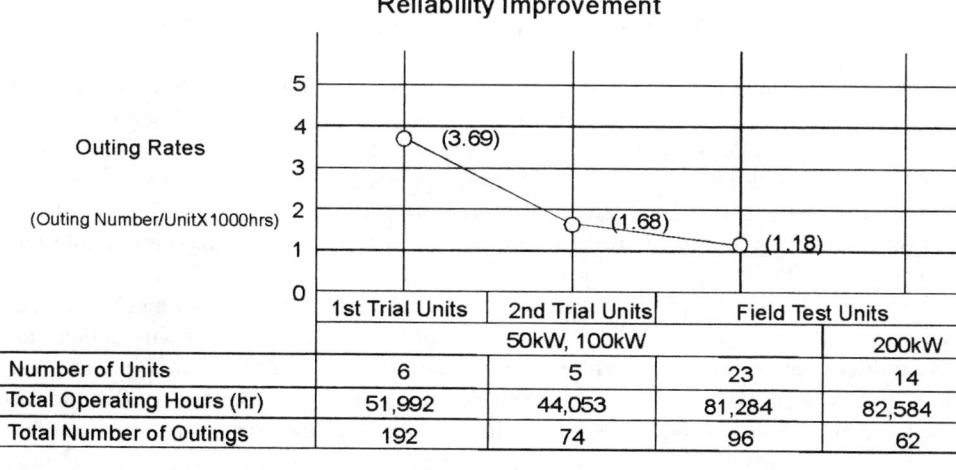

	1st Trial Units	2nd Trial Units	Field Test Units	
		50kW, 100kW		200kW
Number of Units	6	5	23	14
Total Operating Hours (hr)	51,992	44,053	81,284	82,584
Total Number of Outings	192	74	96	62

(as of Feb. 1994)

Figure 9-2. The reliability improvement of PAFC power plants operated in Japan [4]

The situation is quite different for transportation applications. Many electric vehicles propelled by a fuel cell power plant have been constructed, but no lifetimes have been reported in a manner which would allow a comparison with the lifetime of combustion engines. The engine of an automobile does not have to show an operational life beyond 3000–5000 hours (which are equivalent to about 200000 km). However, the calendar life must be at least 5 years.

For fuel cell systems to be competitive with engine propulsion systems a cumulative counting of the "operational working hours" would be sufficient, but it must then be sure that the fuel cell does not deteriorate on standby.

This way of looking at fuel cell electric vehicle power plant life is unconventional. There are just not enough published data available about such life expectancy, except some estimates by the authors of this book, considering the intermittent operation of low temperature hydrogen-air systems. PEM, AFC and also some acidic systems are principal candiates for such a manner of usage. More about that will be said in connection with circulating (flowing) electrolyte batteries which can be completely shut down when not in use.

- **Markets**

Central power generation provides a long range market for fuel cells, based on fuel cells integrated with coal gasifiers (IGFC) for base load central stations. However, performance, life, and cost estimates must be met in competition with existing systems. Fortunately, the higher efficiency and better environmental compatibility is strongly in favour of fuel cells.

In the USA, electricity consumption is expected to rise from 2.7 trillion kWh in 1990 to 4.5 trillion kWh in 2010, a 1.7% yearly increase (this includes imports). More than 900 GW (Giga = 10^9) of generating capacity will be needed in the year 2010, perhaps 1300 GW in 2030 (see Fig. 9-3).

Any technology capable of producing clean electric power at a reasonable price will be needed. Means for electricity storage must be improved, advanced utility storage batteries are looked for.

Fuel cells are able to fulfill both requirements in one: direct generation of electricity from chemical storage. In connection with this, dispersed power is an essential point of consideration. Fuel cells are ideally suited for power plants in the 50 kW to 1 MW sizes where turbines and gas motors do not perform efficiently. Cogeneration systems, combining turbines with PAFC systems can represent a very favourable symbiosis [6]. The combination of high temperature fuel cell systems with a hydrogen-oxygen double loop steam cycle (1500 °C) is a distinct possibility [7].

Commercial on-site applications are considered to be the "consumer range" of dispersed power. These applications include office buildings, hospitals, shopping centers and also military bases. It also means thermal cogeneration of energy (room and water heating, low grade steam-supply and also refrigeration). In these cases, over 80% fuel use can be achieved, compared to ca. 35% with conventional methods. With the use of reformers, natural gas, easily available from pipelines, becomes a preferred fuel. Presently, the PAFC is the most suitable system for on-site applications. Conservative estimates speak of a demand of 20000 MW around the year 2000 (see Fig. 9-4).

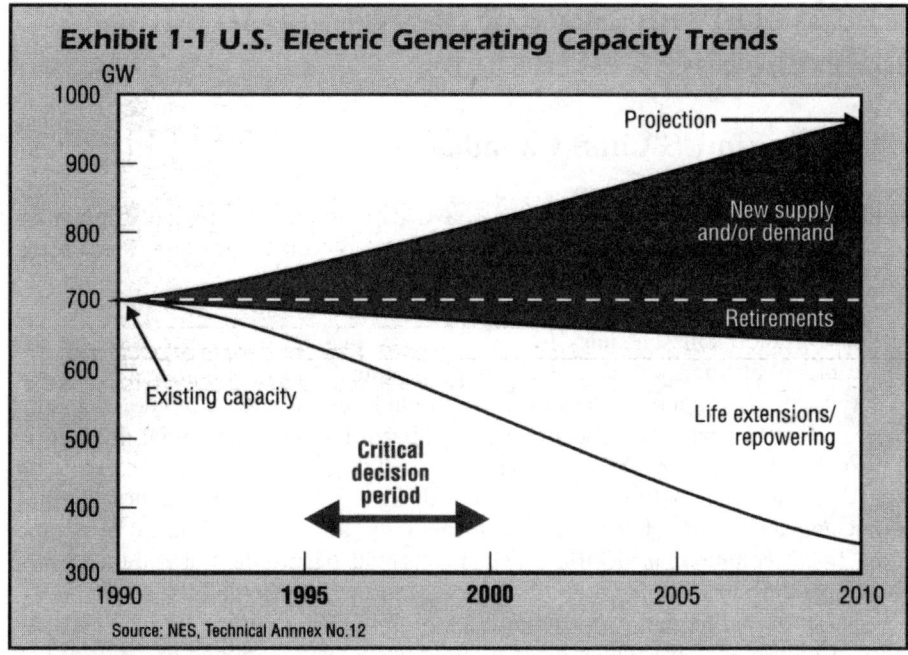

Figure 9-3. The expected trend in US.Electric Generating Capacity [5]

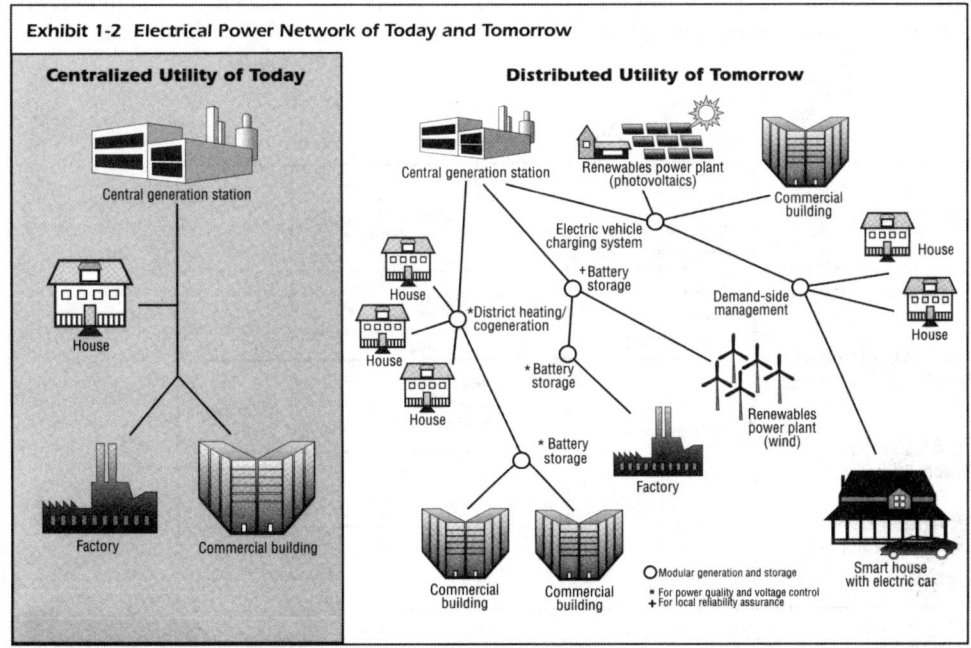

Figure 9-4. Comparison of the Electric Power Networks of today and tomorrow [9]

9.2 The Worldwide State of the Art and Readiness of the Different Fuel Cell Systems

9.2.1 Programs in USA and Canada

Most of the current projects and applications have already been mentioned in previous Chapters. More recently announced or future goals and objectives are described in the following survey:

The Department of Energy Program is shown in an overall DOE fuel cell systems plan (1994) [9] and in the DOE schedules for fossil energy fuel cell programs issued at the 1995 Office of Fossil Energy's FY 1996 budget briefing [10] (see Figures 9-6 to 9-9). The project strategy is to aid in the development and demonstration of the necessary technologies to ensure that national benefits are realized through commercial production and use of fuel cells. It covers the use of natural gas during the 1990s and the operation of integrated coal gasifiers before 2010. DOE considers the PAFC as the furthest developed hydrogen/air and reformate/air fuel cell technology, already in advanced prototype production. Local commercial electric power plants will be installed at military bases with low-temperature heat applications. The DOE is further developing the MCFC technology because of its higher efficiency (50–60%), its insensitivity to CO and the availability of high-grade heat which leads to effective cogeneration and to efficient use of coal-derived systems. The SOFC is expected to provide additional benefits. (see Fig. 9-5) The alkaline fuel cell system is not anymore supported by the DOE as it is only used as on-board power plant in the NASA Space Shuttle Program.

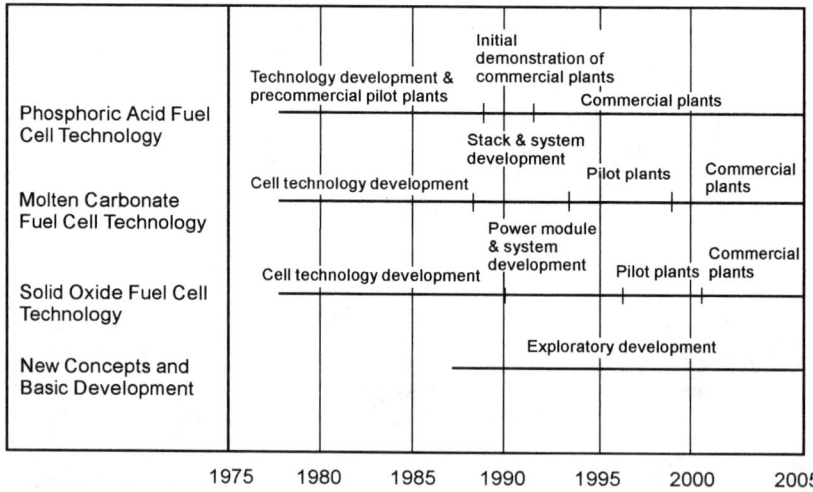

Figure 9-5. The DOE Overall Fuel Cell Program Activities

The DOE supports the progress of MCFC and SOFC systems by cost sharing programs. Energy Research Corp. (ERC) specializes in the internal reforming, externally manifolded system and Molten Carbonate Power Corp. (M-C Power) in the external reforming, internally manifolded type of the MCFC. Development contracts for SOFC exist with Westinghouse for tubular stacks and Allied Signal Aerospace Corp. for a monolithic concept (also with Argonne Natl. Laboratory). Planar cell configurations are under development and the proof of concepts is part of the advanced research effort. Among other projects with environmental goals, interest is shown in ways to dispose of CO_2 which can be removed from the exhaust gases of fuel cells [11].

DEVELOPMENT TARGETS FOR FUEL CELL SYSTEMS

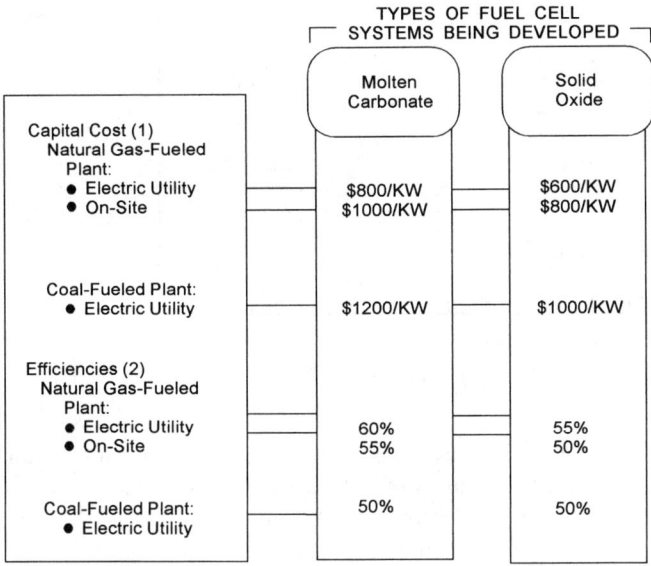

	TYPES OF FUEL CELL SYSTEMS BEING DEVELOPED	
	Molten Carbonate	Solid Oxide
Capital Cost (1) Natural Gas-Fueled Plant:		
• Electric Utility	$800/KW	$600/KW
• On-Site	$1000/KW	$800/KW
Coal-Fueled Plant:		
• Electric Utility	$1200/KW	$1000/KW
Efficiencies (2) Natural Gas-Fueled Plant:		
• Electric Utility	60%	55%
• On-Site	55%	50%
Coal-Fueled Plant: • Electric Utility	50%	50%

Figure 9-6. DOE Program Targets for efficiency and capital cost for MCFC and SOFCs

The U.S. Department of Energy, Office of Fossil Energy states in the Conclusion of the Fuel Cell Handbook (Revision 3), 1994 about MCFC and SOFC's: [12]. *The MCFC has reached a point where cells are being built at commercial size and full height stacks are beginning to be tested. Shorter and smaller stacks have demonstrated life times of about 9000 hours. It is predicted that the materials used in MCFC's can cost as low as $ 300 per kW if stacks can be produced in quantities in the order of 200 MW per year.* Presently the costs are about $ 4400 / kW.

Utilities and manufacturing industries and state authorities are supporting commercialization plans. USA developers of MCFC are Energy Research Corp. (ERC), M-C Power Corp. (MCP) and International Fuel Cells (IFC). ERC has constructed a pilot manufacturing facility in Torrington, Connecticut which manufactures 120 kW stacks. The manufacturing capacity of this plant is 2 to 3 MW per year. In cooperation with Ameri-

FUEL CELL PROGRAM ACTIVITIES

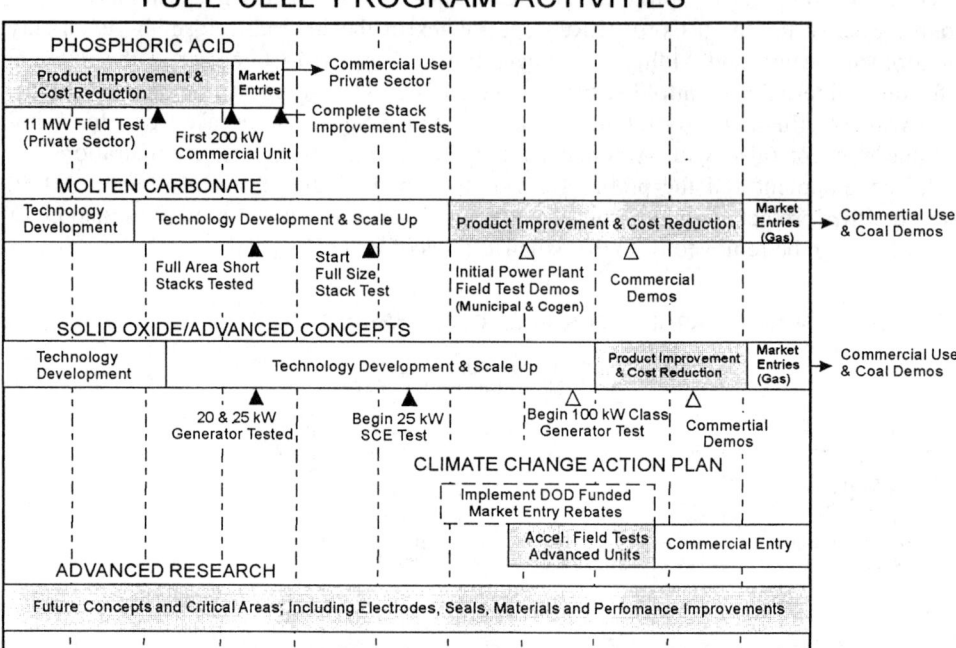

Figure 9-7. The DOE Fuel Cell Program Activities until the year 2000

can Public Power Association (APPA), ERC has developed a commmercialization plan that aims to have orders for over 100 MW of units by 1997. The Fuel Cell Commercialization Group is helping ERC to secure orders for fifty 2 MW plants.

M-C Power produces full size stack components in its pilot manufacturing plant in Burr Ridge, Illinois. Manufacturing capacity is estimated at 2 to 3 MW per year. DOE and M-C Power agreed in 1992 on a 250 kW cogeneration demonstration plant located in San Diego, CA. The tests are scheduled to start in 1995. MC-Power has plans to enter the commercial market around 1998. A number of gas and electric utilities have formed a group known as the Alliance to Commercialize Carbonate Technology (ACCT) to promote MC-Power technology. International Fuel Cells has contracts with DOE to improve MCFC stack performance.

Several solid oxide fuel cell configurations are under development, in the USA. The SOFC tubular concept needs scaling to commercial size and cost reductions too. Recently 25 kW stacks have been operating. The alternate designs are in earlier states of development.

Westinghouse Electric Corp. designs packages in arrays of three by six. Groups of such bundles are then mounted in thermally integrated systems which make up a module. A module consisting of 32 bundles of 50 cm long cells produces 20 kW. Gas utilities are the customers for two 25 kW units. Westinghouse has a pre-pilot manufacturing plant

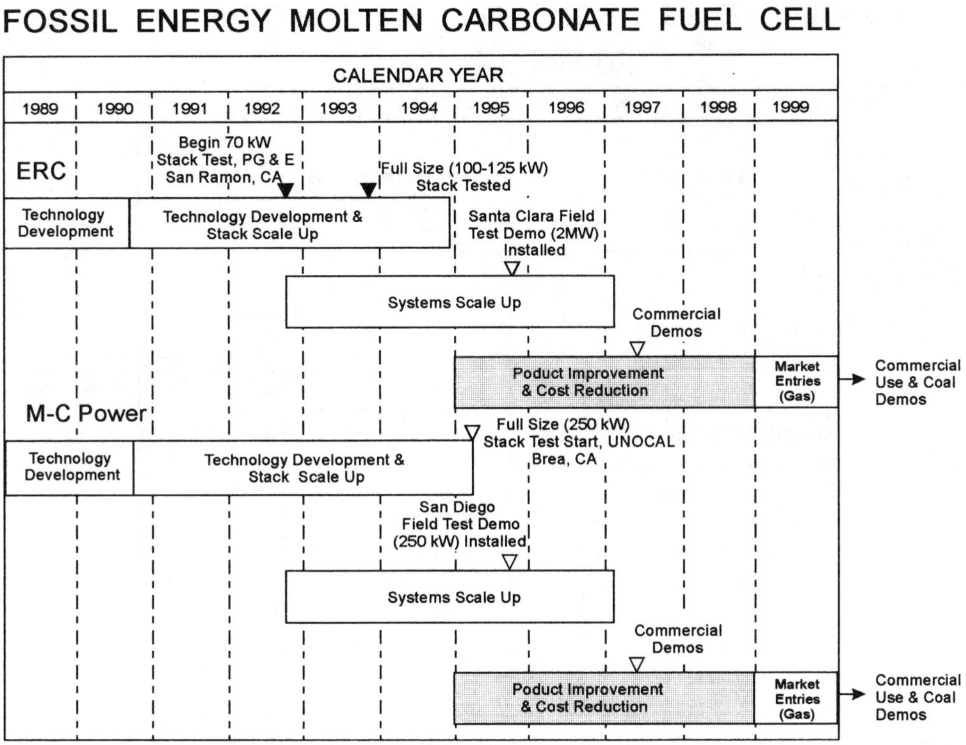

FOSSIL ENERGY MOLTEN CARBONATE FUEL CELL

Figure 9-8. The DOE Fossile Energy Molten Carbonate Fuel Cell. Program [10]

which is capable of producing 300 kW of cells per year. The operating life of cells is predicted to be ultimately near 40000 hours. Integrated gasifier/fuel cell plants have been designed; testing is scheduled for 1995.

Under DOE sponsorship, Allied Signal and Argonne National Laboratory have fabricated stacks of monolithic cells. Alternative designs, like planar cells, are still in the laboratory state. It should be mentioned that the DOE-PAFC program was completed in 1992. The commercial programs have been discussed in Section 4.4. The PC-25™ power plants were the workhorses of the PAFC technology. Built by ONSI, these fuel cell power plants have been sited, permitted, installed, started, operated and maintained by members of the North American Fuel Cell Owners Group (NAFCOG). This group of 13 companies and electricity utilities covers a wide range of applications: hospitals, nursing homes, offices, university, airport, industrial, government and military sites. In the majority of cases the fuel cells provide local hot water and space heating too. The record for an uninterrupted run is held by the South California Gas Co., with 7570 hours [13].

Chapter 4.3 discussed the principles and the decade-old history of PEM fuel cell systems and also their renaissance in the 90's. Since Ballard in Canada again started to develop powerful membrane fuel cells, the development efforts skyrocketed. Ballard's membrane (PEM) fuel cells have been built in 5 kW stacks and systems operating on

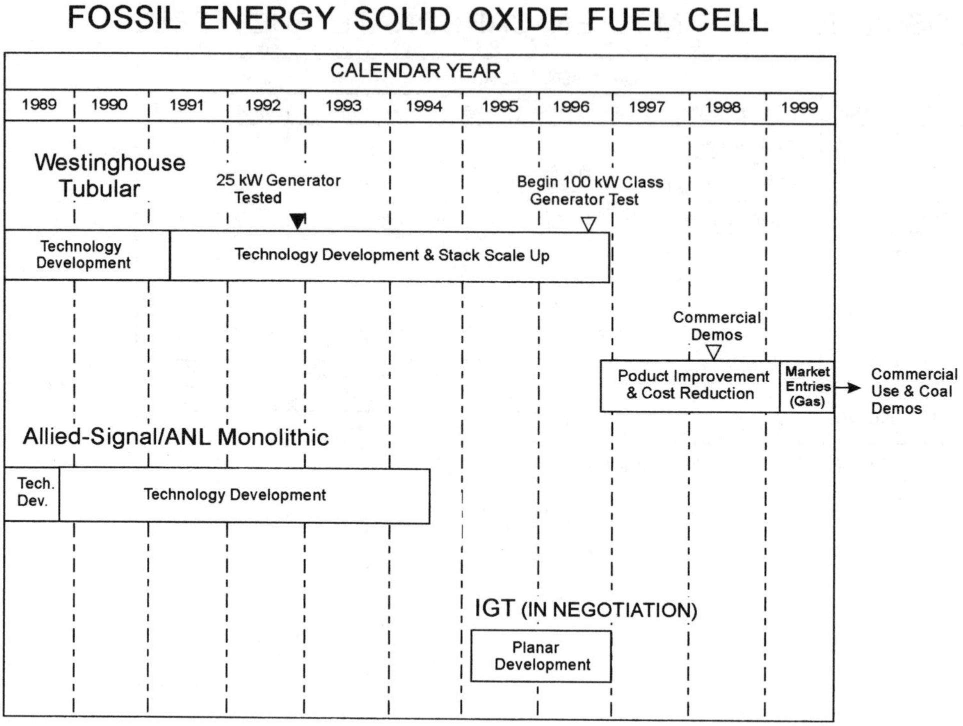

FOSSIL ENERGY SOLID OXIDE FUEL CELL

Figure 9-9. The DOE Fossil Energy Solid Oxide Fuel Cell Program [10]

hydrogen and air are now tested all over the world. Methanol/Air systems are following. The cost question is open. The credibility of the extremely low estimates is low, in spite of the fact that they have been made by large companies. The same must unfortunately be assumed about many claims about the widespread use in electric street automobiles in the near future. Very reputable automobile companies worldwide have started large development efforts with PEM systems. General Motors in the USA, Daimler Benz and BMW in Germany are studying the system. Siemens AG looks at submarine applications. The sources of the stacks are Ballard and International Fuel Cells, these companies are mainly concerned with improvements of the systems.

An example for the widespread belief in the PEM system is the widely advertized "Green Car" by Energy Partners [14], intended to be the test bed for a 125 V, 15 kW PEM hydrogen-air fuel cell operating at 20 psig and with a parallel lead-acid battery as peak power system. Surprisingly the listed performance of the 1300 kg car (range: 100 km) is of the order of advanced lead-acid battery vehicles, which is rather low when compared to demonstrated data of alkaline or phosphoric acid fuel cell systems (see the corresponding sections on AFC and PAFC).

Existing electric vehicles propelled by advanced rechargeable batteries have a range nearly 3 times as large. Examples: The "Wave"™ powered vehicle, which is using a zinc bromide flow battery manufactured by Powercell, USA [15], won the World Clean Air Vehicle Ralley 1994 with a range of 270 km per charge. The zinc-air mechanically rechargeable battery built into postal vehicles presently tested in Germany is supposed to provide about 300 km driving distance for each anodic charge [16].

AFC's with matrix immobilized electrolyte are still produced for NASA. International Fuel Cell Corporation (IFC) designed, developed, and manufactures the AFC power plants for the manned space (orbiter) program. In the space shuttle, three 12 kW power plants fueled by liquid hydrogen and oxygen provide the power on board, including the payloads. Potable water is the reaction product. The efficiency of the AFC system is in the 70% range. It consists of 96 cells arranged in 3 parallel groups of 32 cells each. The powerplant weight is 120 kg and the energy density is ca. 230 Wh/kg. The system is pressurized at about 4 bar and operates around 85 °C. The safety and durability record is exemplary: in 63 missions over 43000 hours have been accumulated with only one failure due to a pump stall caused by contaminations. Afterwards a filter was installed. NASA maintains an inventory of 21 fuel cells, 12 in vehicles, 9 as spares. IFC maintains the units which are qualified for 2400 hours of operation between overhauls. The limit of durability has been found to be the deterioration of the fibre glass-epoxy frames. NASA contractors have improved the framing material and the shuttle fuel cells are now expected to serve for over 10000 hours. The power output has been increased twelve-fold (!) in an SDI-sponsored program (1994 FC 412). The Russian "PHOTON" System, dedicated for space operation was also an AFC system.

A historic note: the same failure mode was realized to be the cause of frame/terminal KOH leakage in the Kordesch Hydrogen-Air electric passenger car with AFC's after 3 years of city driving performance accumulating about 2000 operating hours during that time period. Just for comparison: the lead-acid batteries used in parallel with the 6 kW fuel cell had to be replaced twice over the said 3 years of operation. It shows that low-cost AFC systems are not necessarily out of considerations for electric cars! However, there is no question that more efforts are needed to further improve alkaline fuel cell systems! (See chapter 7, Fuel Cell Powered Electric Vehicles).

9.2.2 Japan

The New Energy and Industrial Technology Development Organisation (NEDO) reported that 87 PAFC plants are in operation in Japan, with a total output of over 20 MW (see Table 9-2).

Table 9-2. PAFC Plants in Operation in Japan (March 1994) [17].

Company	Output (kW)	No.	Manufacturer
NEDO	50	14	Fuji
	200	1	Mitsubishi
Tohoku EPCO	50	2	Fuji
Tokyo EPCO	50	1	Fuji
	200	1	IFC
	200	1	Mitsubishi
	11000	1	IFC

Table 9-2. continued

Company	Output (kW)	No.	Manufacturer
Chubu EPCO	50	2	Fuji
	200	1	Toshiba
Chubu EPCO NEDO	50	1	Fuji
Hokuriku EPCO	50	1	Fuji
Kansai EPCO	200	1	Mitsubishi
Chugoku EPCO	200	1	Mitsubishi
Shicogu EPCO	50	1	Fuji
Kyushu EPCO	50	1	Fuji
Tokyo Gas	50	4	Fuji
	100	2	Fuji
	200	6	ONSI
Tokyo Gas NEDO	50	2	Fuji
	100	2	Fuji
	200	1	OSI
Toho Gas	50	1	Fuji
	200	1	ONSI
Toho Gas NEDO	100	1	Fuji
Osaka Gas	50	10	Fuji
	100	6	Fuji
	200	6	ONSI
	500	1	Fuji
Osaka Gas NEDO	50	5	Fuji
	100	2	Fuji
	200	2	ONSI
	500	2	Fuji
Shikoku Gas	50	1	Fuji
Saibu Gas	50	1	Fuji
Petroleum Energy Center	50	1	Fuji
	170	1	ONSI

In addition, NEDO and PAFC Research Association will start a 5 MW plant for electric utilities and a 1 MW plant for on-site use in 1995.

Table 9-3. Current Operating Status of PAFC of Japanese Gas Industry (as of Feb. 1994)

	50 kW	100 kW	200 kW	500 kW	Total
Number of Units	24	14	17	3	58
Site (Number of Units)	In-House Facilities (22) Offices (6) Works (6) Laboratories (5)	Apartments (4) Hotels (3) Sport Facilities (3) Complex Bldg., DHC (3)	Telecommunica- tion-Building (2) Schools (2) Supermarkets (2)		(58)
Longest Opera- tional (h)	15037	8574	11373	3216	
Longest Run (h)	4865	3576	5476	1564	

Generally, in Japan, fuel cells are considered to be one of the most promising environmentally friendly technologies, mainly because of the high efficiencies achievable and the clean exhaust gases. The PAFCs are presently the most developed and used ones. Table 9-3 gives a summary about the results obtained in major trials runs of fuel cells.

The current (Feb. 1994) operational status of PAFC Fuel Cells in Japan is shown in Table 9-4.

Table 9-4. Results obtained in Major Trial Runs of Fuel Cells (as of Feb. 1994)

Fuel Cell	Manufacturer	Start of Operational	Operational (h)	Cum. generating Power (KWh)	Load Factor (%)	Efficiency (%-LHV) Generation/Total		No$_x$ (ppm)
PC-11 (12.5 kW)	IFC	1973.3	3 509	9 923	22.6	30	30	19
PC-18 (40 kW)	IFC	1984.11	15 588	209 000	33.5	40	80.5	3
Moon Light (200 kW)	Mitsubishi	1989.8	13 038	1 797 000	68.9	36	80.2	6
FP-50 (50 kW)	Fuji	1991.3	15 037	578 000	76.9	40	80	7.5
FP-100 (100 kW)	Fuji	1992.4	8 547	592 000	69.1	38	80	7.5
PC-25 (200 kW)	ONSI (IFC)	1992.10	10 820	1 822 000	84.2	40	84	2
FP-500 (500 kW)	Fuji	1003.5	3 216	985 000	61.3	40	85	7.5

Compared with the PAFC technology, the other fuel cell systems lagged behind somewhat in Japan, but are now rapidly pushed to the pilot plant state.

It has been recognized that MCFC systems are better suited for large scale utility use than PAFC systems. The Japanese 1 MW NEDO pilot plant project was only started in 1993, long after ERC and MC-Power in the USA.

SOFC systems are now considered as the next generation of large power plants. Planar types of 10 to 30 kW size have been developed by NEDO, tubular types together with Westinghouse. While PEM fuel cells were rapidly developing in USA (IFC) and Canada (Ballard), in Japan the Polymer Electrolyte Fuel Cells were in 1994 demonstrated only in the 1 kW-sizes. However, the NEDO project committments have increased rapidly, competing with Ballard and IFC. It should be noted that the "Moonshine Program" was renamed in 1993 as "Sunshine Program". The New Sunshine Program is managed by the Agency of Industrial Science and Technology (AIST). The budget for fuel cell developments for 1994 was 5 billion Yen (about 50 million Dollar), see Table 9-5.

Mitsubishi Heavy Industries have developed SOFC systems since 1984. The design is called MOLB (mono-block-layer built) and the tubular type has reached 3000 hours continuous operation. A 10 kW model is planned to be in operation in 1995. A planar type SOFC consists of flat plates and corrugated support structures. 1 kW stacks have been operated. A 5 kW stack is to be ready in 1996. (1994 FC 642).

Table 9-5. The amounts which the 1994 – programs received (in 10 000 Dollar) [18]

New Sunshine Program: (National Project)	PAFC	MCFC	SOFC	PEMFC	Total
	– 	4.415	400	234	5.049
Urban Demonstration	314*)				
On-Site Use	507*)				
Field Test	525*)				
Petroleum Energy Project	310*)				

Development of Reforming Systems for Hydrogen Production (Natl. Project): 90
Total of all Programs : $ 67 950 000   Note: *) : 50% Subsidy only

9.2.3 Europe

In previous Chapters 1.3.3 we discussed the competing activities in Europe. Since 1994 PEM fuel cell developments are favoured, but mainly concerned with electric vehicles. AEG-Daimler-Benz Industrie is working on the project "NECAR" (New Electric Car) which will be operating on a PEFC system to be further developed in cooperation with Ballard, Canada. Hydrogen will be supplied by a methanol reforming unit [19]. Deutsche Aerospace is also part of the Daimler Benz Group and is mainly concerned with MCFC and SOFC fuel cell research and development.

Siemens AG is developing a PEM fuel cell system for vehicles: initially for the Federal German Office for Defense Technology to be used for the KL212 submarines (with oxygen as oxidant) perhaps later for surface transportation (with air). The principal PEM system is licensed from IFC. NAFION membranes are used in the initial 34 kW size stacks operating on hydrogen and oxygen. Reforming of methanol will be considered in later phases.

The problems to be solved are the same as faced by all PEM systems: high costs, rather poor performance on air compared to oxygen. Using a reformed fuel (methanol) brings up the danger of catalyst poisoning, especially at low Pt-catalyst loadings. Energetically, the air compressor is the energy wasting component. Life questions are connected with the extent of gas purification. Basic research at Siemens (FC 474) looks therefore to the direct methanol fuel cell system (DMFC) as a distant future solution. 100 W models are to be demonstrated first. Within the frame of JOULE II, a 12-cell stack with 250 cm^2 electrodes was constructed and tested. At a current density of 130 mAcm^{-2}, 150 W were produced at 0.45 V per cell (80 °C). Operation at 130 °C increases the current density to 500 mA at a voltage of 0.5 V. The problem is the poor efficiency.

Renault (FC 215) is planning to have a fuel cell powered vehicle (a Renault Clio, mid-sized car) operating in 1996. To achieve this, Renault has joined within the European Commission JOULE II project the following partners: Ansaldo Ricerche (Italy) for FC systems integration, Volvo-AB (Sweden) for safety and system studies, De Nora Permelec (Italy) for the (PEFC) fuel cell stack manufacture, Volvo aeroturbines (Sweden) for the air compressor, Air Liquide (France) for the liquid hydrogen fuel tank and L'Ecole des Mines (France) for fundamental aspects. The basic specifications are: Power: 30 kW; Voltage: 100–150 V; Range: 400–500 km; Top speed: 130 km/h; Cruising speed: 100 km/h.

Netherlands Energy Research Foundation (ECN) is the coordinator of the European "Brass-Board" Project, which is developing a traction system for a 5 ton vehicle, comprising four Ballard 5 kW stacks, batteries, and an electric motor. One task is to build the air compressor system, another is the Polymer Electrolyte fuel Cells (PEFC) stack component development. The objective is to find ways of reducing corrosion problems between membrane materials (e.g. Nafion), stainless steel and carbon. Also glass-fibre containing PTFE material poisoned the Pt catalyst on the membrane by silicon release (at 80 °C). (FC 503).

The Flemish Institute for Technological Research (VITO) in Belgium initiated a test program evaluating fuel cell stack technologies for transportation. The 4 kW stacks were obtained from Energy Partners, Palm Beach, Florida. A rather high spread of cell voltages occurred during air operation. (1994 FC 638).

Alkaline fuel cell (AFC) systems with flowing electrolyte are considered to be "past technology" and the recent demise of ELENCO in Belgium seems to reinforce that belief. ELENCO was involved in transportation (City bus) and niche applications, using the high efficiency as reason for the development and claiming relatively low material costs – features which were not achieved by the other fuel cell systems at the present time. The Russian efforts to build an AFC space fuel cell system "Photon" have been described. The only ongoing efforts are the research studies in Sweden at the Royal Institute of Technology in Stockholm [20,21] and in Korea [22].

The chances to apply alkaline systems to large power plants are dependent on the possibilities of removing carbon dioxide from fuel gases. In this respect, comparisons have been made between AFC and MCFC power plants. The Pressure Swing Adsorption (PSA) method was proposed [20]. The recently proposed "Iron Sponge Process" is very well suited for H_2-cleaning. Iron sponge is manufactured by the reduction of iron ores with reducing gases (hydrogen or crack gases) and for decades has been used in the steel industry in huge quantities. When iron sponge is treated with steam at high temperatures it produces pure hydrogen. This cycle is repeatable many times [23].

The CO_2 in the air is not the obstacle to widespread operation, as was reported in many papers comparing it with other systems. Soda lime is an effective and cheap absorbent and can be used with an indicator to show its pending exhaustion. ELENCO has found that not all carbon dioxide must be removed from the air; about half (0.01%) can remain in the cathodic gas stream without causing structural damage to a properly wetproofed carbon electrode. This is also important to consider when alkaline systems are operated intermittently.

Studies of the CO_2-tolerance of AFC electrodes at the DLR, Germany [24] show that porous PTFE bonded electrodes (anodes and cathodes) can be operated with reactants containing up to 5% carbon dioxide (!!) for several thousands of hours at 150 mA/cm^2 constant loads at 70 °C. CO_2 does not necessarily deposit in the electrode pores either: it can get through the electrodes into the electrolyte. The electrolyte was liquid KOH and the cumulative amount of carbonate was found in it and the carbonate formation in the electrolyte increased only the resistance of the electrolyte. The measured degradation value of the Ag-catalysed cathodes was only 17 microV/h essentially with or without carbon dioxide contamination. Experiences showed also that exposure of Pt/Pd catalized anodes to air improves their performance, the same way as potential reversal.

Intermittent operational life data confirm also the positive experiences made with mechanically rechargeable zinc-air and aluminium-air systems operating with liquid KOH electrolytes. The hydrazine-air electric vehicle fuel cell systems (Kordesch 2 kW Motorcycle and DOE 20 kW Hybrid Van) showed years of life when operated intermittently, with no electrolyte in the system when shut down. The Kordesch Hydrogen-Air 6 kW-Hybrid passenger car allows the same conclusions.

In Europe, the PAFC fuel cells are considered near to their commercial introduction and therefore no larger developments of stacks are planned. The manufacturers in USA and Japan will be the sources of licenses. Application studies are directed mainly to environmental problems with fossil fuels in conventional and in dispersed power plants.

9.3 The Hydrogen Technology (Gaseous-, Liquid- or Solid Fuels?)

The concept of the "Hydrogen Economy" concerns the ultimate replacement of our conventional fossil fuel economy with a fuel system in which hydrogen produced by electrolysis from water (by renewable energies like solar- or wind energy or water power, but also by nuclear power) would play the role of oil, gas and coal. Such a readily available supply of hydrogen would establish a potential role for fuel cells, making it the most attractive way of producing electricity.

The personal mobility of people would switch to hydrogen operated vehicles, maybe first still with combustion engines, later to electric cars with fuel cells. Hydrogen could also be burned in thermal plants or used in catalytic ovens to produce heat; it could also be used in metallurgical processes instead of coal for reduction of ores. In a cyclic process, reduced iron oxides (iron sponge) could be used for the renewed production of hydrogen by the reaction with steam, supplying heat from solar furnaces or any by-product heat available, covering the field from atomic power plant off-heat to conventional cogeneration and even biogas heating. Last, but not least, hydrogen is a feed stock for the chemical industry, leading to ammonia, fertilizers, petroleum and hydrocarbon modification and is used in the production of plastics and pharmaceuticals.

The transportation of hydrogen can be done as gas in a pipe line, in compressed form in steel cylinders, and as a cryogenic liquid. Hydrogen as gas could even be stored in underground caves. Environmentally, hydrogen is not objectionable: its reaction product is pure water. The danger aspects are not worse than those of gasoline or other flammable or explosive gases used commercially, but they are psychologically associated with dramatic and widely publicised events (for example, "The Hindenburg Syndrom").

The ultimate "Hydrogen Technology" may be far in the future. A realistic transfer technology which allows the adaptation to changing conditions of fuel types, industrial demands and environmental restrictions is the "Synthetic Fuel Technology". Some constraints apply to this concept: the fuels must be made from materials which are in abundant supply and the fuel must be able to be burned in air without a noxious effluent. Because we will not be able to collect the products of combustion and transport them back

Table 9-6. The Storage of Hydrogen – A Comparison

Storage	Energy Density				Refilling Time (min)	Notes
	(kWh/kg)	Rel. to Diesel (%)	(kWh/l)	Rel. to Diesel (%)		
High pressure H_2 (300 bar)	1.00	9	0.55	7	10 to 60	Cylinder storage
Liquid H_2 (−253 °C)	6.00	55	1–1.7	12–20	up to 60	Energy loss for liquification vaporisation loss 2%/day
Metal hydride	0.40	4	0.80	10	10	Recovery of heat 10%
Methylcyclo-hexan	0.56	5	0.37	4	10	Additional H_2 needed for dehydration

to the generating station, or deposit them somewhere at low cost, the recycling must be carried out by natural prevailing processes.

The restraint is therefore the demand that the materials involved, are either hydrogen, nitrogen or carbon (or their derivates) and oxygen. Carbon is included in the list; it is available in the form of the fossil deposits, but due to the "green house effect" it is not anymore considered as a preferred fuel constituent. On the other hand, we cannot find other substances: hydrogen itself, ammonia and hydrides are the only choices. Synthetic methane, alcohols (methanol, ethanol) and biomasses are next in line, but the required CO_2 control calls for a recycling mechanism so that they serve only as carriers. In this consideration also the, iron oxide/iron sponge/steam/H_2 process cycle could serve as a transport and storage possibility in a "Fuel distribution network". The use of fuel cells is again the best way of utilizing these synthetic fuels. However, we should not forget that the preparation of all these synthetic fuels will also cost additional energy.

9.4. The Fuel Cell Strategies until and beyond 2000 (Support and Opposition)

In the USA the support of fuel cells by the utilities, industry and by governmental agencies is strong at the present time. It is expected to continue. However, it should be remembered how fast the support of large projects in the USA can disappear when the base conditions for success are in doubt. Examples are the abandonment of energy projects as soon as the oil crisis was over, how fast space efforts dropped in the late 70's, or how sensitive scientific research and developments have shown to be in times of political changes. Last but not least, the competition to fuel cells from other electric power producing systems is considerable. The performance of turbines and heat engines will still be improved and cogeneration is increasing the efficiency not only for fuel cell systems, but also for others. A sad example is the rise and fall of the Stirling Engine Program which was supported extensively by Government Agencies and Automobile Co. in USA, Europe and Japan, ranging into expenditures of hundreds of millions of dollars worldwide [25].

In the USA: Resentments and Lobbying, which may also apply to fuel cells:
Strong opposition to the Californian "Zero Emission Vehicle" mandate was voiced by the California Manufacturers Association (CMA). They called it a "well intentionated but misguided attempt by government to force technology into the market place prematurely [25] CMA, which represents more than 800 companies throughout California would work to stop the mandate before it takes effect in 1998". A recent report from the U.S. General Accounting Office was said "to sound the alarm". The report warns "that batteries are far from being perfected, the production costs are excessively high and the market potential is highly uncertain. Ultra-low emission vehicles, cleaner gasoline and programs to get gross poluting vehicles off the road are more promising at far less cost".

Completely the opposite opinion was released by the Southern California Edison Company on the following pages of the same publication. The California Energy Commission, The South Coast Air Quality Management District, the California Public Utilities Comission and the California Council of Science and Technology supported the Zero Emission Vehicle law, pointing to the advantages of the new technologies, also through creating new jobs.

Of course, the fuel cell supporting groups can also present excellent arguments for the ZEV law. However, Automobile Companies have already concluded that intermediate solutions like compressed gas or improved gasoline or catalysts for diesel engines may provide a "Minimum Emission level" in the near future. "Zero Emission" is not possible in legal language, pointing to analytical limits and energy considerations. The basic argument, that no batteries suitable for adequately performing electric vehicles are in production at affordable costs anywhere, must even be conceded by electrochemists. The reasons for this dilemma may have different causes: scientific, technical, commercial, political, etc. Fuel cells operated vehicles may even benefit from these poor judgements about secondary batteries, because they alone can provide a power plant rivaling the present combustion engine.

A cautious voice from Japan:
After all the tremendous efforts made over the last decade in Japan, the concluding assessment expressed by T. Homma of NEDO at the end of the Fuel Cell Seminar in San Diego (Nov. 1994) was indeed surprising: "Fuel cell technology has not yet attained a high level of development at this stage. We have to clear many technical barriers before approaching commercial feasibility, such as: life span, system reliability and cost. In finding solutions to these obstacles, worlwide technical information exchange and cooperative development are of great importance".

Suggestions and Expectations in Europe:
The European Commission has suggested a ten year fuel cell research, development and demonstration strategy for Europe [27]. The essential points for support were:

- R & D on low temperature fuel cells (especialy low-cost PEMFC). AFC were ignored.
- R & D on high temperature fuel cells (MCFC and SOFC)
- Demonstration of a fuel cell network for clean cogeneration and transport
- Demonstration of fuel cell driven electric vehicles

The European Planning must be verified:
The implementation of the various fuel cell programs should be carried out in close collaboration with different services in the Commission and with national and international programs. The EU services involved in fuel cell activities are: JOULE (basic R & D), BRITE (manufacturing) and THERMIE (demonstration). Funding estimates varied from 30–40 Million ECU for each of the PEMFC, MCFC and SOFCs. However, the money for PEMFC systems should be spent in the next 5 years, for the other systems evenly distributed over 10 years. For the development of fuel cell driven vehicles about 30 million ECU are envisioned. Note : 1 ECU equals about 1.5 US $.

Fears of Competition in an open market:
The European Fuel Cell Group Ltd. [28] criticises that the EU funding in the past has focused on R & D and not on commercial objectives. The members of the EFCG are afraid that Europe will not be able to compete with the USA and Japan, which spend about 50 million ECU each, per year (1994), and the DOE fossil energy fuel cell budget requests for 1995 and 1996 are $ 50 million and $ 55 million, respectively. The US Defense Department may add 20–30 million $ per year (Source: DOE fuel cell Workshop). Adding to these amounts the 50% support by industry and utilities allows a strategic planning which is rather difficult in Europe.

Key Points for a Fuel Cell Strategy in Europe, were expressed at the ECU Fuel Cell Technology Workshop, held in Brussels, 22–23 November 1994:

- Achievement of competitive commercialization of fuel cells.
- Currently only the PAFC system is near maturity; cost in 1998 should be $ 1500 per kW and at that price a return on investments is expected.
- Most European technologies are licensed from USA or Japan, "Europeanisation" is needed.
- An initial cost sharing program (support grants) for fuel cell users may be decisive for success.
- Fuel cells need success, they have the attitude of promising and not delivering since 1960.
- Fuel cells could be a growing industry providing jobs for thousands of people in 2002, the public funding of fuel cells is an investment. It should have priority in the fuel cell strategy.

9.5 My Outlook on Fuel Cells (A personal post scriptum by the senior author)

At fuel cell meetings in the nineties, one could clearly recognize the enthusiasm of the relatively few basic research and the successful technical people, presenting their papers or posters. It may have concerned small steps in the laboratory or large steps in the technology; the pride in their achievements was obvious. On the other side we noticed the more careful management and commercial representatives of the companies which were supposed to implement the next steps. They rightly considered the alternatives, the competitive situations and the cost questions, showing their concerns often more critically than necessary. However, in many cases progress was questioned just to look experienced and responsible. If something goes wrong (which is more probable) they predicted it; if it succeeds, they knew it, too. This was disheartening for young people.

Fuel cells since the times of Grove went through many cycles of high expectations and failures. Also, the different types followed each other as a result of new demands, but also due to the advent of new materials. The efficiency advantage of avoiding Carnot's Law was known to the prominent electrochemists at the begin of this century and high temperature "coal fired" fuel cells were proposed to power ships not long after the electric motor and the propeller were invented!

From these rudimentary beginnings the technology has made tremendous strides, space vehicles, MW size power stations and experimental electric cars are now a reality. "Not profitable", the doubting banker will say, but his adventurous colleague will invest in shares of new and old companies involved in the technology, hoping that fuel cells will deliver what they promised.

The bets that fuel cells will propel future electric vehicles are rather safe in the long run. There is no other immaginable solution to the mobility problem once the fossil fuels diminish. Hydrogen or some hydrogen carrier will replace the liquid or gaseous carbon derivates of today and the energy needed by mankind and produced by renewable sources will produce less CO_2. Some prophets say it will be fusion power which will do it ultimately, but that does not remove the need for the most efficient electric energy producer, the fuel cell. What fuel ultimately? I believe that besides hydrogen and alcohols, ammonia has good chances: it can be stored at low pressure and delivers 75% hydrogen in a simple catalytic converter. It smells, people say, but that is good: any leakage will be detected! It also plays a major role in the hydrogen-nitrogen-oxygen scenario which is basic to life as we know it. Even now, as starting compound for fertilizers it is among the most produced chemicals in all our chemistry. We can also store hydrogen that way for dispersed power generation plants.

When I first planned my hydrogen-air car, it was supposed to use liquid ammonia as fuel. Unfortunately, at that time only 2 kW-size NH_3 converters were available [29]. Some years before, a huge military program including liquid ammonia as logistic (energy) fuel was abandoned in favour of jet fuels, petrol and diesel. It may be interesting to note that heat and combustion engines do operate on ammonia, with no nitrogen oxides produced. Just the starting is difficult; a little propane gas injection helps.

Alkaline hydrogen-air fuel cells with circulating KOH-electrolyte could operate directly on the catalytically dissociated ammonia and the effluent gas, containing a few percent hydrogen used to heat the converter. Filters soaked with Cu-salt solution remove any smell from the exhaust gas. Air-operation of carbon electrodes is not difficult, a soda-lime cleaner is sufficient and the old belief that complete removal of CO_2 is required has been proven wrong. It depends on the repellancy of the electrode to what extent the electrolyte is penetrating the electrode structure; small amounts of CO_2 diffuse right through the pores system and end up as carbonate in the KOH. With a circulating electrolyte, periodic exchange (like an oil change) and the removal of many other impurities is easy; the heat exchange and the water removal is no problem. There is no gas-cross-over – a big advantage compared with matrices, which were declared to be more convenient and simpler to use (and nearly everyone believes it until today).

I have mentioned the interrupted use of fuel cells as a possible way to increase life expectancy of a fuel cell. The comparison was made with an automobile engine which works only a few thousand hours in its many years of life. There are good indications that this is true if the optimal way of emptying the liquid electrolyte and the corresponding gas-shut off mechanisms are employed. Fuel cells on open circuit do deteriorate faster than on load! Also, with such a completely shut-down system one can sleep without any nightmares of an accident. This was the reason why I selected this shut-down operational mode for my City Car [30, 31]. After the starting of the circulating pump the battery was operational within one minute and the hydrogen gas could be admitted by the control system. The car was used in public traffic for over three years and the electrodes were supposed to have a life time of only 2000 hours. Only far later did we realize that the intermittent operation had such a large effect on the life time. A detailed description of the electrodes [32] and the fuel cell – battery hybrid system [33] is available.

Presently members of my work group are contemplating such electrode life extending possibilities again, with low-cost alkaline, but also with other electrolyte systems It is also important to remember, that the repeated (or programmed) exposure of Pt-catalyzed carbon – hydrogen electrodes to air is a means to recover damaged or poisoned catalyst systems. This may be another point for studying interrupted operation of low temperature air-fuel cell systems of different types, including liquid acidic and membrane systems. In respect to circulating electrolyte management, considerable experience has been collected in other electrochemical systems, like in the zinc-bromine rechargeable battery. Especially, the stack design and the reduction or the elimination of parasitic currents in bipolar stacks has made considerable progress [34].

Therefore, the outlook for the fuel cell is, in my opinion, still good and it is worthwhile to make it attractive for scientists and engineers to try their talents on these admittedly old, but still "young at heart" energy converting technologies. The plural is correct: the systems have been spread out into many areas of physics and chemistry, not imaginable a few years ago. Coupled with the progress of materials and processes and of the controls by electronic means, the future is bright. The only sad thing is that the technical specialist may lose himself easily in details and may miss the major (in hindsight "obvious") steps of improvements. This danger exists for the modern man, and it is not limited to scientists.

References to Chapter 9

[1] T. N. Veziroglue, F. Barbier, Int. J. Hydrogen Energy **17** (7) (1992) 527
[2] EPRI Journal, April/May 1993, pp. 7–17
[3] L. A. Wittrup et al. 1994 Fuel Cell Seminar, San Diego, Proceedings, pp. 567
[4] Fuel Cell Development Information Center, "Fuel Cells Now" Oct.1994
[5] U.S. Department of Energy, Utility Storage Systems Program Plan FY 1994 – FY 1998
[6] R. Hendriks, H. J. Ankersmit, L. J. M. J. Blomen, "Heron Turbine Integrated Systems with Hydrogen Technology and Fuel Cells", 21st Int. Congress on Combustion Engines, Interloaken, 1995

[7] H. Jericha, W. Sanz, J. Woisetschläger, M. Fesharaki, CO_2-Retention Capability of CH_4/O_2-Fired Graz Cycle, 21st Int. Congress on Combustion Engines, Interlaken, 1995

[8] U.S. Department of Energy, Utility Storage Systems Program Plan FY 1994 – FY 1998

[9] Fuel Cell Systems Program Plan, Fiscal Year 1994, U.S. Department of Energy, DOE/FE-0311P

[10] "DOE-Workshop on Fuel Cells, March 2, 1995", Fuel Cell News, Vol. XII, No.1. Spring Issue 1995, pp.10–17, The Fuel Cell Institute, P.O.B. 65481, Washington, DC, 20035-5481, USA

[11] D. B. Stauffer et al., 1994 Fuel Cell Seminar, San Diego, Proceedings, pp. 562

[12] DOE Fuel Cells, A Handbook (Revision 3), DOE/METC-94/1006 (DE94004072) by the Office of Fossil Energy, Morgentown Energy Technology Center, Morgantown, West Virginia, 26507-0880, Jan. 1994

[13] D. McClelland, "First Commercial Fuel Cell Flut: Experience, Lessons learned, and Future Perspective, 1994 – Fuel Cell Seminar, San Diego (CA), Nov. 28 – Dec. 1, 1994, pp. 11–14

[14] "The EP Green Car"[TM], Energy Partners, Inc., 1501 Northpoint Parkway, Suite 102, WPB FL, 33407

[15] Powercell Corp., Cambridge, Mass, 02142, Developer of the Zn-Br Flow Battery, previously: S.E.A., Austria

[16] Electric Fuel Ltd. (Lutz Industries) Mechanically rechargeable Zinc-Air Battery, Special Report, Batteries International, October 1994, pp. 10–13. Batteries International Ltd., Aberdeen House, Surrey, GU26 6La, UK

[17] T. Homma et al., "Current Status of Fuel Cells in Japan", 1994 – Fuel Cell Seminar, San Diego (CA), pp. 675

[18] Fuel Cell Development Information Center, "Fuel Cells Now" Oct. 1994, p. 5, (R. Anahara, Secretary General) Fuel Cell Development Information Center (FCDIC) 2-1-7, Kanda-ogawamachi, Chiyoda-ku, Tokyo 101, Japan

[19] Daimler Benz, "Neue Antriebe", High Tech Report 3/1994, pp. 14–21

[20] T. Kivisaari, "CO_2 Management – A Critical Issue for Two Fuel Cell Options", Fuel Cell Seminar 1994, San Diego (CA), pp. 535–538

[21] Y.Kiros, "Tolerance of Gases by the Anode for the Alkaline Fuel Cell", 1994 – Fuel Cell Seminar, pp. 364–367

[22] Hong-Ki-Li et al., "Development of Electrode performance and Stack Designs for Alkaline Fuel Cells", 1994 – Fuel Cell Seminar, Nov. 28–Dec. 1, San Diego (CA), pp. 585–588

[23] M. Selan et al., "The Process Cycle: Sponge Iron/Hydrogen/Iron Oxide Used for Fuel Conditioning in Fuel Cells", VDI-Bericht 1168, 1994, pp. 191–200

[24] E. Guelzow et al., German Aerospace Res. Establishment, Stuttgart, 1994 – Fuel Cell Seminar, pp. 319–322.

[25] G. Walker et al., "The Stirling Alternative", University of Calgary, Alberta, Canada, Gordon and Breach Science Publishers, Yverdon, Switzerland, 1994

[26] Advanced Battery Technology, June 1995, pp. 8–10, Robert Morey Associates, Cooperstown, NY, 13326, USA

[27] European Commission, Directorate General XII: Science Research and Development. Also Directorate XVII, Energy, Direction Energy Technologies, Draft Document, Nov. 1994. To be edited and released 1995.

[28] European Fuel Cell Group, Ltd., (EFCG) Main Office: P.O. Box 104, 2810 AC Reeuwijk, The Netherlands, EFCG's 1995 Summer Workshop was held June 8–9, Genoa, Italy

[29] "Ammonia-Air Fuel Cell System for Vehicle Propulsion", Union Carbide Corp., Techn. Proposal to US. Army, Ft. Belvoir, Virginia, RFP 64-3730-C, May 15, 1964

[30] K. V. Kordesch, "City Car with Hydrogen-Air Fuel Cell and Lead-acid Battery", Proceedings of the Intersociety Energy Conversion Engineering Conf., Boston 1971, Paper No. 719016

[31] K. V. Kordesch, "Power Sources for Electric Vehicles", In: Modern Aspects of Electrochemistry, Vol. 10, (Bockris, J'OM, Conway, B. E., eds.) pp. 339–443, NY., Plenum Press 1975

[32] K. Kordesch, J. Gsellmann, S. Jahangir, M. Schautz, "The Technology of PTFE-bonded Carbon Electrodes", Proceedings of the Symposium on Porous Electrodes, (Maru, H. et al., eds.) The Electrochem. Soc., 1984–1988

[33] K. Kordesch, "Brennstoffbatterien", Springer Verlag, Wien, New York, 1984 (In German)

[34] The WAVE^{TM}, Electric Vehicle Power Plants, Powercell Corporation, Cambridge Mass, 02142, USA, the former Studiengesellschaft für Energiespeicher und Antriebssysteme (S.E.A.) Mürzzuschlag, A-8680, Austria

Index

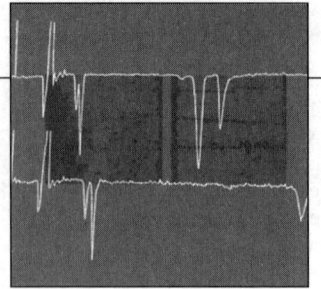

Electric Field Applications

in Chromatography, Industrial and Chemical Processes

Edited by Takao Tsuda

1995. XVI, 311 pages with 213 figures and 17 tables. Hardcover.
DM 298.00. ISBN 3-527-28687-X

This authoritative review brings scientists up-to-date with the exciting recent developments in modern electric field applications and highlights their benefits compared with other methods.

In Part 1 the book opens with a complete account of electrochromatography - a state-of-the-art technique that combines chromatography and electrophoresis. It reveals how you can achieve first-class separations in numerous analytical and biochemical applications. Part 2 focuses on the unique characteristics of electro-processes in industry, and several examples, such as electroosmotic dewatering, new electro-rheological fluid technologies and demulsification processes in the car and oil industries, are given. The role of the electric field in chemical processes is discussed in Part 3. The chapters explore its use in concentration processes, immunoassay and molecular orientation methods, and important examples are presented in each case.

This book is essential reading for analytical chemists, applied chemists and chemical engineers working in research and development wishing to keep up with this dynamic field.

Date of Information:
January 1996

Prices ar subject to change without notice.

VCH, P.O.Box 10 11 61,
D-69451 Weinheim,
Fax 0 62 01 - 60 61 84

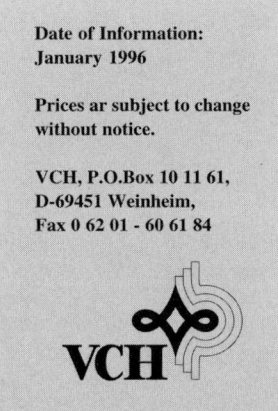